FLUCTUATION THEORY
OF SOLUTIONS

Applications in Chemistry, Chemical
Engineering, and Biophysics

FLUCTUATION THEORY OF SOLUTIONS

Applications in Chemistry, Chemical Engineering, and Biophysics

EDITED BY

Paul E. Smith ■ **Enrico Matteoli** ■ **John P. O'Connell**

CRC Press
Taylor & Francis Group
Boca Raton London New York

CRC Press is an imprint of the
Taylor & Francis Group, an **informa** business

CRC Press
Taylor & Francis Group
6000 Broken Sound Parkway NW, Suite 300
Boca Raton, FL 33487-2742

First issued in paperback 2019

ISBN-13: 978-1-4398-9922-9 (hbk)
ISBN-13: 978-0-367-38034-2 (pbk)

Library of Congress Cataloging-in-Publication Data

Fluctuation theory of solutions : applications in chemistry, chemical engineering and biophysics / editors, Paul E. Smith, Enrico Matteoli, John P. O'Connell.
    p. cm.
Includes bibliographical references and index.
ISBN 978-1-4398-9922-9 (hardback)
    1. Solution (Chemistry) 2. Fluctuations (Physics) I. Smith, Paul E. II. Matteoli, Enrico. III. O'Connell, John P.

QD543.F626 2013
541'.3413--dc23   2012037609

Visit the Taylor & Francis Web site at
http://www.taylorandfrancis.com

and the CRC Press Web site at
http://www.crcpress.com

Contents

Preface

Many, if not most, processes of interest occur in solutions. It is therefore somewhat unfortunate that our understanding of solutions and their properties remains rather limited. There are essentially two theories of solutions that can be considered exact. These are the McMillan–Mayer theory of solutions and Fluctuation Solution Theory (FST), or the Kirkwood–Buff (KB) theory of solutions. The former has practical issues, which limit most applications to solutes at low concentrations. The latter has no such issues. Nevertheless, the general acceptance and appreciation of FST remains limited. It is the intention of this book to outline and promote the considerable advantages of using FST/KB theory to study a wide range of solution properties.

Fluctuation solution theory is an exact theory that can be applied to any stable solution containing any number of components at any concentration involving any type of molecules of any size. The theory is primarily used to relate thermodynamic properties of solutions to the underlying molecular distributions, and vice versa. This collection has been developed to outline the general concepts and theoretical basis of FST, and to provide a range of applications relevant to the areas of chemistry, chemical engineering, and biophysics, as described by experts in each field. It serves as an update to a previous compilation published over two decades ago (Matteoli and Mansoori 1990). Many substantial advances have been made since the previous compilation was published, and these are included in the present edition. In particular, the application of FST to study biological systems is now well established and promises to be even more fruitful in the near future. In addition, continuing developments in computer simulation hardware and software have increased the range of potential applications, helping to improve our understanding of solution properties, and providing access to the required integrals that form the basis of the theory.

This book includes a historical perspective (Prolegomenon) and an introductory section (Chapter 1) outlining the basic theory, including the underlying concepts and a basic derivation that is aimed at the casual reader. Additional chapters then provide applications of FST to help rationalize and understand simple model (Chapter 2), binary (Chapter 3), and ternary (Chapter 4) systems with a focus on their thermodynamic properties and the concept of preferential solvation. The use of FST to help develop more accurate potential functions for simulation is illustrated (Chapter 5), followed by a detailed outline of the problems and possible solutions for determining the integrals over molecular distribution functions from simulation as required by the theory (Chapter 6). New approaches to help understand microheterogeneities in solutions are then described (Chapter 7), together with an overview of solvation in real and model systems including systems under critical conditions (Chapter 8). The use of FST to describe and model solute solubility in a variety of systems is then discussed (Chapters 9 and 10). Finally, a series of biological applications are provided which illustrate the use of FST to the study of cosolvent effects on proteins (Chapter 11), and the implications for protein folding (Chapter 12). Where possible, we have attempted to maintain the same notation (established in Chapter 1) and a set of symbol descriptions is provided for reference.

However, the number of possible applications of FST extends beyond those presented here. Indeed, there are many additional applications that deserve attention, but have not been included due to either space limitations, or because they represent newly emerging areas, which are not yet fully mature. A reasonably comprehensive list of the currently available applications of FST includes: thermodynamic properties of binary and ternary solutions; transfer free energies; osmotic systems; solute solubility and Henry's constant (including critical regions); descriptions of preferential solvation; preferential interactions in biological systems including osmotic stress and volumetric studies; density fluctuations provided by light scattering; evaluation of force fields for computer simulations; chemical equilibria, and the effects of pressure and composition on molecular crowding and protein denaturation; and the effects of cosolvents on surface tension, crystal morphology, and micelle formation. More recently, one has also been able to move beyond isothermal conditions, which provide molecular level interpretations of additional thermodynamic quantities.

It is hoped that the efforts described here help to convey the beauty and simplicity of FST to a range of researchers in a variety of fields. We are confident that FST provides the most rigorous and useful approach for understanding and rationalizing a wide range of solution properties, especially when used in conjunction with computer simulation data.

Paul E. Smith
Enrico Matteoli
John P. O'Connell

Acknowledgments

The editors would like to thank some of the many people who have made this collection possible. PES expresses his gratitude to current group members—Yuanfang Jiao, Shu Dai, Sadish Karunaweera, Elizabeth Ploetz, Gayani Pallewela, Nawavi Naleem, and Jacob Mercer—for their help proofreading the manuscript. JPO'C is grateful for the stimulation of his many colleagues and students, whose works are cited in Chapters 6 and 9, especially Jens Abildskov of the Danish Technical University; John M. Prausnitz, now retired from the University of California at Berkeley; and Peter T. Cummings, now at Vanderbilt University. The editors would like to give a very special thank you to Elizabeth Ploetz, who was an integral part of the organization and execution of this project. Quite simply, the project would not have been realized without her help, commitment, and dedication.

Contributor List

Jens Abildskov
CAPEC
Department of Chemical and
    Biochemical Engineering
Technical University of Denmark
Kongens Lyngby, Denmark

Matthew Auton
Cardiovascular Sciences and
    Thrombosis Research Section
Department of Medicine
Baylor College of Medicine
Houston, Texas

Arieh Ben-Naim
Department of Physical Chemistry
The Hebrew University of Jerusalem
Jerusalem, Israel

Ariel A. Chialvo
Chemical Sciences Division
Oak Ridge National Laboratory
Oak Ridge, Tennessee

Paolo Gianni
IPCF-CNR
Istituto per i Processi Chimico-Fisici
Pisa, Italy

Myungshim Kang
Department of Chemistry
University of California
Riverside, California

Luciano Lepori
IPCF-CNR
Istituto per i Processi Chimico-Fisici
Pisa, Italy

Yizhak Marcus
Institute of Chemistry
The Hebrew University of Jerusalem
Jerusalem, Israel

Enrico Matteoli
IPCF-CNR
Istituto per i Processi Chimico-Fisici
Pisa, Italy

Robert M. Mazo
Institute of Theoretical Science and
    Department of Chemistry
University of Oregon
Eugene, Oregon

John P. O'Connell
Department of Chemical Engineering
University of Virginia
Charlottesville, Virginia

Aurélien Perera
Laboratoire de Physique Théorique de
    la Matière Condensée
Université Pierre et Marie Curie
Paris, France

B. Montgomery Pettitt
Center for Structural Biology and
    Molecular Biophysics
University of Texas Medical Branch
Galveston, Texas

Elizabeth A. Ploetz
Department of Chemistry
Kansas State University
Manhattan, Kansas

Eli Ruckenstein
Department of Chemical and Biological
   Engineering
State University of New York at Buffalo
Amherst, New York

Seishi Shimizu
York Structural Biology Laboratory
Department of Chemistry
University of York
Heslington, York, United Kingdom

Ivan L. Shulgin
Department of Chemical and Biological
   Engineering
State University of New York at Buffalo
Amherst, New York

Paul E. Smith
Department of Chemistry
Kansas State University
Manhattan, Kansas

Rasmus Wedberg
FOI–Swedish Defense Research
   Agency
Division of Defense and Security,
   Systems and Technology
Tumba, Sweden

Samantha Weerasinghe
Department of Chemistry
University of Colombo
Colombo, Sri Lanka

Prolegomenon to the Fluctuation Theory of Solutions

Robert M. Mazo

Solutions are mixtures that are homogeneous on any macroscopic scale. One may argue that this definition excludes colloidal systems. So it does, but not because of any matter of principle; it is just that we shall not consider colloidal solutions in this chapter, so it is a matter of convenience to exclude them from the terminology also. Solutions can be gaseous, liquid, or solid; here we shall be concerned almost exclusively with liquid solutions. Furthermore, we shall restrict ourselves primarily to solutions of nonelectrolytes. Electrolytes require some special attention stemming from the long range of interionic forces, although they pose no fundamental problems for fluctuation theory.

Solutions are by far much more common than pure substances. Just think of the effort needed to separate a solution into its components: distillation, crystallization, zone melting, and so forth. It is no wonder that since ancient times humans have been interested in the properties of solutions and how they are modified from those of the pure constituents. The scientific study of these matters, however, dates from the early part of the 19th century and the first period of discovery may be said to have culminated in the 1880s with the discovery of Raoult's law.

Perhaps the first quantitative law governing the properties of solutions was published by William Henry in 1803. Henry was studying the solubility of gases in liquids and found that this solubility was proportional to the gas partial pressure (Henry 1803). He did not express his results as an equation, but published tables of data from which the proportionality could be extracted. An interesting review of the current status of Henry's law has been given by Rosenberg and Peticolas (Rosenberg and Peticolas 2004).

The next major step was the enunciation of Raoult's law (Raoult 1887, 1888). In 1887, Francois Raoult published his investigations on the vapor pressure of the solvent in dilute solutions. He studied five solutes in water and 14 solutes in each of 11 organic solvents and found that the diminution of the vapor pressure of the solvent upon addition of a given (small) amount of solute was proportionally the same for all cases. The proportionality factor is the mole fraction of the solute. This may be expressed in the currently accepted notation as $p_1^0 - p_1 = p_1^0 x_2$; this is known as *Raoult's law*. Raoult had previously discovered the laws of freezing point depression and boiling point elevation (Raoult 1878, 1882), three of the so-called colligative[*] properties of dilute solutions.

---

[*] From the Latin *colligare*, to bind together (*Oxford English Dictionary*). However, some authors derive it from *colligere*, to gather.

Another property of solutions, the osmotic pressure, was studied by Jacobus van't Hoff in 1887 (Van't Hoff 1887, 1894). Here, Van't Hoff was trying to understand the properties of liquids in terms of those of gases, which were considered fairly well understood. He seized on the phenomenon of osmotic pressure as the analog of the pressure of a gas, and derived the eponymous equation $\pi = c_2 RT$, where c_2 is the molar concentration of the solute and π the osmotic pressure. The osmotic pressure was more talked about than measured in those days because the available semipermeable membranes were not very good. They very often leaked and accurate experiments were very hard to do. Nevertheless, it was a favorite function for discussing the properties of solutions.

Van't Hoff also showed, thermodynamically, the relations between the colligative properties found by Raoult and the osmotic pressure, namely, that they were all alternative ways of counting molecules.

Not long after these developments, the subject of statistical mechanics began to be developed (Gibbs 1902). Statistical mechanics had brilliant success in the calculation of the properties of gases, especially after the advent of quantum theory permitted a proper description of the internal states of molecules, but its application to condensed phases was less successful. A survey of the state of the molecular theory in 1939 can be found in the textbook of Fowler and Guggenheim (Fowler and Guggenheim 1939). The theory at that time was based on the cell model of liquids, which overestimates the correlation between molecular positions.

During World War II, little or no work was done on solution theory, but after the war, activity began again. Now, the emphasis of many theories began to fall on the properties and usefulness of molecular distribution functions, in particular the pair correlation function. This was due, in part, I believe, to the thesis of Jan de Boer (De Boer 1940, 1949). As an aside, I once asked J. E. Mayer why he used the canonical ensemble in his early work on statistical mechanics and the grand ensemble in his later works. He replied, "Oh, I switched after I read de Boer's thesis and saw how easy the grand ensemble made things." De Boer's work was for pure fluids, not solutions, and other authors, in particular John G. Kirkwood (Kirkwood 1935), also developed the correlation function method.

This work on correlation functions, when generalized to mixtures, led to two equivalent, though superficially different, formally *exact* theories of solutions, due to Joseph Mayer and William McMillan (McMillan and Mayer 1945) and to John Kirkwood and Frank Buff (Kirkwood and Buff 1951). These theories and their experimental consequences form the bulk of the material in the remainder of this book. Before discussing them, however, let us describe several approximate theories, which had a considerable vogue in the 1950s and 1960s but which are not much used nowadays.

The first of these developments is perturbation theory. Its application to solution theory was perhaps first made by H. C. Longuet-Higgins in his conformal solution theory (Longuet-Higgins 1951). The formal theory of statistical mechanical perturbation theory is very simple in the canonical ensemble. If V_N denotes the intermolecular potential energy of a classical N-body system (not necessarily the sum of pair potentials), the central problem is to evaluate the partition function,

$$Q_N = \frac{1}{N!} \int \exp(-\beta V_N) d\{r\} \tag{P.1}$$

where Q_N is the partition function for an N particle system. The symbol $d\{r\}$ indicates integration over the positions of all particles.

We suppose that the potential energy V_N can be written in the form,

$$V_N = V_N^0 + \Delta V_N \tag{P.2}$$

then,

$$Q_N = Q_N^0 \left(1 + \sum_{n=1}^{\infty} (-\beta)^n \left\langle (\Delta V_N)^n \right\rangle_0 / n! \right) \tag{P.3}$$

The symbol $<\ldots>_0$ means average over the unperturbed Boltzmann factor,

$$\langle X \rangle_0 = \frac{1}{N! Q_N^0} \int X \exp(-\beta V_N^0) d\{r\} \tag{P.4}$$

Thus, the partition function has been expressed as a series in β whose coefficients are powers of the perturbation ΔV. Note that this is not strictly a series expansion in β, since the coefficients themselves will generally have some temperature dependence.

However, the quantity of main interest is not the partition function but the free energy, $A_N = -k_B T \ln Q_N$. But, since the partition function is expressible as a series, so is its logarithm. There is a standard procedure for passing from the coefficients of one series to those of the other. In mathematical statistics, this is called *passing from the moments of a distribution to its cumulants*. We record here only the leading term in the cumulant expansion, since that is the only one that has ever been used in solution theory. The actual computation of higher-order terms involves molecular correlation functions of third order and higher, about which essentially nothing is known. So, to lowest order in ΔV,

$$A_N \approx A_N^0 + \left\langle \Delta V_N \right\rangle_0 \tag{P.5}$$

This is the entire formal structure of classical statistical mechanical perturbation theory. The reader will note how much simpler it is than quantum perturbation theory. But the devil lies in the details. How does one choose the *unperturbed* potential, V_N^0? How does one evaluate the first-order perturbation? It is quite difficult to compute the quantities in Equation P.5 from first principles. Most progress has been made by some clever application of the law of corresponding states. It is not the aim of this chapter to follow this road to solution theory any further.

A second strand of approximate theory that held interest for a while was variational theory. This idea is based on the rigorous inequality,

$$A_N \leq A_N^0 + \left\langle V_N - V_N^0 \right\rangle_0 \qquad (P.6)$$

This is almost the same as Equation P.5 except that the $\approx$ sign has been replaced by a $\leq$ sign. This is known as the *Gibbs–Bogoliubov variational principle*. Gibbs proved it for classical statistics (Gibbs 1948), and Bogoliubov for quantum statistics (Tolmachev 1960). The idea here is to choose an unperturbed, or reference, potential with a certain amount of flexibility of form (adjustable parameters, functional form, etc.) and vary the right-hand side of Equation P.6 to make it as small as possible.

The problems here are twofold. First, V_N^0 should be simple enough to make effective computation possible, yet complex enough to accommodate the inherent complexities of V_N. These two criteria are often incompatible. Second, some of the quantities we want are derivatives of the free energy, for example, the pressure or chemical potential. An upper bound on a function, even a close upper bound, does not guarantee that the derivative of the function serving as the bound is close to the derivative of the original function. For example, consider the function $f(x)$ and $f(x) +$ $\varepsilon \sin(\varepsilon^{-1} x)$. When ε is very small, the functions can be very close but their derivatives (with respect to x) can be very different because of the high frequency *ripple*. This is not likely to occur when using smooth trial functions, but its possibility should always be kept in mind. Again, we shall go no further into the details of variational theory. A review of the subject has been given by Girardeau and Mazo (1973).

The last of the approximate theories that we wish to mention is that of Prigogine and collaborators (Prigogine with contributions from A. Bellemans and V. Mathot. 1957). This theory combined ideas from the cell theory of solutions and from perturbation theory, both mentioned above. This approach was qualitatively quite successful especially insofar as it correctly predicted the relative signs of the various excess functions of mixing; these were incorrectly predicted by most other approximate theories in a number of cases.

Now we want to leave our discussion of what might be called the ancient and early modern periods of solution theory history and concentrate on the modern period, characterized by the theories of Mayer and McMillan (McMillan and Mayer 1945) and of Kirkwood and Buff (Kirkwood and Buff 1951). The McMillan–Mayer theory was the earlier of the two, by some 6 years, and had already captured the attention of the experimental community by the time the Kirkwood–Buff theory appeared.

J. E. Mayer and his students had, in the 1930s, developed the theory of the equation of state of gases in terms of the intermolecular potential. Essentially, they derived the virial equation of state from first principles with explicit expressions for the virial coefficients in terms of certain integrals, called *irreducible cluster integrals*, of certain functions of the intermolecular potential. What McMillan and Mayer did was to perform an analogous task for the osmotic pressure of a solution. They obtained an expansion for the osmotic pressure in powers of the solute concentration (for short-range forces) completely analogous to the gas case. The coefficients in this series are called *osmotic virial coefficients*. They are even expressible in terms of integrals analogous to cluster integrals. However, and this is an important caveat, the integrands do not depend directly on the free space potential between solute molecules, but on the *potential of mean force* at infinite dilution between solute molecules in the solution. These functions are different. For example, the intermolecular potential between a pair of molecules usually has a single minimum, whereas the potential of mean force usually oscillates. Furthermore, even if the intermolecular potential is pairwise additive, the potential of mean force between n molecules is not; the presence of the solvent generates nonadditivity.

We begin our discussion with a bit of notation. The n body distribution function in an open system described by the grand canonical ensemble is,

$$\rho_{\mathbf{n}}(\{n\},\lambda) = \exp(-\beta pV) \sum_{\mathbf{N} \geq \mathbf{n}} \lambda^{\mathbf{N}}/(\mathbf{N} - \mathbf{n})! \int \exp(-\beta V_N) d\{\mathbf{N} - \mathbf{n}\} \qquad (\text{P.7})$$

Here $\{n\}$ means $(n_1, n_2, \ldots)$, $\lambda^N = \lambda_1^{N_1} \lambda_2^{N_2} \ldots$, and so forth. $\lambda_i = \Lambda_i^{-1} \exp(\beta \mu_i)$ where Λ_i is the internal and translational partition function of species i, and λ_i is called the *absolute activity of species i*: this activity should not be confused with the thermodynamic, or Lewis, activity defined in Chapter 1, Section 1.1.3.

The starting point of McMillan–Mayer theory is a relationship between distribution functions at different activity sets. The derivation of this relationship is the difficult part of the theory. But once obtained, the relation leads to an expression for the osmotic pressure of a solution, since the components permeable to the osmotic membrane have the same chemical potential on both sides of the membrane while those impermeable have differing chemical potentials. A lengthy computation then leads to an expansion for the osmotic pressure, completely analogous to the activity expansion of the pressure in the theory of imperfect gases. Indeed, for the purpose of comparing gas theory with solution theory, it helps to regard the gas as a solute in a very special and very simple solvent—vacuum. The λ expansion is,

$$\beta \pi = \sum_{j \geq 1} b_j \lambda^j \qquad (\text{P.8})$$

where the λ_js refer to solute species only. The coefficients, b_j are given explicitly in terms of the partition functions of small numbers of solute molecules (i.e., infinite dilution) in the solvent. For example,

$$1Vb_1 = Q_1$$

$$2Vb_2 = Q_2 - Q_1^2 \qquad (\text{P.9})$$

$$\cdots$$

But λ is not a convenient experimental variable, so the final step, as in gas theory, is to convert Equation P.8 into a series in the density, the virial series. This is done by using the thermodynamic relationship,

$$c_i = \lambda_i \left(\frac{\partial \beta \pi}{\partial \lambda_i} \right)_{\beta, \{\lambda\}'} \qquad \text{(P.10)}$$

Inserting Equation P.8 in this relation, one obtains a series for **c** in terms of λ that can be inverted to give $\{\lambda\}$ as a series in $\{c\}$. Finally, inserting this series in the λ series for π, one obtains the so-called osmotic virial expansion,

$$\beta \pi = c_2 \left(1 + B_2 c_2 + B_3 c_2^2 + \cdots \right) \qquad \text{(P.11)}$$

the coefficients in this expansion are called *osmotic virial coefficients*. The second virial coefficient is given by $B_2 = -b_2$. Higher order Bs are more complicated algebraic combinations of the bs and thereby the Qs.

Note that these look just like the corresponding expansion coefficients in gas theory except for one important difference: the potential of mean force takes the place of the intermolecular potential. Since the potential of mean force is not, in general, pairwise additive, the familiar technology of Mayer f functions and cluster diagrams are not available to the solution theorist. It is interesting to note that the emphasis on osmotic pressure in McMillan–Mayer theory seems to bring one back to the ideas of van't Hoff.

The results of McMillan–Mayer theory have been used primarily in the area of solutions of macromolecules in low molecular weight solvents. The osmotic second virial coefficient, which can be measured either by osmometry or light scattering, gives information on the size of the solute molecules. We shall see why in more detail later when we discuss fluctuation theory.

The theory of McMillan and Mayer is exact, but only useful in dilute solutions. It delivers thermodynamic functions as a power series in the solute concentrations and it is quite difficult to compute, or even to interpret the coefficients higher than the second virial coefficient, B_2. About 6 years after the McMillan–Mayer theory was developed a new solution theory appeared, not subject to this difficulty, that of Kirkwood and Buff (Kirkwood and Buff 1951), of course this new theory had computational problems of its own. KB (Kirkwood–Buff) theory is also known as *fluctuation theory* for reasons that will become obvious below. It is the basis for the rest of this volume and therefore will occupy the remainder of this chapter.

The paper by Kirkwood and Buff is quite remarkable. It is only four pages long and only two of those four contain the important parts of the theory. It does this by giving only definitions and results and leaving the reader to fill in all of the intermediate steps. Most of these are straightforward, but there are several tricky points, which we shall discuss below.

The basis of KB theory is the relation between number fluctuations in an open system and the thermodynamic properties of that system. This relation is usually ascribed to Einstein (1910), but many of the results can be found in Gibbs (1948). Kirkwood had long been interested in fluctuations. He discussed them extensively in lectures given at Princeton University in 1947 (Kirkwood 1947, privately circulated). I regard these notes as a precursor to KB theory.

The relation of fluctuations in concentration to thermodynamic functions was also previously recognized in the theory of light scattering from solutions (Brinkman and Hermans 1949; Kirkwood and Goldberg 1950; Stockmayer 1950) since light is scattered by inhomogeneities in refractive index, which, in turn, arise in part from concentration fluctuations.

The theory proceeds by deriving two different formulas for the concentration fluctuations, and then equates the results. On the one hand, the probability that the system contains exactly N particles is,

$$P_N = \lambda^N Q_N / \Xi \qquad (P.12)$$

where Ξ is the grand partition function. Since $\Sigma P_N = 1$, differentiating Equation P.12 twice with respect to any of the λs yields,

$$V^{-1}\left[\langle N_i N_j \rangle - \langle N_i \rangle \langle N_j \rangle \right] = A_{ij}^{-1}$$

$$A_{ij} = V\left(\partial \beta \mu_i / \partial N_j \right)_{T,V,\{N\}'} \qquad (P.13)$$

where the inverse symbol is meant in the matrix sense. Note that this is a slightly different definition of the **A** matrix than used in the original paper.

The second formula alluded to is the relation between the pair correlation function between pairs of species in the solution and the number fluctuations,

$$G_{ij} = 4\pi \int_0^\infty \left[g_{ij}(r) - 1 \right] r^2 dr = V\langle \delta N_i \delta N_j \rangle / \langle N_i \rangle \langle N_j \rangle - \delta_{ij}/\rho_i \qquad (P.14)$$

where ρ_i is the number density of species i. This is derived from the definition of the pair correlation function in the grand ensemble as,

$$\exp[\beta p V] = \sum_{N \geq 0} \lambda^N Q_N / N! \qquad (P.15)$$

This connects the fluctuations to thermodynamics.

Equating these two expressions for the number fluctuations, one arrives at an expression for the composition derivatives of the chemical potential,

$$A_{ij} = B_{ij}^{-1}$$

$$B_{ij} = \rho_i \left(\delta_{ij} + \rho_j G_{ij} \right) \qquad (P.16)$$

This is the cornerstone of Kirkwood–Buff theory (see Chapter 1, Section 1.2.1). Most of what follows is just modification of Equation P.16 using thermodynamic identities.

There are three remarks to be made here. The first is that Equation P.12 of KB (hereafter called *KB12*), an equation for $(\partial\mu_i/\partial N_j)_{P,T,\{N\}'}$ treating one component, *the solvent*, in an unsymmetrical way, does not follow in an obvious way from substituting the definition of the G_{ij} in terms of the B_{ij}. This has led some to suspect that Equation *KB12* is not correct. I have been able to go backward, so to speak. By using the properties of partitioned matrices I have gone from Equation *KB12* to Equation *KB11*, but I have found a derivation of Equation *KB12* directly from the other equations of KB in only one place (Münster 1969) to which the reader is referred. The interested reader may also consider Chapter 1, Section 1.2.1 and other literature (O'Connell 1971b).

Second, one might ask, since McMillan–Mayer and Kirkwood–Buff theories are both exact, what is the relation between them? McMillan–Mayer theory is formulated in terms of potentials of mean force at infinite dilution, albeit of increasing numbers of particles. Kirkwood–Buff theory is formulated in terms of the potential of mean force between pairs only, but at the actual concentration of the solution. The answer to this question is given by Equation *KB23*, written down without derivation. A future publication with a derivation is promised but, as far as I know, now 60 years later, none has appeared. This is an unsatisfactory state of affairs.

The third comment concerns a passing remark in the Kirkwood–Buff paper (1951): "The preceding relations are completely general, and it is of interest to remark that in electrolyte theory Eq. (20) provides an alternative to the usual charging process." However, the matrix B is singular for ionic solutions because of the constraint of (average) electroneutrality imposed by the high-energy cost of a fluctuation with an imbalance of charge. This implies that, for charged systems,

$$\sum_i z_i B_{ij} = 0 \tag{P.17}$$

that is, the rows of B are linearly dependent. It is not clear whether Kirkwood and Buff were aware of this problem. To the best of my knowledge the problem was first explicitly pointed out by Friedman and Ramanathan (1970). The problem with the B matrix arises fundamentally from the nonmeasurable, that is, nonphysical, nature of single-ion chemical potentials. This suggests several methods of avoiding the problem. One is to treat the solution as a mixture of independent ions and solvent. This method was introduced by Friedman and Ramanathan (1970) and has been exploited by Smith and coworkers (Chitra and Smith 2000). Other methods for avoiding the problem have been suggested by Kusalik and Patey (1987) and Behera (1998). The first of these is particularly elegant; it involves working in Fourier transform space where the $B(k)$ matrices are nonsingular for $k > 0$, and then going to the limit of $k = 0$ at the very end of the computation when only the chemical potentials of neutral salts enter. This is achieved through a series of Fourier transforms of the KB integrals,

$$\hat{G}_{ij}(k) = 4\pi k^{-1}\int_0^\infty \left[g_{ij}(r)-1\right]r\sin(kr)dr \tag{P.18}$$

However, in keeping with the general historical nature of this introduction, we shall not treat electrolyte solutions any further here. More details can be found in Chapter 1, Section 1.3.6 and Chapter 9, Section 9.4.

The Kirkwood–Buff paper did not make much of an impression on those working in solution chemistry in the years immediately following its publication. I entered Kirkwood's department as a graduate student in 1952, and Kirkwood's research group in 1953. According to my best recollection, no one in the theoretical group was working on solution theory at that time. Kirkwood did have an experimental group working on protein solutions, staffed primarily by postdoctoral students, but my memory is that the experimental and theoretical groups did not interact, except socially.

There was, however, one important follow-up paper, by Buff and Brout (1955). The reader may have noticed that the Kirkwood–Buff paper concerns exclusively those properties of solutions that can be obtained from the grand potential by differentiation with respect to pressure or particle number. Those such as partial molar energies, entropies, heat capacities, and so forth, are completely ignored. The original KB theory is an isothermal theory. The Buff–Brout paper completes the story by extending the theory to those properties derivable by differentiation with respect to the temperature. Because these functions can involve molecular distribution functions of higher order than the second, they are not as useful as the original KB theory. Yet they do provide a coherent framework for a complete theory of solution thermodynamics and not just the isothermal part.

In the middle to late 1950s, perturbation theory was very popular, and two early applications of KB theory were to perturbation theory (Buff and Schindler 1958; Mazo 1958). Since these did not appear to be any more useful than perturbation theory based directly on the partition function, they were never followed up.

The problem in application of KB theory in the manner in which it was originally presented in those early days was twofold. First, given intermolecular potentials for the species involved, it was quite difficult to determine the molecular distribution functions needed to compute the G_{ij} integrals. Recall that at that time even pure fluid distribution functions were only available through use of the superposition approximation. Second, even had accurate methods for obtaining the Gs been available, intermolecular potentials were not as well studied at that time as they are now.

This situation changed suddenly in 1977. A paper by A. Ben-Naim pointed out that the Kirkwood–Buff equations could be inverted (Ben-Naim 1977). That is, instead of regarding the formulas of the Kirkwood–Buff paper as enabling one to compute the macroscopic thermodynamic properties of a solution from the $g(r)$ functions, one could equally logically think of the equations as enabling computations of the particle number fluctuations $\langle N_i N_j \rangle - \langle N_i \rangle \langle N_j \rangle$ in terms of the laboratory measurable thermodynamic functions. This is important because the definition of G_{ij} in Equation P.14 means that $\rho_j G_{ij}$ is the mean excess number of j particles in the neighborhood of an i particle in the solution. The excess is reckoned with respect to a random distribution, $\langle N_j/V \rangle$. Thus, one has a direct measurement of the local clustering properties of solutions by making thermodynamic measurements. True, it is an overall gross measure. It does not give detailed distance information; this is contained in the gs. Nevertheless, it is *local* information because the gs are short-ranged functions. The contribution of a given g to its G comes from a range of r of only a few intermolecular distances (except near critical phases).

This was a, "Why didn't I think of that?" idea, exceedingly simple, but very powerful. It very rapidly changed the status of the theory from an elegant, but hard to apply, formal theory to a useful tool of solution chemistry. It was a very important paper but, in my opinion, the importance was not so much technical as it was psychological. In a sense, it was a paradigm shift.

There were two other lines of attack that also led to appreciable progress in the use of KB theory. The first of these is experimental. The G_{ij} values can be considered as the zero wave number limit of the transformed quantity $G_{ij}(k)$ defined in Equation P.18. In the field of radiation scattering from molecular systems, this is sometimes called the *structure factor* and given the symbol $S(k)$. For small molecules, X-ray or neutron scattering are useful techniques. Small angle neutron scattering (SANS) is a preferred approach because of the contrast in scattering powers between various nuclear species. Many studies evaluating KB integrals using SANS have been carried out. We quote only one example here (Almasy, Jancso, and Cser 2002), although small angle X-ray scattering (SAXS) has also been used. See also Chapter 7.

The second line of inquiry alluded to was the use of modern computational power, both hardware and software, for the evaluation of pair distribution functions. When the Kirkwood–Buff paper was published, the use of computers for this kind of scientific computation was in its infancy. Indeed, one can say that it was in its prenatal stage. It is difficult to put a date on the time when computers became powerful enough to compute pair correlation functions and, consequently, KB integrals with sufficient accuracy for application to real systems. They have certainly reached that stage at the time of the writing of these words. The computational method of choice in carrying out these calculations is the molecular dynamics method. Since this kind of calculation is discussed in detail in several of the later chapters of this work, we eschew discussion here.

The remaining source of imprecision in the calculation of KB integrals from molecular theory is the imperfection of our knowledge of intermolecular potentials. Here also, KB theory comes to our aid in a way reminiscent of the inversion of KB theory discussed previously. For example, Weerasinghe and Smith refined the interaction potentials between Na^+ ions, Cl^- ions, and water molecules by calculating the relevant correlation functions by molecular dynamics for an assumed potential (Weerasinghe and Smith 2003d). They then adjusted the potential and redid the calculation, checking the results against the known experimental thermodynamic properties of the system, and repeated this procedure until no discernible difference was noted. This adjusted potential can then be used in other calculations with some confidence. See Chapter 5 for more details.

Subsequent chapters in this treatise describe modern applications of KB theory in detail. That is why the last several paragraphs of this historical introduction have been somewhat brief and relatively devoid of references to the literature. So, let us end this brief background survey by merely mentioning some other applications of KB theory that have been made: applications to systems containing solutes of biological interest, solubility (both under normal and supercritical conditions), as well as to salting out, mixed solvents, solvent effects on equilibria, and the framing of hypotheses on the local structure of solutions. This list is not intended to be exhaustive.

It is rather remarkable that a theory as simple as that of Kirkwood and Buff (recall that the essential part of the theory is given in two pages of a four-page paper) has turned out to be so powerful. The remainder of this treatise is an affirmation of how widespread the appreciation of its power has become. It is firmly based in statistical mechanics and classical physical chemistry. We have tried to illustrate here the organic nature of its concepts and how they came to be appreciated. The subsequent chapters of this volume will show in technical detail how they are used.

GREEK SYMBOLS

α_p	Isobaric thermal expansion coefficient (Equation 1.6)
Γ_{23}	Preferential binding parameter (Equation 1.86)
$\gamma_\pm$	Mean ion molal activity coefficient (Equation 1.92)
γ_i	Lewis–Randall/rational/mole fraction activity coefficient (Equation 1.19)
γ_i^c	Molar activity coefficient
γ_i^m	Molal activity coefficient
Δ	Isothermal–isobaric partition function (Equation 1.28)
ΔG_{ij}	$G_{ii} + G_{jj} - 2G_{ij}$ (Equation 1.93)
ζ_2	$1 + c_i G_{ii} + c_j G_{jj} + c_i c_j (G_{ii} G_{jj} - G_{ij}^2)$ (Equation 1.66)
η_{12}	$c_i + c_j + c_i c_j (G_{ii} + G_{jj} - 2G_{ij})$ (Equation 1.66)
κ_T	Isothermal compressibility (Equation 1.5)
Λ_i	Thermal de Broglie wavelength of species i
λ_i	Absolute activity of i
μ_i	Chemical potential of component i
μ_{ij}	Chemical potential derivative (Equation 1.1)
ν	Number of cations/anions, $\nu = \nu_+ + \nu_-$
Ξ	Grand canonical partition function (Equation 1.28)
π	Osmotic pressure
ρ	Mass or total number density
ρ_i	Number density of $i = N_i/V$, see also c_i
ϕ_i	Volume fraction of $i = \rho_i \bar{V}_i$
φ_i	Fugacity coefficient of i (Equation 1.23)
Ω	Microcanonical partition function (Equation 1.28)

MATHEMATICAL

$\langle \ \rangle$	Ensemble or time average
$\{X\}$	Set notation, $\{X_1, X_2, \ldots\}$
$\mathbf{1}$	Unit matrix

$|\mathbf{A}|$	Determinant of matrix $\mathbf{A}$
$\mathbf{A}^{ij}$	Cofactor of matrix $\mathbf{A}$
$\mathbf{A}$	Matrix with elements A_{ij}
$\mathbf{I}$	Identity matrix
δ_{ij}	Kronecker delta function
δX_i	Instantaneous fluctuation, $X_i - \langle X_i \rangle$
$\langle (\delta X)^2 \rangle$	Mean square fluctuation of a property X

LATIN

$+$	Cation
$-$	Anion
o	Pure
∞	Infinitely dilute (limiting)
1	Solvent
$1/k_B T$	β
2	Solute
3	Cosolvent/cosolute/additive
$\Delta_r G$	Reaction Gibbs energy (function)
$\Delta_r H$	Reaction enthalpy
$\Delta_r S$	Reaction entropy
δx_{ji}	Preferential solvation parameter (j surrounding i)
$\delta x'_{ji}$	Corrected preferential solvation parameter
$\mu V T$	Grand canonical ensemble
$\hat{X}$	Fourier transformed X
$\bar{X}_i$	Partial molar property of X
$a_\pm$	Mean activity of electrolyte in solution
A	Helmholtz energy (function)
A_i	Aggregate/multimer of i monomers
a_i	Activity of i
aq	Aqueous solution
c_i	Molarity of i, see also number density, ρ_i
C_{ij}	$\rho \int c_{ij}(r)dr$, DCFI, elements of the $\mathbf{C}$ matrix (Equation 1.39)
$c_{ij}(r)$	Direct correlation function
C_p	Constant pressure heat capacity (Equation 1.7)
D	Activity derivative, concentration fluctuation term (Equation 1.73)
f_i	Fugacity of a substance i in a gaseous mixture (Equation 1.21)
G	Gibbs energy (function)
g_{ij}	Radial (pair) distribution function, RDF
G_{ij}	Kirkwood–Buff integral, KBI
H	Enthalpy

h	Planck's constant
H_{ij}	$\rho \int h_{ij}(r)dr = \rho G_{ij}$, TCFI (see below Equation 1.38), or Henry's law constant
$h_{ij}(r)$	Total correlation function, TCF, $g_{ij}(r)$-1
id	Ideal (mole fraction scale)
K	Equilibrium constant
k	Rate constant
k_B	Boltzmann constant
k_H	Henry's law constant, see also H_{ij}
M	Monomer
m	m-value for protein denaturation (see Equation 1.99)
$m_i = c_i/c_1$	(Dimensionless) molality
mix	Mixing process
n	Number of monomers in an aggregate
n_c	Number of components in the system
N_A	Avogadro's number
N_i	Number of entities (usually molecules, atoms, or ions)
N_{ij}	Excess coordination number
NpT	Isothermal–isobaric (Gibbs) ensemble
NVE	Microcanonical ensemble
NVT	Canonical ensemble
p	Pressure
Q	Canonical partition function (Equation 1.28)
r	$|r_1\text{-}r_2|$, distance between COM of molecules
R	Gas constant
R_{cor}	Correlation radius (see V_{cor})
S	Entropy
T	Temperature (thermodynamic)
T_m	Melting temperature
trs	Transfer between two phases
U	Internal energy
V	Volume
V_{cor}	Correlation volume (see Equation 1.81)
X^*	Reduced or characteristic quantity X
X_c	Critical X (X is pressure or temperature)
X^E	Excess of X
X^r	Residual of quantity X
x_i	Liquid phase mole fraction composition
X_m	Molar quantity
y_i	Gas phase mole fraction composition, or solute solubility
$z_{+/-}$	Charge of cation/anion

ACRONYMS

COM	Center of mass
DCF	Direct correlation function
DCFI	Direct correlation function integral
EOS	Equation of state
FF	Force field
FST	Fluctuation solution theory
FT	Fluctuation theory
GD	Gibbs–Duhem
IG	Ideal gas
KB	Kirkwood–Buff
KBFF	Kirkwood–Buff Force Field
KBI	Kirkwood–Buff Integral
LJ	Lennard–Jones
MC	Monte Carlo
MD	Molecular dynamics
MDF	Molecular distribution function
MM	McMillan–Mayer
MW	Molecular weight
NRTL	Non-random two liquid
OSA	Osmotic stress analysis
OZ	Ornstein–Zernike
PF	Partition function
PI	Preferential interaction
PMF	Potential of mean force
PS	Preferential solvation
PY	Percus–Yevick
RDF	Radial distribution function
RK	Redlich–Kister
RISM	Reference interaction-site model
SAFT	Statistical associated-fluid theory
SANS	Small-angle neutron scattering
SAXS	Small-angle X-ray scattering
SI	Symmetric ideal
SPT	Scaled particle theory
TCF	Total correlation function
TCFI	Total correlation function integral
UNIFAC	UNIversal Functional Activity Coefficient
UNIQUAC	UNIversal QUAsiChemical
VDW	van der Waals

1 Fluctuation Solution Theory
A Primer

Paul E. Smith, Enrico Matteoli,
and John P. O'Connell

CONTENTS

Abstract: Fluctuation Theory of Solutions or Fluctuation Solution Theory (FST) combines aspects of statistical mechanics and solution thermodynamics, with an emphasis on the grand canonical ensemble of the former. To understand the most common applications of FST one needs to relate fluctuations observed for a grand canonical system, on which FST is based, to properties of an isothermal–isobaric system, which is the most common type of system studied experimentally. Alternatively, one can invert the whole process to provide experimental information concerning particle number (density) fluctuations, or the local composition, from the available thermodynamic data. In this chapter, we provide the basic background material required to formulate and apply FST to a variety of applications. The major aims of this section are: (i) to provide a brief introduction or recap of the relevant thermodynamics and statistical thermodynamics behind the formulation and primary uses of the Fluctuation Theory of Solutions; (ii) to establish a consistent notation which helps to emphasize the similarities between apparently different applications of FST; and (iii) to provide the working expressions for some of the potential applications of FST.

1.1   BACKGROUND AND THEORY

The Fluctuation Theory of Solutions—also known as *Fluctuation Solution Theory*, *Kirkwood–Buff Theory*, or simply *Fluctuation Theory*—provides an elegant approach relating solution thermodynamics to the underlying molecular distributions or particle number fluctuations. Here, we provide the background material required to develop the basic theory. More details can be found in standard texts on thermodynamics and statistical mechanics (Hill 1956; Münster 1970). Indeed, the experienced reader may skip this chapter completely, or jump to Section 1.2. A list of standard symbols is also provided in the Prolegomenon to aid the reader, and we have attempted to use the same set of symbols and notations in all subsequent chapters. Throughout this work we refer to a collection of species (1, 2, 3,...) in a system of interest. We consider this to represent a primary solvent (1), a solute of interest (2), and a series of additional cosolutes or cosolvents (3, 4,...) which may also be present in the solution. However, other notations such as A/B or u/v is also used in the various chapters. All summations appearing here refer to the set of thermodynamically independent components (n_c) in the mixture unless stated otherwise. Derivatives of the chemical potentials with respect to composition form a central component of the theory. The primary derivative of interest here is defined as

$$\mu_{ij} = \left(\frac{\partial \beta \mu_i}{\partial x_j} \right)_{p,T} \tag{1.1}$$

although the most convenient derivative often depends on the exact application. A general fluctuation in a property X is written as $\delta X = X - \langle X \rangle$, where the angular brackets denote an ensemble or time average in the grand canonical ensemble unless stated otherwise. The Kronecker delta function (δ_{ij}) is used consistently during the theoretical development and is equal to unity when $i = j$, and zero otherwise.

Finally, throughout the various chapters the contributors have generally used the terms *species*, *component*, and *molecule type* interchangeably. This is perfectly acceptable for most applications. However, there are some applications for which a distinction between species and components must be made, and extra care should be taken to avoid confusion in these cases. Examples include the treatment of electrolytes provided in Section 1.3.6 and in Chapter 8, the treatment of solute association in Section 1.3.8, as well as the general treatment of reactive equilibria outlined in Chapter 9.

1.1.1 BASIC THERMODYNAMICS

Let us consider the fundamental equation of Gibbs describing the dependence of the internal energy on entropy, volume, and composition, that is, $U(S,V,\{N\})$ such that

$$dU = TdS - pdV + \sum_i \mu_i dN_i \tag{1.2}$$

for systems in equilibrium with only pV work (Münster 1970). More convenient relationships, which express the above relationship in terms of alternative sets of variables, can be obtained from the above equation by a series of Legendre transformations. Consequently, a series of thermodynamic potentials are obtained with the most common being

$$dH = TdS + Vdp + \sum_i \mu_i dN_i$$

$$dA = -SdT - pdV + \sum_i \mu_i dN_i$$

$$dG = -SdT + Vdp + \sum_i \mu_i dN_i \tag{1.3}$$

$$dpV = SdT + pdV + \sum_i N_i d\mu_i$$

where we have used the usual definitions; $H = U + pV$, $A = U - TS$, and $G = H - TS$. One is then free to choose the most appropriate variables of interest for a specific application. The set of variables of most interest in this work are $\{N\}$, p, and T for which the characteristic function is the Gibbs free energy, and $\{\mu\}$, V, and T for which the characteristic function is pV. The primary aim is to use statistical thermodynamics for an open system to describe the properties of an equivalent closed system that are amenable to experiment.

An alternative formulation of the above expressions can be obtained by treating the entropy in terms of the internal energy, volume, and composition, that is, $S(U,V,\{N\})$. This leads to the entropy formulation of Gibbs' fundamental equation and, while not commonly used in thermodynamics, provides certain advantages in statistic thermodynamics. The following expressions are then obtained:

$$d\beta A = U d\beta - \beta p dV + \sum_i \beta\mu_i dN_i$$

$$d\beta G = H d\beta + \beta V dp + \sum_i \beta\mu_i dN_i \qquad (1.4)$$

$$d\beta pV = -U d\beta + \beta p dV + \sum_i N_i d\beta\mu_i$$

and also provide a series of expressions for thermodynamic properties in terms of partial derivatives. These relationships are particularly useful when investigating changes in temperature (or β).

The above relationships provide expressions for many thermodynamic variables (U, H, S, p, V, etc.) in terms of first derivatives of the thermodynamic potentials. We will see that, in general, fluctuations are related to second derivatives of the thermodynamic potentials. Second derivatives of the Gibbs free energy provide three very important properties of solutions. These are the isothermal compressibility,

$$\kappa_T = -\frac{1}{V}\left(\frac{\partial V}{\partial p}\right)_{T,\{N\}} = -\frac{1}{V_m}\left(\frac{\partial V_m}{\partial p}\right)_T = \frac{1}{\rho}\left(\frac{\partial \rho}{\partial p}\right)_T \qquad (1.5)$$

the isobaric thermal expansion coefficient,

$$\alpha_p = \frac{1}{V}\left(\frac{\partial V}{\partial T}\right)_{p,\{N\}} = \frac{1}{V_m}\left(\frac{\partial V_m}{\partial T}\right)_p = -\frac{1}{\rho}\left(\frac{\partial \rho}{\partial T}\right)_p \qquad (1.6)$$

and the constant pressure molar heat capacity,

$$C_{p,m} = \left(\frac{\partial H_m}{\partial T}\right)_p \qquad (1.7)$$

all of which can be obtained directly from experiment.

The application of Euler's theorem to a series of extensive properties ($X = U$, S, V, H, G, and C_p) of a system at constant T and p indicates that (Davidson 1962)

$$X = \sum_i N_i\left(\frac{\partial X}{\partial N_i}\right)_{p,T,\{N\}'} = \sum_i N_i \bar{X}_i \qquad X_m = \sum_i x_i \bar{X}_i \qquad (1.8)$$

where $\bar{X}_i$ is known as a partial molar quantity. All the extensive quantities depend on T, p, and composition, and hence, one can write the following differential:

$$dX = \left(\frac{\partial X}{\partial p}\right)_{T,\{N\}} dp + \left(\frac{\partial X}{\partial T}\right)_{p,\{N\}} dT + \sum_i \bar{X}_i dN_i \qquad (1.9)$$

Comparing this relationship with the expression for dX obtained from Equation 1.8 then provides a series of relationships, one for each extensive quantity, such that

$$-\left(\frac{\partial X}{\partial p}\right)_{T,\{N\}} dp - \left(\frac{\partial X}{\partial T}\right)_{p,\{N\}} dT + \sum_i N_i d\bar{X}_i = 0 \qquad (1.10)$$

This is known as the generalized Gibbs–Duhem equation. The most common and useful of these expressions involves the Gibbs free energy ($X = G$) for which one obtains the Gibbs–Duhem (GD) equation,

$$-Vdp + SdT + \sum_i N_i d\mu_i = 0 \qquad (1.11)$$

or equivalently,

$$-\beta Vdp - Hd\beta + \sum_i N_i d\beta\mu_i = 0 \qquad (1.12)$$

At constant T and p, the above expressions indicate that changes in the partial molar quantities are not independent. Alternatively, the GD expression can be viewed along the lines of the other thermodynamic potentials where the independent variables are pressure, temperature, and the chemical potentials, and the thermodynamic potential is zero. We note that derivatives of the GD expression with respect to T or p simply generate the expressions provided in Equation 1.8 for $X = S$, V and H, but derivatives with respect to composition at constant T and p provide new relationships, which are extremely useful (see Section 4.3.1 in Chapter 4, for instance).

1.1.2 SOLUTION THERMODYNAMICS

In practice, many of the extensive functions (U, S, H, G), and their corresponding partial molar quantities, can only be determined up to an additive constant—absolute values of U, H, and μ_i can be obtained from simulation, but these then depend on the model of choice. Hence, their values are typically expressed with respect to a set of defined reference or standard states. These are usually taken as the pure solutions of each component at the same T and p. One can then define a series of mixing quantities such that

$$\Delta_{\text{mix}} X_{\text{m}} = X_{\text{m}} - \sum_i x_i X_{\text{m}}^{\circ} = \sum_i x_i (\bar{X}_i - X_{\text{m}}^{\circ}) \qquad (1.13)$$

This concept is also often applied to the volume of the solution even though absolute molar volumes of mixing can be determined quite easily. Other standard states are possible. The most common alternative involves an infinitely dilute solute, also known as the Henry's law standard state.

A series of excess quantities may then be defined using the properties of ideal solutions such that

$$X_m^E = \Delta_{mix} X_m - \Delta_{mix} X_m^{id} = \sum_i x_i \overline{X}_i^E \qquad (1.14)$$

where the excess mixing quantities are expressed in terms of the corresponding excess partial molar quantities. These are the quantities ($X = G, V, H, S$) that are normally available experimentally. The first law properties (U, V, H) can be obtained directly and the corresponding mixing properties are zero for ideal solutions. Properties that relate to the second law (S, G) have to be determined indirectly—from phase equilibria, for example—and their mixing properties are not zero even for ideal solutions. Expressions for the excess partial molar quantities and their derivatives with respect to composition can then be expressed in terms of derivatives of the excess mixing quantities. One finds

$$\overline{X}_i^E = X_m^E + (1 - x_i) \left(\frac{\partial X_m^E}{\partial x_i} \right)_{p,T} \qquad (1.15)$$

and for derivatives of the chemical potentials in binary systems (see Section 4.3 in Chapter 4 for ternary systems),

$$\left(\frac{\partial \mu_i^E}{\partial x_i} \right)_{p,T} = (1 - x_i) \left(\frac{\partial^2 G_m^E}{\partial x_i^2} \right)_{p,T} \qquad (1.16)$$

Obtaining the derivatives on the right-hand side requires a fitting equation for the excess mixing quantities. The Wilson, Redlich–Kister, and nonrandom two liquid (NRTL) model equations are some of the most commonly used (Poling, Praunitz, and O'Connell 2000). Some additional practical considerations are also provided in Section 1.3.9 and Section 4.2 in Chapter 4.

1.1.3 MORE ON CHEMICAL POTENTIALS

Chemical potentials are central for an understanding of material/phase equilibrium and phase stability. FST can be used to study metastable phase and phase instabilities. However, the vast majority of the studies using the FST of solutions involve a single stable phase with multiple components. Here, we are concerned with the relationships among the chemical potentials, and their derivatives, and the local solution distributions. Thermodynamically, from Equations 1.2 and 1.3, we have:

$$\mu_i = \left(\frac{\partial U}{\partial N_i}\right)_{S,V,\{N\}'} = \left(\frac{\partial H}{\partial N_i}\right)_{S,p,\{N\}'} = \left(\frac{\partial A}{\partial N_i}\right)_{T,V,\{N\}'} = \left(\frac{\partial G}{\partial N_i}\right)_{T,p,\{N\}'} = \bar{G}_i \quad (1.17)$$

the latter definition being the most useful for practical applications. Hence, the chemical potential quantifies the change in the Gibbs free energy on addition of a single particle of species i while keeping p, T, and the number of all other species constant. There are additional useful relationships among the partial molar quantities, which can be obtained from derivatives of the chemical potentials. These include

$$\left(\frac{\partial \mu_i}{\partial N_j}\right)_{T,p,\{N\}'} = \left(\frac{\partial \mu_j}{\partial N_i}\right)_{T,p,\{N\}'} \qquad \left(\frac{\partial \mu_i}{\partial p}\right)_{T,\{N\}} = \bar{V}_i$$

$$\left(\frac{\partial \mu_i}{\partial T}\right)_{p,\{N\}} = -\bar{S}_i \qquad \left(\frac{\partial \beta \mu_i}{\partial \beta}\right)_{p,\{N\}} = \bar{H}_i \qquad (1.18)$$

many of which will be used later.

In practice, the relative chemical potential of a species in solution obtained from experiment can be expressed in many closely related forms. The primary differences relate to the choice of the reference (or standard) states, together with the concentration scale adopted for quantifying changes in composition. The most common choice was proposed by Lewis and Randall (LR) and adopts the pure liquids at the same T and p for the reference states and mole fractions for the composition variables (O'Connell and Haile 2005). Consequently, the chemical potentials are expressed in the form

$$\mu_i(p,T,\{x\}) = \mu_i^\circ(p,T) + RT \ln a_i \qquad (1.19)$$

where the Lewis activity, $a_i = \gamma_i x_i$ is the activity of i, and γ_i is known as either the LR, rational, or mole fraction activity coefficient. An ideal solution can then be defined as one in which all the activity coefficients are unity and independent of composition, and where the excess enthalpy and volume of mixing are zero. Hence, the chemical potentials for ideal solutions are described by

$$\mu_i^{id} = \mu_i^\circ(p,T) + RT \ln x_i$$

$$\left(\frac{\partial \beta \mu_i^{id}}{\partial x_i}\right)_{p,T} = \frac{1}{x_i} \qquad (1.20)$$

There are, however, no perfectly ideal solutions—although many systems may approach this behavior. Ideal behavior is observed in the limit of a pure component, where the activity coefficient approaches unity and its composition derivative is then zero. Nevertheless, ideal solutions provide very useful reference points for real solution behavior (see Section 1.3.7).

Unfortunately, the LR scale is not always the most convenient for practical use. In many cases, one would like to be able to use molalities or molarities as concentration variables, and to use different reference states. This does not change the form of Equation 1.19 nor the value of the chemical potential, but it does alter the activities and the activity coefficients. It might seem strange that one can have a variety of activity coefficients. However, we shall see that it is only derivatives of the activities, which enter into fluctuation theory. Hence, the exact choice of reference state is often irrelevant.

The fugacity was defined by G. N. Lewis as a substitute for the chemical potential to more directly relate a component's mixture properties to measurable properties and to avoid the divergence of the chemical potential at the limit of infinite dilution (Lewis 1900a, 1900b, 1901). The definition is an isothermal differential,

$$d\beta\mu_i \equiv d\ln f_i \text{ constant } T \tag{1.21}$$

with the boundary condition for an ideal gas that

$$\lim_{IG} f_i/x_i p = 1 \tag{1.22}$$

For nonideal gases, the fugacity is obtained from

$$f_i = x_i\varphi_i p \tag{1.23}$$

where the fugacity coefficient, φ_i, is obtained from an equation of state. For condensed phases where an equation of state would not be reliable, the Lewis–Randall standard state is used, and the fugacity is expressed as

$$f_i = x_i\gamma_i(T,p,\{x\})f_i^\circ(T,p) \tag{1.24}$$

where f_i° is the fugacity of the component in its pure standard state at the system T and p. It can be found from the chemical potential difference,

$$\beta\mu_i^\circ(T,p) - \beta\mu_i^{IG}(T,p) = \ln\left(f_i^\circ/p\right) \tag{1.25}$$

Finally, composition derivatives of chemical potentials are related to the activity coefficient derivative by

$$\left(\frac{\partial\beta\mu_i}{\partial x_j}\right)_{T,p} = \frac{\delta_{ij} - x_j}{x_j(1-x_j)} + \left(\frac{\partial\ln\gamma_i}{\partial x_j}\right)_{T,p} \tag{1.26}$$

This gives a direct connection of the derivative to G^E models as mentioned above.

In many cases, it is necessary to transform derivatives (of the chemical potential) between different concentration units. This can be achieved by use of the composition variables and volume fractions according to the following relationships:

$$\left(\frac{\partial \ln c_i}{\partial \ln m_j}\right)_{p,T} = \delta_{ij} - \phi_j \qquad \left(\frac{\partial \ln x_i}{\partial \ln m_j}\right)_{p,T} = \delta_{ij} - x_j \qquad \left(\frac{\partial \ln c_i}{\partial \ln x_j}\right)_{p,T} = \frac{\delta_{ij} - \phi_j}{1 - x_j}$$

$$\left(\frac{\partial \ln c_i}{\partial \ln c_j}\right)_{p,T} = \frac{\delta_{ij} - \phi_j}{1 - \phi_j} \qquad \left(\frac{\partial \ln x_i}{\partial \ln x_j}\right)_{p,T} = \frac{\delta_{ij} - x_j}{1 - x_j} \qquad \left(\frac{\partial \ln m_i}{\partial \ln m_j}\right)_{p,T} = \delta_{ij}$$

(1.27)

which apply for solutions with any number of components.

1.1.4 STATISTICAL THERMODYNAMICS

The previous summary provides the basic relationships, derived from the first and second laws, used for the manipulation of available experimental data. However, statistical thermodynamics is then required to develop expressions for the thermodynamic properties in terms of the fluctuating quantities of interest here. First, we will use statistical thermodynamics to provide the characteristic thermodynamic potentials in terms of the appropriate partition function, which will involve a sum over the microscopic states available to the system. Second, we will provide relevant expressions for the fluctuations under one set of variables, which can then be used to rationalize the thermodynamic properties of a system characterized by a different set of variables.

There are four main ensembles in statistical thermodynamics for which the independent variables are NVE (microcanonical), NVT (canonical), NpT (Gibbs or isothermal isobaric), and μVT (grand canonical). The characteristic functions provided in Equations 1.2 and 1.3 can be expressed in terms of a series of partition functions such that (Hill 1956)

$$S(\{N\},V,E) = k_B \ln \Omega(\{N\},V,E)$$

$$A(\{N\},V,T) = -k_B T \ln Q(\{N\},V,T) \qquad Q(\{N\},V,T) = \sum_j e^{-\beta E_j}$$

$$G(\{N\},p,T) = -k_B T \ln \Delta(\{N\},p,T) \qquad \Delta(\{N\},p,T) = \sum_V \sum_j e^{-\beta E_j} e^{-\beta pV} \quad (1.28)$$

$$pV(\{\mu\},V,T) = k_B T \ln \Xi(\{\mu\},V,T) \qquad \Xi(\{\mu\},V,T) = \sum_{\{N\}} \sum_j e^{-\beta E_j} e^{\beta \mu \cdot N}$$

where we have used the shorthand $\mu \cdot N = \mu_1 N_1 + \mu_2 N_2 \cdots$ for simplicity, and the sums are over all members of the ensemble with different energies $E_j(\{N\},V)$, volumes, and number of molecules. Our focus will be the grand canonical ensemble, where we shall develop expressions for the fluctuations, and the isothermal isobaric ensemble, which is representative of most experimental conditions. We note that one does not have to assume classical behavior for FST to be valid.

Using Equation 1.4 and Equation 1.28 it can be shown that the volume and enthalpy in the Gibbs ensemble are given by

$$V = \left(\frac{\partial G}{\partial p}\right)_{\{N\},T} = -k_B T \left(\frac{\partial \ln \Delta}{\partial p}\right)_{\{N\},T} = \frac{1}{\Delta}\sum_V\sum_j V e^{-\beta E_j} e^{-\beta p V} = \langle V \rangle \quad (1.29)$$

$$H = \left(\frac{\partial \beta G}{\partial \beta}\right)_{\{N\},p} = -\left(\frac{\partial \ln \Delta}{\partial \beta}\right)_{\{N\},p} = \frac{1}{\Delta}\sum_V\sum_j (E_j + pV) e^{-\beta E_j} e^{-\beta p V} = \langle E \rangle + p\langle V \rangle$$

Furthermore, using the above expressions in Equation 1.5 through Equation 1.7, and Equation 1.29, and then evaluating the derivatives leads to expressions for the compressibility, thermal expansion, and heat capacity. The results expressed in terms of fluctuations in the isothermal–isobaric ensemble are

$$\kappa_T = \beta \frac{\langle \delta V \delta V \rangle}{\langle V \rangle}$$

$$\alpha_p = \frac{1}{k_B T^2} \frac{\langle \delta H \delta V \rangle}{\langle V \rangle} \quad (1.30)$$

$$C_{p,m} = \frac{1}{k_B T^2} \frac{\langle \delta H \delta H \rangle}{N}$$

It is important to note that the above formulas represent fluctuations ($\delta X = X - \langle X \rangle$) in the properties of the whole system, that is, bulk fluctuations. They are useful expressions but provide no information concerning fluctuations in the local vicinity of atoms or molecules. These latter quantities will prove to be most useful and informative. One can also derive expressions for partial molar quantities by taking appropriate first (to give the chemical potential) and second (to give partial molar volume and enthalpy) derivatives of the expressions presented in Equation 1.28. However, these do not typically lead to useful simple formulas that can be applied directly to theory or simulation. For instance, while it is straightforward to calculate the compressibility, thermal expansion, and heat capacity from simulation, the determination of chemical potentials is much more involved (especially for large molecules and high densities).

Turning now to the grand canonical ensemble, one finds the following expressions for the number of particles of each species and the internal energy,

$$N_i = \left(\frac{\partial \beta p V}{\partial \beta \mu_i}\right)_{\{\beta\mu\}',V,\beta} = \left(\frac{\partial \ln \Xi}{\partial \beta \mu_i}\right)_{\{\beta\mu\}',V,\beta} = \frac{1}{\Xi}\sum_{\{N\}}\sum_j N_i e^{-\beta E_j} e^{\beta\mu\cdot N} = \langle N_i \rangle$$

$$\hspace{8cm} (1.31)$$

$$U = -\left(\frac{\partial \beta p V}{\partial \beta}\right)_{\{\beta\mu\},V} = -\left(\frac{\partial \ln \Xi}{\partial \beta}\right)_{\{\beta\mu\},V} = \frac{1}{\Xi}\sum_{\{N\}}\sum_j E_j e^{-\beta E_j} e^{\beta\mu\cdot N} = \langle E \rangle$$

Second derivatives in the grand canonical ensemble play a central role in FST. These can be obtained on taking derivatives of the two above expressions to give

$$\left(\frac{\partial \langle N_i \rangle}{\partial \beta \mu_j}\right)_{\{\beta\mu\}',V,\beta} = \langle \delta N_i \delta N_j \rangle \hspace{2cm} \left(\frac{\partial \langle E \rangle}{\partial \beta}\right)_{\{\beta\mu\},V} = -\langle \delta E \delta E \rangle$$

$$\left(\frac{\partial \langle E \rangle}{\partial \beta \mu_i}\right)_{\{\beta\mu\}',V,\beta} = -\left(\frac{\partial \langle N_i \rangle}{\partial \beta}\right)_{\{\beta\mu\},V} = \langle \delta E \delta N_i \rangle \hspace{2cm} (1.32)$$

$$\left(\frac{\partial \langle E \rangle}{\partial V}\right)_{\{\beta\mu\},\beta} = \frac{\langle E \rangle}{V} = \frac{U}{V} \hspace{2cm} \left(\frac{\partial \langle N_i \rangle}{\partial V}\right)_{\{\beta\mu\},\beta} = \frac{\langle N_i \rangle}{V} = \rho_i$$

The first of these expressions involves particle–particle fluctuations and will be the major properties of interest in the most common uses of FST.

Before leaving this section it is important to note that the above expressions can be used to describe the properties of a collection of systems characterizing the grand canonical ensemble, in which case they correspond to bulk properties. Alternatively, they can be used to describe small regions of any bulk system in any ensemble at constant *T*, in which case they represent local fluctuations within the much larger bulk system. This is the key aspect to the large number of applications of FST.

1.1.5   DISTRIBUTION FUNCTIONS AND KIRKWOOD–BUFF INTEGRALS

The previous expressions involve particle number (and energy) fluctuations. It is more common, and totally equivalent, to use correlation/distribution functions to replace the number fluctuations. In many cases this can help to clarify the significance of the number fluctuations (correlations) as we indicate in this section. However, in doing so one has to remember that these distributions correspond to a system volume that is open to all species.

In the grand canonical ensemble, the probability that any N_1 molecules of species 1, and N_2 molecules of species 2, and so forth, are within $d\{r\}$ at $\{r\}$ is given by $\rho^{(n)}\{(r)\}d\{r\}$ where

$$\int \rho^{(n)}(\{r\})d\{r\} = \left\langle \prod_s \frac{N_s!}{(N_s - n_s)!} \right\rangle \hspace{2cm} (1.33)$$

the integral is over all space and the product is over all types of species (s) in the mixture (Hill 1956). Here, n_s is the number of molecules of type s in the n particle probability distribution. Hence, we can evaluate the following integrals related to the singlet and doublet distributions,

$$\int \rho_i^{(1)}(r_1)dr_1 = \langle N_i \rangle$$

$$\iint \rho_{ij}^{(2)}(r_1,r_2)dr_1 dr_2 = \langle N_i N_j \rangle - \delta_{ij}\langle N_i \rangle$$

(1.34)

Various combinations of the above integrals are commonly encountered in statistical thermodynamic theories of solutions. The most relevant is given by,

$$\iint \left[\rho_{ij}^{(2)}(r_1,r_2) - \rho_i^{(1)}(r_1)\rho_j^{(1)}(r_2) \right]dr_1 dr_2 = \langle \delta N_i \delta N_j \rangle - \delta_{ij}\langle N_i \rangle \qquad (1.35)$$

and is identical in form to the integrals appearing in the theory of imperfect gases and the McMillan–Mayer theory of solutions (McMillan and Mayer 1945).

A set of grand canonical distribution functions $g_{ij}^{(n)}$ can then be defined for species i and j by,

$$\rho_i^{(1)}(r_1) = c_i g_i^{(1)}(r_1) = c_i \qquad \rho_{ij}^{(2)}(r_1,r_2) = c_i c_j g_{ij}^{(2)}(r_1,r_2) \qquad (1.36)$$

which then provide expressions for integrals over the pair ($n = 2$) distribution functions,

$$G_{ij} = G_{ji} = 4\pi \int_0^\infty \left[g_{ij}^{(2)}(r) - 1 \right]r^2 dr = V \frac{\langle \delta N_i \delta N_j \rangle}{\langle N_i \rangle \langle N_j \rangle} - \frac{\delta_{ij}}{c_i} \qquad (1.37)$$

where we have integrated over the position of the central particle and only consider the scalar interparticle distance, $r = |r_2 - r_1|$. At this point we have related the particle number fluctuations to integrals over radial distribution functions (RDFs) in the grand canonical ensemble. FST does not require information on the angular distributions for pairs of molecules—these are averaged out in the above expressions (see Chapter 6). The RDFs correspond to distributions obtained in a solution at the composition of interest, after averaging over all the remaining molecular degrees of freedom. A typical RDF and the corresponding integral are displayed in Figure 1.1. The G_{ij}s are known as Kirkwood–Buff integrals (KBIs) and are the central components of FST (Kirkwood and Buff 1951).

The KBIs quantify the average deviation, from a random distribution, in the distribution of j molecules surrounding a central i molecule summed over all space. In this respect they are more informative than the particle number fluctuations as they can then be decomposed and interpreted in terms of spatial contributions—using computer simulation data, for example. They clearly resemble the integrals

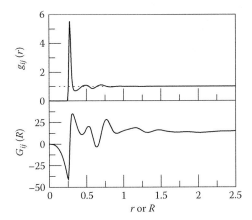

FIGURE 1.1 A typical center of mass-based pair radial distribution function as a function of molecule separation, r (in nm) (top), and the corresponding Kirkwood–Buff integral (in cm³/mol) as a function of integration distance, R (in nm) (bottom).

one encounters in the study of imperfect gases and the McMillan–Mayer theory of solutions. However, the FST approach does not suffer from convergence limitations in typical series expansions, and the KBIs are relevant at any solution composition—not just infinitely dilute solutes in the gas phase or a primary solvent.

Multiplying the KBIs by number densities (concentrations) one obtains an alternative picture of the integrals. The quantities, $N_{ij} = \rho_j G_{ij}$, have been referred to as *excess coordination numbers* and quantify the change in the number of j molecules observed in an open volume of solution on introduction of a central i molecule from that observed for the same volume of bulk solution in the absence of the i molecule (Newman 1994).

Multiple combinations of KBIs and number densities have appeared in the literature. Some of the more common variations include,

$$B_{ij} = B_{ji} = \rho_i \left(\delta_{ij} + \rho_j G_{ij} \right) = \frac{\langle \delta N_i \delta N_j \rangle}{V}$$

$$\delta_{ij} + N_{ij} = \frac{\langle \delta N_i \delta N_j \rangle}{\langle N_i \rangle}$$

(1.38)

The most convenient form often depends upon the application and/or the derivation used. In many applications, total correlation function integrals (TCFIs) are often used in place of the KBIs (O'Connell 1971b). These are typically defined and denoted as $H_{ij} = \rho G_{ij} = x_j^{-1} N_{ij}$, but care should always be taken to ensure other closely related definitions are not being used.

Finally, we note that particle number fluctuations increase, and many properties diverge, as one approaches a critical point. O'Connell and coworkers have shown that one can still apply FST under these circumstances by using integrals over direct correlation functions, and not integrals over the total correlation function as adopted in the standard KBIs (see Section 1.2.3 and Chapter 9). The direct correlation function integrals (DCFIs) are related to the direct correlation function (DCF) by

$$C_{ij} = 4\pi\rho \int_0^\infty c_{ij}(r)r^2 dr \qquad (1.39)$$

and are defined by the relationship between the TCFIs and the DCFIs provided by the matrix equation,

$$\mathbf{I} - \mathbf{C} = (\mathbf{I} + \mathbf{H})^{-1} \qquad (1.40)$$

The advantages of the direct correlation function approach include the ability to treat systems close to critical points, together with a general simplification of the expressions such as those in Section 1.2.3. The disadvantages are that the DCF or DCFI cannot be calculated directly from simulation, only through the defining expression in Equation 1.40, and their molecular interpretation is less clear compared to the KBIs.

1.2   FLUCTUATION THEORY OF SOLUTIONS

The previous background material covered many aspects of thermodynamics, statistical thermodynamics, and solution thermodynamics. At this point, we have all we need to derive the main expressions provided by FST. There are many derivations of the principal expressions in the literature, including a matrix approach that is general for any number of components, all of which involve a series of thermodynamic manipulations (Kirkwood and Buff 1951; Hall 1971; O'Connell 1971b; Valdeavella, Perkyns, and Pettitt 1994; Ben-Naim 2006; Kang and Smith 2008; Nichols, Moore, and Wheeler 2009). We will not use that type of approach here as our primary concern is binary and ternary solutions for which a more transparent and, in our opinion, simpler approach is available.

1.2.1   GENERAL EXPRESSIONS

The basic idea is to develop expressions for common thermodynamic quantities in terms of fluctuations in a system open to all species. The key lies in the fact that the fluctuating quantities characteristic of the grand canonical ensemble can then be transformed into expressions, which provide properties representative of the isothermal isobaric ensemble. Using an equivalence of ensembles argument, one can then consider these fluctuations to represent properties of small microscopic local regions of the solution of interest. This approach can be used to understand many properties of isobaric, isochoric, or osmotic systems in terms of particle number (and energy) fluctuations.

The average number of particles of species i in the grand canonical ensemble is a function of the temperature, volume, and set of chemical potentials. Consequently, one can write the following differential,

$$d\langle N_i\rangle = \left(\frac{\partial\langle N_i\rangle}{\partial\beta}\right)_{V,\{\beta\mu\}} d\beta + \left(\frac{\partial\langle N_i\rangle}{\partial V}\right)_{\beta,\{\beta\mu\}} dV + \sum_j \left(\frac{\partial\langle N_i\rangle}{\partial\beta\mu_j}\right)_{\beta,V,\{\beta\mu\}'} d\beta\mu_j \quad (1.41)$$

valid for each i species and any number of solution components. All the partial derivatives can be expressed in terms of ensemble averages in the grand canonical ensemble. Using the expressions provided in Equations 1.32, the above differential simplifies to

$$d\langle N_i\rangle = -\langle\delta N_i\delta E\rangle d\beta + \frac{\langle N_i\rangle}{V} dV + \sum_j \langle\delta N_i\delta N_j\rangle d\beta\mu_j \quad (1.42)$$

Derivatives of the above differential can be used to develop expressions valid for a variety of ensembles. Some applications where changes in T are important are provided in the literature (Buff and Brout 1955; Debenedetti 1988; Jiao and Smith 2011; Ploetz and Smith 2011b). However, the vast majority of applications of FST are restricted to isothermal systems. Hence, the first term on the right-hand side of Equation 1.42 disappears and the above expression can be rewritten, after dividing throughout by $\langle N_i\rangle$ and rearranging as

$$d\ln\rho_i = \sum_j (\delta_{ij} + N_{ij})d\beta\mu_j \quad (1.43)$$

This set of simultaneous equations, coupled with the Gibbs–Duhem equation at constant T, are all we need to derive the most common forms of the basic theory providing a range of relationships between fluctuating quantities and thermodynamic properties for multicomponent systems.

In the original Kirkwood and Buff paper (Kirkwood and Buff 1951), the starting point for their derivation is the **A** matrix (see also the Prolegomenon) which has elements of the form $V(\partial\beta\mu_i/\partial N_j)_{T,V,\{N\}'}$. When used in combination with the **B** matrix, with elements B_{ij} given by Equation 1.38, one finds the matrix relationship $\mathbf{I} = \mathbf{A}\,\mathbf{B}$, which is obtained directly from Equation 1.43 by taking derivatives with respect to particle numbers with volume and T constant. Others have adopted similar approaches (Ben-Naim 2006). This is usually followed by a series of thermodynamic transformations to convert the isochoric chemical potential derivatives to provide the more common and useful isobaric expressions. Here, we wish to eliminate the majority of these transformations. In fact, the results for a small number of components can be obtained directly from Equation 1.43, as we shall see in the next section.

Before leaving this section, we note that Equation 1.43 also provides several additional relationships that are of general use and/or might be more convenient for specific applications (Smith and Mazo 2008). An expression for changes in the species molality ($m_i = \rho_i/\rho_1$) can be obtained by subtracting the differential for the primary solvent ($i = 1$) to give

$$d\ln m_i = \sum_j (\delta_{ij} + N_{ij} - \delta_{1j} - N_{1j})d\beta\mu_j \quad (1.44)$$

This can, however, cause problems if the solvent concentration disappears. A similar differential for changes in the mole fractions can also be developed by noting that $d \ln x_i = d \ln \rho_i - \sum_k x_k d \ln \rho_k$ and therefore,

$$d \ln x_i = \sum_j \left[\delta_{ij} + N_{ij} - \sum_k x_k (\delta_{kj} + N_{kj}) \right] d\beta\mu_j \qquad (1.45)$$

The three previous differentials are valid under isothermal conditions. Taking the derivative of Equation 1.43 with respect to p with composition and T constant one finds,

$$k_B T \kappa_T = \sum_j (\delta_{ij} + N_{ij}) \bar{V}_j \qquad (1.46)$$

for any i species. Alternatively, taking derivatives with respect to composition with p and T constant provides,

$$\delta_{ik} - \rho_i \bar{V}_k = \sum_j (\delta_{ij} + N_{ij}) \left(\frac{\partial \beta \mu_j}{\partial \ln N_k} \right)_{T,p,\{N\}'} \qquad (1.47)$$

The last two expressions involve a mixture of fluctuating quantities and thermodynamic properties. The thermodynamic properties on the right-hand side will ultimately be expressed in terms of fluctuating quantities. However, the above mixed expressions are often extremely useful in themselves.

If desired, the relationships in Equation 1.43 through Equation 1.45 can be further manipulated by eliminating one of the chemical potentials (usually the primary solvent) using the GD expression at constant T and p. The results for components other than the solvent are then given by

$$d \ln \rho_i = \sum_{j>1} (\delta_{ij} + N_{ij} - m_j N_{i1}) d\beta\mu_j$$

$$d \ln m_i = \sum_{j>1} (\delta_{ij} + N_{ij}^+) d\beta\mu_j \qquad (1.48)$$

where $N_{ij}^+ = N_{ij} + m_j (1 + N_{11} - N_{i1} - N_{j1})$ has been used as a shorthand (Hall 1971), and represents essentially the matrix elements used in the original paper (Kirkwood and Buff 1951). The expressions are now only valid under constant T and p conditions. The mole fraction differential is significantly more complicated and is rarely used (Smith and Mazo 2008).

1.2.2 FLUCTUATION THEORY OF BINARY SOLUTIONS

Isothermal–isobaric binary solutions are by far the most common system of interest for FST. Since there are only two components, the number of simultaneous equations and the number terms in these equations are minimal, which means that matrices are unnecessary. Here, we derive the FST-based equations for binary mixtures as a simple illustration of the general approach. If one takes the derivative of Equation 1.48 ($i = 2$) with respect to the natural log of the number density of component 2 with p and T constant, then one immediately finds

$$\left(\frac{\partial \beta \mu_2}{\partial \ln \rho_2} \right)_{T,p} = \frac{1}{1 + N_{22} - N_{12}} \tag{1.49}$$

Hence, the derivative of the solute chemical potential (or activity) with respect to solute concentration can be expressed in terms of a combination of number densities and particle number fluctuations or KBIs. The ability to express thermodynamic properties in terms of KBIs is the major strength of FST. This has been achieved without approximation and the relationship holds for any stable binary solution at any composition involving any type of components. Derivatives of other chemical potentials can be obtained by application of the GD equation, or by a simple interchange of indices. The same approach can be applied to the second expression in Equation 1.48, with a subsequent application of Equation 1.27, to provide chemical potential derivatives with respect to other concentration scales,

$$\left(\frac{\partial \beta \mu_2}{\partial \ln x_2} \right)_{T,p} = \frac{\rho}{\rho_1 (1 + N_{22} - N_{12}) + \rho_2 (1 + N_{11} - N_{21})}$$

$$\left(\frac{\partial \beta \mu_2}{\partial \ln N_2} \right)_{T,p,N_1} = \left(\frac{\partial \beta \mu_2}{\partial \ln m_2} \right)_{T,p} = \frac{\rho_1}{\rho_1 (1 + N_{22} - N_{12}) + \rho_2 (1 + N_{11} - N_{21})} \tag{1.50}$$

The most appropriate derivative of interest will depend on the ultimate application. Cabezas and O'Connell (1993) show connections among derivatives with different properties held constant.

After expressions for the chemical potential derivatives have been obtained, one can use them to determine corresponding expressions for the partial molar volumes and isothermal compressibility. Using Equation 1.50 in Equation 1.47 with $i = k = 2$ followed by some rearrangement using the GD expression provides

$$\bar{V}_2 = \frac{1 + N_{11} - N_{21}}{\rho_1 (1 + N_{22} - N_{12}) + \rho_2 (1 + N_{11} - N_{21})} \tag{1.51}$$

The solvent partial molar volume is then provided by a simple index change. Using both partial molar volumes in Equation 1.46 and rearranging provides

$$k_B T \kappa_T = \frac{1 + N_{11} + N_{22} + N_{11}N_{22} - N_{12}N_{21}}{\rho_1 (1 + N_{22} - N_{12}) + \rho_2 (1 + N_{11} - N_{21})} \qquad (1.52)$$

which is the FST expression for the isothermal compressibility of a binary solution. This completes the application of FST to binary solutions. Equation 1.49 through Equation 1.50, and Equations 1.51 and 1.52 involve three thermodynamic properties of a binary mixture expressed in terms of composition and three KBIs (G_{11}, G_{22}, $G_{12} = G_{21}$).

1.2.3 GENERAL MATRIX FORMULATION OF FLUCTUATION THEORY

The number of simultaneous equations provided by Equation 1.43 increases as the number of components increases. It then becomes more efficient to express the results in terms of matrices. Again, there are many equivalent matrix formulations that have been presented (Kirkwood and Buff 1951; O'Connell 1971b; Ben-Naim 2006; Nichols, Moore, and Wheeler 2009). Here, we present one of the simplest. A general formulation is easiest starting from the first expression in Equation 1.44. Writing the number fluctuations in matrix form for an n_c component system where we also include the number densities (GD expression at constant T) in the first row provides,

$$\mathbf{M} = \begin{pmatrix} \rho_1 & \rho_2 & \cdots & \rho_{n_c} \\ N_{21} - (1 + N_{11}) & 1 + N_{22} - N_{12} & \cdots & N_{2n_c} - N_{1n_c} \\ \vdots & \vdots & \ddots & \vdots \\ N_{n_c 1} - (1 + N_{11}) & N_{n_c 2} - N_{12} & \cdots & 1 + N_{n_c n_c} - N_{1n_c} \end{pmatrix} \qquad (1.53)$$

By taking derivatives of the GD expression with respect to pressure with all molalities (and T) constant, together with a series of derivatives of Equation 1.44 with respect to each species molality at constant p (and T), the results can be expressed in the general form,

$$\begin{pmatrix} \bar{V}_1 & \mu'_{12} & \cdots & \mu'_{1n} \\ \bar{V}_2 & \mu'_{22} & \cdots & \mu'_{2n} \\ \vdots & \vdots & \ddots & \vdots \\ \bar{V}_n & \mu'_{n2} & \cdots & \mu'_{nn} \end{pmatrix} = \mathbf{M}^{-1} \qquad \bar{V}_i = \frac{\mathbf{M}^{i1}}{|\mathbf{M}|} \qquad \mu'_{ij>1} = \left(\frac{\partial \beta \mu_i}{\partial \ln m_j} \right)_{T,p,\{m\}'} = \frac{\mathbf{M}^{ij}}{|\mathbf{M}|} \qquad (1.54)$$

that is, the expressions or values for the partial molar volumes and activity derivatives are just the elements of the inverse of the $\mathbf{M}$ matrix, which can also be expressed in terms of cofactors of the $\mathbf{M}$ matrix. A simple expression for the compressibility is then

$$k_B T \kappa_T = \frac{|\mathbf{NN}|}{|\mathbf{M}|} \qquad (1.55)$$

where **NN** is the matrix of $\delta_{ij} + N_{ij}$ elements. At this point, we have a matrix formulation for the chemical potential derivatives, partial molar volumes, and isothermal compressibility. This is the theory of Kirkwood and Buff.

A general formulation of KB/FST theory in terms of direct correlation functions is also possible. Because the DCFIs are related to TCFIs through a matrix inversion, their general multicomponent formulas do not involve matrix cofactors (O'Connell 1971b). The equivalents of Equations 1.54 and 1.55, which apply over the entire composition range are then

$$1/\rho k_B T \kappa_T = \sum_{i=1} \sum_{j=1} x_i x_j \left(1 - C_{ij}\right) \tag{1.56}$$

$$\bar{V}_i / k_B T \kappa_T = \sum_{j=1} x_j \left(1 - C_{ij}\right) \tag{1.57}$$

$$\frac{N_i}{\rho k_B T \kappa_T} \left(\frac{\partial \beta \mu_i}{\partial N_j}\right)_{T,V,N_{k \neq j}} = \delta_{ij} - C_{ij} \tag{1.58}$$

and

$$\frac{N_i}{\rho k_B T \kappa_T} \left(\frac{\partial \beta \mu_i}{\partial N_j}\right)_{T,p,N_{k \neq j}} = \delta_{ij} - C_{ij} - \frac{\left[\sum_{k=1} x_k \left(1 - C_{ik}\right)\right]\left[\sum_{k=1} x_k \left(1 - C_{jk}\right)\right]}{\left[\sum_{m=1} \sum_{n=1} x_m x_n \left(1 - C_{mn}\right)\right]} \tag{1.59}$$

for any number of components.

1.2.4 INVERSION OF FLUCTUATION THEORY

The KB/FST inversion procedure is the process of obtaining expressions for the particle number fluctuations or KBIs in terms of experimentally available (isothermal–isobaric) data. Again, there are multiple approaches to the inversion procedure (Ben-Naim 1977; O'Connell 1994; Smith 2008). Arguably, the simplest approach involves the pseudo chemical potential and partial molar volumes (Ben-Naim 2006). First, we note that combining Equations 1.46 and 1.47 provides

$$\delta_{ik} - \rho_i V_k^* = \sum_j (\delta_{ij} + N_{ji})(\delta_{kj} + \mu_{kj}^*)$$

$$V_k^* = \bar{V}_k - k_B T \kappa_T \tag{1.60}$$

$$\left(\frac{\partial \beta \mu_k}{\partial \ln N_j}\right)_{p,T,\{N\}'} = \mu_{kj}^* + \delta_{kj} - \phi_j$$

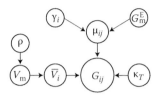

FIGURE 1.2 The Kirkwood–Buff inversion approach for obtaining KBIs from the available experimental data. See Prolegomenon for symbol definitions. (Ploetz, E. A., and Smith, P. E., 2011, Local Fluctuations in Solution Mixtures, *Journal of Chemical Physics*, 4, 135. With permission of the American Institute of Physics.)

The above equations can then be expressed in matrix form,

$$
\begin{pmatrix}
1+N_{11} & \cdots & N_{n1} \\
\vdots & \ddots & \vdots \\
N_{1n} & \cdots & 1+N_{nn}
\end{pmatrix}
\begin{pmatrix}
1+\mu^{*}_{11} & \cdots & \mu^{*}_{n1} \\
\vdots & \ddots & \vdots \\
\mu^{*}_{1n} & \cdots & 1+\mu^{*}_{nn}
\end{pmatrix}
=
\begin{pmatrix}
1-\rho_{1}V^{*}_{1} & \cdots & -\rho_{1}V^{*}_{n} \\
\vdots & \ddots & \vdots \\
-\rho_{n}V^{*}_{1} & \cdots & 1-\rho_{n}V^{*}_{n}
\end{pmatrix}
\tag{1.61}
$$

which provides a simple matrix expression for the elements of the transposed **NN** matrix (and thereby the KBIs) in terms of a matrix containing just volumes and a matrix containing just chemical potentials, both of which can be expressed in terms of experimental data.

Alternatively, one can simply use the matrix relation between the **A** and **B** matrices, and a thermodynamic relation between the isochoric and isobaric chemical potential derivatives to provide the elements of the **A** matrix. The relevant expressions are

$$
B_{ij} = \frac{\mathbf{A}^{ij}}{|\mathbf{A}|}
$$

$$
A_{ij} = V\left(\frac{\partial \beta \mu_i}{\partial N_j}\right)_{T,p,\{N\}'} + \frac{\bar{V}_i \bar{V}_j}{k_B T \kappa_T}
\tag{1.62}
$$

Explicit expressions for up to four components have also been provided (Kang and Smith 2008). The overall inversion approach is summarized schematically in Figure 1.2.

1.2.5 SUMMARY OF FUNDAMENTAL RELATIONS

At this point it is worth pausing and reminding ourselves what we have accomplished here. The above expressions correspond to thermodynamic properties of solutions that are amenable to experiment. The expressions involve fluctuating quantities for an equivalent open system in which the average particle numbers are equal to the fixed number of particles used in the experiment. The same argument holds for the chemical potentials, except in the opposite direction, and the volume and the pressure, that is, we are invoking the equivalence of ensembles approach. In addition, the open regions can be thought of as local finite regions in an essentially infinite bulk solvent. Hence, the approach provides information on local structure and fluctuations within the system. These are different from the bulk fluctuations characteristic of isothermal isobaric systems. The role of FST can be summarized schematically as

$$\left\{ \left\langle \delta N_i \delta N_j \right\rangle \right\} \qquad \leftrightarrow \qquad \left\{ \mu_{ij} \right\}, \left\{ \bar{V}_i \right\}, \kappa_T \tag{1.63}$$

where FST provides the link, in either direction, between local fluctuating quantities and global thermodynamic properties.

1.3   APPLICATIONS OF FLUCTUATION THEORY

In the previous section we outlined the basic equations of FST or KB theory. A general matrix formulation was provided for several thermodynamic properties of solutions in terms of either particle number fluctuations and/or integrals over pair radial distribution functions. While the matrix formulation has the advantage of being totally general, it is also less transparent and the underlying relationships between the integrals are often hidden. Furthermore, it is often informative to examine the expanded expressions for many applications, especially as one or more species approach infinite dilution. In this section we explore many of these relationships and present the fundamental expressions behind a series of applications of the theory. Many practical examples of these applications are found in the accompanying chapters.

1.3.1   PURE LIQUIDS

The application of FST to pure liquids is rather trivial. However, we include it here for the sake of completeness. We find from Equation 1.46 that for a single component (1),

$$\rho_1^o k_B T \kappa_{T,1}^o = 1 + N_{11}^o$$

$$\bar{V}_1^o = V_m^o = \frac{1}{\rho_1^o} \tag{1.64}$$

which is just the well-known compressibility equation. The inverse of the first expression provides the bulk modulus. In situations where the solution approaches a critical point, the number fluctuations become macroscopic in nature and the compressibility diverges. In contrast, for the near-critical region, the DCFI is convergent since,

$$\lim_{\text{critical}} \left(1 - \rho_1^o \right) = \lim_{\text{critical}} \left(\rho_1^o k_B T \kappa_{T,1}^o \right)^{-1} = 0 \tag{1.65}$$

for any system.

1.3.2   CLOSED BINARY SYSTEMS

Binary systems have been the most common system of interest when applying FST. Most investigations of closed binary systems have involved the determination and examination of the three KBIs, possibly followed by some description of local composition or preferential solvation (see Section 1.3.4), all of which is composition dependent. The equations provided in the previous section are more commonly expressed and simplified by the definition of two additional variables (Ben-Naim 1977),

$$\eta_{12} = |\mathbf{M}| = \rho_1(1 + N_{22} - N_{12}) + \rho_2(1 + N_{11} - N_{21})$$

$$\zeta_2 = |\mathbf{NN}| = (1 + N_{11})(1 + N_{22}) - N_{12}N_{21} \tag{1.66}$$

If we consider a binary mixture of a solvent (1) and a solute (2), then FST provides the following expressions,

$$\left(\frac{\partial \beta \mu_2}{\partial \ln m_2}\right)_{T,p} = \frac{\rho_1}{\eta_{12}} \qquad \left(\frac{\partial \beta \mu_2}{\partial \ln x_2}\right)_{T,p} = \frac{\rho}{\eta_{12}} \qquad \left(\frac{\partial \beta \mu_2}{\partial \ln \rho_2}\right)_{T,p} = \frac{1}{1 + N_{22} - N_{12}}$$

$$\bar{V}_1 = \frac{1 + N_{22} - N_{12}}{\eta_{12}} \qquad \bar{V}_2 = \frac{1 + N_{11} - N_{21}}{\eta_{12}} \qquad k_B T \kappa_T = \frac{\zeta_2}{\eta_{12}} \tag{1.67}$$

All the quantities on the left-hand side of the expressions are second derivatives of the Gibbs free energy, while all the quantities on the right-hand side involve second derivatives of pV—the characteristic thermodynamic potential of the grand canonical ensemble. Application of the stability conditions for stable (miscible) solutions indicates that we must have $\eta_{12} > 0$ and $\zeta_2 > 0$ (Prigogine and Dufay 1954).

The above expressions can be manipulated further to provide relationships for the various activity coefficients,

$$\left(\frac{\partial \ln \gamma_2}{\partial \ln x_2}\right)_{T,p} = -\frac{\rho_1(N_{22} - N_{12}) + \rho_2(N_{11} - N_{21})}{\eta_{12}}$$

$$\left(\frac{\partial \ln \gamma_2^m}{\partial \ln m_2}\right)_{T,p} = -\frac{\rho_1(N_{22} - N_{12}) + \rho_2(1 + N_{11} - N_{21})}{\eta_{12}} \tag{1.68}$$

$$\left(\frac{\partial \ln \gamma_2^c}{\partial \ln \rho_2}\right)_{T,p} = -\frac{N_{22} - N_{12}}{1 + N_{22} - N_{12}}$$

which are often of more practical interest. It is well known that the activity coefficient provides an indication of deviations from ideal behavior. Indeed, for dilute solute standard states on the molarity scale, a positive deviation from unity suggests a dominance of favorable solute–solvent interactions ($G_{12} > G_{22}$), while negative deviations suggest a significant contribution from favorable solute–solute associations $(G_{22} > G_{12})$. FST provides a way to quantify these effects.

In many cases, the solute may appear at low concentrations. Limiting values of the above expressions are then

$$\left(\frac{\partial \beta \mu_1}{\partial m_2}\right)_{T,p}^{\infty} = -1 \qquad \left(\frac{\partial \beta \mu_1}{\partial x_2}\right)_{T,p}^{\infty} = -1 \qquad \left(\frac{\partial \beta \mu_1}{\partial \rho_2}\right)_{T,p}^{\infty} = -\frac{1}{\rho_1^0}$$

$$\left(\frac{\partial \beta \mu_2}{\partial m_2}\right)_{T,p}^{\infty} = \frac{1}{m_2} \qquad \left(\frac{\partial \beta \mu_2}{\partial x_2}\right)_{T,p}^{\infty} = \frac{1}{x_2} \qquad \left(\frac{\partial \beta \mu_2}{\partial \rho_2}\right)_{T,p}^{\infty} = \frac{1}{\rho_2} \qquad (1.69)$$

$$\bar{V}_1^{\infty} = \frac{1}{\rho_1^0} \qquad \bar{V}_2^{\infty} = \frac{1 + N_{11}^0 - N_{21}^{\infty}}{\rho_1^0} = k_B T \kappa_{T,1}^0 - G_{12}^{\infty}$$

$$k_B T \kappa_{T,1}^0 = \frac{1 + N_{11}^0}{\rho_1^0}$$

with

$$\left(\frac{\partial \ln \gamma_2}{\partial m_2}\right)_{T,p}^{\infty} = -\left[1 + \rho_1^0 \left(G_{11}^0 + G_{22}^{\infty} - 2G_{12}^{\infty}\right)\right]$$

$$\left(\frac{\partial \ln f_2}{\partial x_2}\right)_{T,p}^{\infty} = -\rho_1^0 \left(G_{11}^0 + G_{22}^{\infty} - 2G_{12}^{\infty}\right) \qquad (1.70)$$

$$\left(\frac{\partial \ln y_2}{\partial \rho_2}\right)_{T,p}^{\infty} = -(G_{22}^{\infty} - G_{12}^{\infty})$$

and can be used to provide information concerning solute–solute and solute–solvent affinity at infinite dilution (Henry's law). See Chapters 9 and 10 for more details. The corresponding DCFI expressions are provided in the literature (O'Connell 1971a).

The inversion of KB theory to provide fluctuating quantities or KBIs in terms of thermodynamic quantities leads to the following expression for the excess coordination numbers,

$$\delta_{ij} + N_{ij} = \rho_j k_B T \kappa_T + \frac{1}{x_1 x_2} \frac{\rho_j}{\rho_i} \frac{(1 - \phi_i)(1 - \phi_j)}{\mu_{ij}} \qquad (1.71)$$

Other equivalent expressions involving a single chemical potential derivative (usually μ_{22}) have also appeared. The most common operative form of the inversion equations for binary mixtures is

$$G_{11} = k_B T \kappa_T - \frac{1}{\rho_1} + \frac{\rho_2 \bar{V}_2^2}{x_1 D}$$

$$G_{12} = k_B T \kappa_T - \frac{\bar{V}_1 \bar{V}_2 (\rho_1 + \rho_2)}{D} \tag{1.72}$$

$$G_{22} = k_B T \kappa_T - \frac{1}{\rho_2} + \frac{\rho_1 \bar{V}_1^2}{x_2 D}$$

where D is defined by

$$D = 1 + x_1 x_2 \left(\frac{\partial^2 \beta G_m^E}{\partial x_1^2} \right)_{p,T} \tag{1.73}$$

and is positive for stable (miscible) solutions.

Similarly, the inversion of Equations 1.56, 1.57, and Equation 1.59 provides the DCFIs in terms of measurable properties. The expressions for binaries are

$$C_{11} = 1 - \frac{\rho \bar{V}_1^2}{k_B T \kappa_T} - N \left(\frac{\partial \ln \gamma_1}{\partial N_1} \right)_{T,p,N_2}$$

$$C_{12} = 1 - \frac{\rho \bar{V}_1 \bar{V}_2}{k_B T \kappa_T} - N \left(\frac{\partial \ln \gamma_1}{\partial N_2} \right)_{T,p,N_1} \tag{1.74}$$

$$C_{22} = 1 - \frac{\rho \bar{V}_2^2}{k_B T \kappa_T} - N \left(\frac{\partial \ln \gamma_2}{\partial N_2} \right)_{T,p,N_1}$$

where γ_i is the activity coefficient of component i.

1.3.3 CLOSED TERNARY SYSTEMS

The explicit expressions for the chemical potential derivatives, partial molar volumes, and isothermal compressibility become rather cumbersome for ternary systems. Experimental data are also much less common. However, there are many interesting effects that involve ternary systems (see Chapter 4). Also, we shall see that considerable simplification is obtained when one of the components is at infinite dilution (see Chapters 10 and 11). If one requires specific expressions for the various properties, it will prove convenient to define the following set of variables (Smith 2006a),

$$\eta_{123} = |\mathbf{M}| = \rho_1 A_2 A_3 + \rho_2 A_1 A_3 + \rho_3 A_1 A_2$$

$$\zeta_3 = |\mathbf{NN}| = (1 + N_{11})(1 + N_{22})(1 + N_{33}) - (1 + N_{11})N_{23}N_{32} - (1 + N_{22})N_{13}N_{31} \tag{1.75}$$

$$- (1 + N_{33})N_{12}N_{21} + 2N_{12}N_{23}N_{31}$$

with

$$A_i = 1 + \rho_i(G_{ii} + G_{jk} - G_{ij} - G_{ik}) \tag{1.76}$$

where i, j, and k can be 1, 2, or 3 and therefore,

$$\eta_{ij} = \rho_i A_j + \rho_j A_i \tag{1.77}$$

which is a generalization of Equation 1.66. The following expressions are then obtained for ternary systems (Smith 2006a):

$$\left(\frac{\partial \beta \mu_1}{\partial \ln m_3}\right)_{T,p,m_2} = -\frac{\rho_3 A_2}{\eta_{123}} \qquad \left(\frac{\partial \beta \mu_2}{\partial \ln m_3}\right)_{T,p,m_2} = -\frac{\rho_3 A_1}{\eta_{123}} \qquad \left(\frac{\partial \beta \mu_3}{\partial \ln m_3}\right)_{T,p,m_2} = \frac{\eta_{12}}{\eta_{123}}$$

$$\bar{V}_1 = \frac{(1 + N_{22} - N_{12})(1 + N_{33} - N_{13}) - (N_{32} - N_{12})(N_{23} - N_{13})}{\eta_{123}}$$

$$k_B T \kappa_T = \frac{\zeta_3}{\eta_{123}} \tag{1.78}$$

Other derivatives and partial molar volumes can be obtained by a simple index change.

When the solute appears at infinite dilution the following limiting expressions apply:

$$\left(\frac{\partial \beta \mu_2}{\partial \ln m_3}\right)_{T,p,m_2}^{\infty} = \left(\frac{\partial \ln \gamma_2^m}{\partial \ln m_3}\right)_{T,p,m_2}^{\infty} = -\frac{\rho_1 \rho_3 (G_{23}^{\infty} - G_{21}^{\infty})}{\eta_{13}} - \phi_3 \qquad \left(\frac{\partial \beta \mu_2}{\partial \ln m_2}\right)_{T,p,m_3}^{\infty} = 1$$

$$\left(\frac{\partial \ln \gamma_2^m}{\partial m_2}\right)_{T,p,m_3}^{\infty} = -\rho_1 \left(G_{22}^{\infty} + G_{31} - G_{21}^{\infty} - G_{23}^{\infty}\right) - \frac{\rho_1 A_1 A_3}{\eta_{13}} \tag{1.79}$$

$$\bar{V}_2^{\infty} = \frac{(1 + N_{11} - N_{21}^{\infty})(1 + N_{33} - N_{13}) - (N_{31} - N_{21}^{\infty})(N_{13} - N_{23}^{\infty})}{\eta_{13}} = k_B T \kappa_T - N_{21}^{\infty} \bar{V}_1 - N_{23}^{\infty} \bar{V}_3$$

and all other expressions reduce to those of the corresponding binary mixture of 1 and 3.

We do not provide expressions for the KBIs in a general form. This is a result of the many equivalent forms for the expressions that can be obtained when the chemical potential derivatives are interchanged using the expressions provided by the GD relationship. However, a useful set of expressions is (Smith 2008)

$$1 + N_{11} = \rho_1 k_B T \kappa_T + \rho_1 \frac{\left(\mu_{33}' \bar{V}_2 - \mu_{23}' \bar{V}_3\right)\phi_2 + \left(\mu_{22}' \bar{V}_3 - \mu_{32}' \bar{V}_2\right)\phi_3}{\mu_{22}' \mu_{33}' - \mu_{23}' \mu_{32}'}$$

$$N_{12} = \rho_2 k_B T \kappa_T + \rho_2 \frac{\left(\mu_{13}' \bar{V}_3 - \mu_{33}' \bar{V}_1\right)\phi_2 + \left(\mu_{32}' \bar{V}_1 - \mu_{12}' \bar{V}_3\right)\phi_3}{\mu_{32}' \mu_{13}' - \mu_{12}' \mu_{33}'} \tag{1.80}$$

where other KBIs are provided by a suitable index change.

1.3.4 Preferential Solvation in Binary and Ternary Systems

As mentioned previously, one of the major advantages of FST is that it provides information on local density fluctuations. An alternative approach to this problem involves the concepts of local composition and preferential solvation (Ben-Naim 2006). To do this we first note that the average number of j particles in a region of spherical volume V_{cor} surrounding a central particle i is

$$\left\langle N_j \right\rangle = N_{ij} + \rho_j V_{cor} \tag{1.81}$$

as long as the volume is large enough that all $g_{ij} = 1$. Using the above expression one can define a local mole fraction of j particles in this volume as

$$x_j^{L} = \frac{\left\langle N_j \right\rangle}{\sum_k \left\langle N_k \right\rangle} = \frac{N_{ij} + \rho_j V_{cor}}{\sum_k (N_{ik} + \rho_k V_{cor})} \tag{1.82}$$

and then define the preferential solvation of i by j as

$$\delta x_{ji} = x_j^{L} - x_j = \frac{x_j \sum_k x_k (G_{ij} - G_{ik})}{V_{cor} + \sum_k x_k G_{ik}} \tag{1.83}$$

which applies for a solution containing any number of molecules. This is the approach used in Chapter 3. Alternative definitions of preferential solvation can also be used and are provided in Chapters 3 and 4. In addition, it has become usual to correct the KBIs by subtraction of the ideal solution values (see Section 1.3.7) in an effort to help interpret the experimental data. A full discussion is provided in Chapter 4.

The above expressions depend explicitly on the spherical volume of interest. This volume must be large enough to ensure all RDFs are unity and therefore that Equation 1.81 is valid. Unfortunately, the distance beyond which the RDFs are unity is generally unknown. It will certainly depend on the species and specific composition, as well as the values of T and p. There are two solutions to this problem. First, one can attempt to estimate the volume and thereby provide quantitative values for the preferential solvation. This is described in Chapters 3 and 4. Alternatively, one can take the approach of Ben-Naim and expand the above expression in powers of V_{cor}^{-1} to give (Ben-Naim 2006)

$$\delta_{ji}^{0} = x_j^{L} - x_j = \frac{x_j \sum_k x_k \left(G_{ij} - G_{ik} \right)}{V_{cor}} + \cdots \tag{1.84}$$

The numerator then provides the sign of the preferential solvation as we decrease the value of V_{cor} from that of an infinite region in a closed system to that of an open local region of the solution. The magnitude of the effect is therefore unimportant, but the direction provides information on the relative distribution of species in solution. For binary solutions this reduces to the difference between two KBIs.

1.3.5   OPEN SYSTEMS

Kirkwood and Buff also provided expressions for an osmotic system where the chemical potential of the primary solvent is held constant (Kirkwood and Buff 1951). Again, this was achieved by a thermodynamic transformation. Alternatively, this can be treated easily in Equation 1.43 by simply setting $d\beta\mu_1 = 0$ and $dp = d\pi$, and noting that the pressure derivatives of the chemical potentials are no longer the partial molar volumes. When there are only two components, the expression for the change in the osmotic pressure due to a nondiffusible solute becomes

$$\beta\left(\frac{\partial\pi}{\partial\rho_2}\right)_{\beta,\mu_1} = \frac{1}{1+N_{22}} \tag{1.85}$$

The above expression holds at any solute concentration. The analysis of osmotic pressure data for solutes therefore provides an indication of solute–solute affinity at the concentration of interest (Karunaweera et al. 2012). One can also show that MM theory is obtained from FST in the limiting case of an infinitely dilute solute, where both approaches then provide a series expansion in concentration (McMillan and Mayer 1945; Kirkwood and Buff 1951). A more detailed discussion is provided in the literature (Cabezas and O'Connell 1993).

The other main application of FST to open systems has involved the study of preferential interactions. This is particularly important for understanding the effects of a cosolvent on the properties of biomolecules—the urea-induced denaturation of proteins, for example. The major source for thermodynamic data on small molecule *binding* to proteins has been equilibrium dialysis studies (Timasheff 1998a). More recently, this has been complemented by isopiestic distillation studies (Anderson, Courtenay, and Record 2002). Both approaches provide thermodynamic descriptions of cosolvent interactions with proteins that can be used to rationalize the effects of cosolvents on protein stability obtained from isothermal–isobaric studies (Smith 2004).

The preferential interaction of a cosolvent (3) with a biomolecular solute (2) in the presence of a solvent (1) is most simply provided by equilibrium dialysis experiments, which provide preferential binding parameters such that (Smith 2006b)

$$\Gamma_{23} = \left(\frac{\partial m_3}{\partial m_2}\right)_{T,\mu_1,\mu_3} = \frac{N_{23} - m_3 N_{21}}{1 + N_{22} - N_{12}} \tag{1.86}$$

The above FST-based expression can be obtained directly from Equation 1.44 after imposing the appropriate thermodynamic constraints. When the solute appears at infinite dilution, the expression reduces to

$$\Gamma_{23}^{\infty} = \rho_3\left(G_{23}^{\infty} - G_{21}^{\infty}\right) = \langle N_3\rangle_2 - m_3\langle N_1\rangle_2 \tag{1.87}$$

where the subscript 2 on the final ensemble averages indicate the average number of cosolvent or solvent molecules in the local vicinity of the solute. The value of the preferential binding parameter is therefore a measure of changes in the local distribution of the cosolvent and water surrounding the protein solute. Positive values occur when the cosolvent/solvent ratio in the vicinity of the protein exceeds that of the bulk solvent (m_3), and vice versa. This includes both direct binding of the cosolvent to the protein, together with any possible changes in the cosolvent and solvent distributions at larger distances from the protein surface. In Section 1.3.8 and Chapter 11, the preferential binding parameter will be used to rationalize the thermodynamics of cosolvent-induced protein denaturation. Other measures of preferential interactions, in a variety of semiopen ensembles, have been used to help rationalize cosolvent effects on biomolecules, and are summarized by Smith (Smith 2006a).

A third illustration of the use of FST for open systems involves the effects of cosolvents or additives on the solubility of a solute in a solvent. If one follows the solute solubility curve, at a fixed temperature and pressure, then the chemical potential of the solute at saturation remains constant as it is in equilibrium with the solid solute. Hence, the effect of an additive on the molar solute solubility (S_2) can be expressed in terms of derivatives of this curve taken at constant T, p, and μ_2. Using these constraints in Equation 1.43 and taking the appropriate derivatives, one immediately finds (Smith and Mazo 2008)

$$\left(\frac{\partial \ln S_2}{\partial \rho_3} \right)_{T,p,\mu_2} = \frac{G_{23} - G_{21}}{1 + N_{33} - N_{13}} \tag{1.88}$$

for any concentration of solute, solvent, and additive. We note, however, that other expressions in closed ensembles have also been used (see Chapters 9 and 10).

1.3.6 ELECTROLYTE SOLUTIONS

The application of FST to electrolyte solutions has received considerable interest. It is generally assumed that the average number of cations surrounding an anion in solution must be such that charge neutrality is obeyed for the local region (Kusalik and Patey 1987). The resulting relationships are known as the *electroneutrality conditions* and can be written in terms of the KBIs as follows:

$$z_+ + z_+\rho_+G_{++} + z_-\rho_-G_{+-} = 0$$

$$z_- + z_-\rho_-G_{--} + z_+\rho_+G_{+-} = 0$$

$$z_+\rho_+G_{i+} + z_-\rho_-G_{i-} = 0 \tag{1.89}$$

$$v_+z_+ + v_-z_- = 0$$

for a salt ($M_{v_+}^{z_+} X_{v_-}^{z_-}$) that generates a total of v ions containing v_+ and v_- cations and anions in solution with charges of z_+ and z_-, respectively, and where the index i refers to any nonelectrolyte component. Unfortunately, the application of the electroneutrality constraints immediately leads to a singular **B** matrix for an electrolyte system if the cation, anion, and solvent are treated as independent variables (Friedman and Ramanathan 1970). Recently, Ben-Naim has argued that there is no fundamental reason why KB theory cannot be applied to study a ternary system of anions, cations, and a solvent (Ben-Naim 2006). He argues that the application of the electroneutrality constraints is incorrect for the KBIs as these are defined for open systems. The simplest way to avoid this problem is to not treat the anion and cation concentrations as independent variables. Indeed, there is no experiment we know of that provides such data, although some electrochemical approaches may have achieved this goal (Wilczek-Vera and Vera 2011). In the absence of general data for anion or cation chemical potentials, it seems reasonable to invoke the electroneutrality expressions and to treat the anion and cation concentrations as dependent variables. This leads to a treatment of electrolytes in terms of either salt *molecules*, or as a collection of indistinguishable ions (Chitra and Smith 2002). The latter approach appears to be more convenient for comparison with simulation data. In addition, the expressions obtained from the latter approach also agree with the more formal treatment of Kusalik and Patey (1987), while individual ions can be distinguished in the more complete analysis by O'Connell and coworkers as mentioned in Chapter 9 (Perry and O'Connell 1984). Hence, we will focus on the indistinguishable ions approach.

A simple substitution of $vN_s = N_2 = N_+ + N_-$ then provides relationships between the distributions obtained using the salt (s), the indistinguishable ion (2), and the individual cation and anion notations (Gee et al. 2011),

$$G_{ss} = \frac{1-v}{\rho_2} + G_{22} \qquad G_{si} = G_{2i}$$

$$G_{22} = \left(\frac{v_+}{v}\right)^2 G_{++} + \left(\frac{v_-}{v}\right)^2 G_{--} + \frac{v_+ v_-}{v}(G_{+-} + G_{-+}) \qquad (1.90)$$

$$G_{2i} = G_{i2} = \frac{v_+}{v} G_{i+} + \frac{v_-}{v} G_{i-}$$

The above expressions do not assume electroneutrality and simply represent an index change. Subsequent application of the electroneutrality conditions leads to a series of relationships between the fluctuating quantities,

$$G_{22} = -\frac{1}{\rho_2} + G_{+-} \qquad\qquad G_{+-} = \frac{1}{\rho_+} + G_{++}$$

$$\frac{1}{\rho_+} + G_{++} = \frac{1}{\rho_-} + G_{--} \qquad\qquad G_{i2} = G_{i+} = G_{i-} \qquad (1.91)$$

Consequently, there is only one unique or independent fluctuating quantity involving the salt species, the others being related by the above expressions. In practical applications one simply has to account for the number of ions generated by the salt when applying the FST or inversion expressions. This results in the following expressions being used to analyze the experimental data,

$$m_2 = \nu m_s \qquad\qquad \rho_2 = \nu \rho_s \qquad\qquad \nu \overline{V}_2 = \overline{V}_s$$

$$\left(\frac{\partial \beta \mu_2}{\partial \ln m_2} \right)_{p,T,\{m\}'} = 1 + \left(\frac{\partial \ln \gamma_{\pm}^m}{\partial \ln m_s} \right)_{p,T,\{m\}'} \qquad\qquad \mu_{ss} = \nu \mu_{22} \qquad (1.92)$$

where $\gamma_{\pm}^m$ is the mean ion molal activity coefficient. See also the approaches outlined in Chapters 8 and 9.

1.3.7 IDEAL SOLUTIONS

Ideal solutions have proved to be a very useful reference point for the analysis of real data on solutions obtained by inversion of FST (Ben-Naim 2006). They can also be used to provide approximate behavior in situations where the required experimental data are not available. In principle, ideal behavior will depend on the concentration scale used. The most common type of ideal solution involves species that are fully miscible and where the composition is described by mole fractions. Ideal solutions of this type are characterized by chemical potentials of the form $\beta d\mu_i = d\ln x_i$, together with zero excess enthalpies and volumes of solution for the whole composition range. This is also referred to as *symmetric ideal* (SI) behavior. Application of the above conditions in Equations 1.54 provides the following general relationships for any number of components (Ploetz, Bentenitis, and Smith 2010b):

$$G_{ij} = k_B T \kappa_T - V_i^\circ - V_j^\circ + S_{n_c} \qquad\qquad S_{n_c} = \sum_k \rho_k (V_k^\circ)^2$$

$$\Delta G_{ij} = G_{ii} + G_{jj} - 2G_{ij} = 0 \qquad\qquad\qquad\qquad (1.93)$$

$$\rho_i = \frac{x_i}{V_m} \qquad\qquad V_m = \sum_k x_k V_k^\circ \qquad\qquad \kappa_T = \sum_k \phi_k \kappa_{T,k}^\circ$$

where the sums are over all the components and the expressions are valid for all i,j combinations.

In many applications, the concentration scale of choice is often molalities or molarities. The above expressions do not lead to similar ideal behavior using these alternative concentration scales. For ideal behavior on the molality scale, one requires that $\beta d\mu_i = d\ln m_i$ for all species except for the primary solvent, which is provided when (Ploetz, Bentenitis, and Smith 2010b)

$$G_{ij} = k_B T \kappa_T - V_i^\circ - V_j^\circ + S_{n_c} - \delta_{1i} \delta_{1j} \rho_1^{-1} \qquad\qquad (1.94)$$

if we set V_1° to zero. Alternatively, for ideal behavior on the molarity scale, one requires that $\beta d\mu_i = d\ln \rho_i$ for all species, for which one must therefore have

$$G_{ij} = k_B T \kappa_T - V_o \tag{1.95}$$

Here, all the volumes of the pure components must be equal. Finally, we note that ideal osmotic solution behavior, using any concentration scale, is observed when all G_{ij}s are zero for all compositions.

1.3.8 CHEMICAL AND ASSOCIATION EQUILIBRIA

FST has also been applied to study changes in chemical and association equilibria. To our knowledge, Ben-Naim was the first to investigate molecular association using KB theory (Ben-Naim 1975). O'Connell and coworkers provided the most general and rigorous application of FST to chemical equilibria (Perry and O'Connell 1984) (see also Section 9.4 in Chapter 9). Hall also applied FST to study micelle formation (Hall 1983). More recently, Smith and coworkers have provided a simple approach for describing conformational and association equilibria (Gee and Smith 2009), which has been extended to include the effects of temperature on equilibria (Jiao and Smith 2011). This is the approach we take here.

Let us consider a simple solute association equilibrium between n monomers (M) and an aggregate (A) such that nM → A, which is described by an equilibrium constant $K = \rho_A / \rho_M^n$. Conformational equilibria can also be investigated by simply setting n = 1. A differential for changes in the equilibrium constant at constant T can then be obtained from Equation 1.43 applied to the monomer and aggregate number densities after application of the equilibrium condition $n d\mu_M = d\mu_A$,

$$d \ln K = \sum_i (N_{Ai} - n N_{Mi}) d\beta \mu_i \tag{1.96}$$

The summation is over all thermodynamically independent species, including the solute, and we have used the notation $N_{M2} = 1 + N_M + n N_{MA}$ and $N_{A2} = n + n N_{AA} + N_{AM}$ for the solute of interest, that is, for the slightly unusual case of when $i = 2$ (Gee and Smith 2009). We note that the equilibrium constant used here is not dimensionless (unless n = 1), but could be made dimensionless by the inclusion of standard concentrations. However, this is unnecessary, as we will only consider changes to the equilibrium constant. We also note that the above definition of the equilibrium constant uses concentrations and not activities. This differs from the traditional thermodynamic approach, but is simpler for following many equilibria studied by spectroscopic techniques—as performed for most biological equilibria.

The above differential can be used directly to obtain the effects of pressure on the equilibrium constant,

$$\left(\frac{\partial \ln K}{\partial p} \right)_{\beta, \{N\}} = \beta \sum_i (N_{Ai} - n N_{Mi}) \bar{V}_i \tag{1.97}$$

together with the effects of a change in composition on the equilibrium constant,

$$\left(\frac{\partial \ln K}{\partial \rho_j}\right)_{\beta,p,\{m\}'} = \sum_i (N_{Ai} - nN_{Mi})\left(\frac{\partial \beta \mu_i}{\partial \rho_j}\right)_{\beta,p,\{m\}'} \quad (1.98)$$

Both expressions are valid for systems containing any number of components at any composition. The chemical potential derivatives are provided by the expressions for an n_c component system. For low solute concentrations, the former equation also provides an expression for the standard volume change for the process. A particularly common situation involves the effect of a single cosolvent on the conformational equilibrium ($n = 1$, $D \rightarrow N$) of an infinitely dilute solute. In this case, Equation 1.98 then reduces to

$$\left(\frac{\partial \ln K}{\partial \rho_3}\right)_{\beta,p}^{\infty} = \frac{1}{\rho_3} \frac{\Gamma_{D3}^{\infty} - \Gamma_{N3}^{\infty}}{1 + N_{33} - N_{13}} \quad (1.99)$$

after application of the GD expression to eliminate one of the chemical potential derivatives, and using the definitions provided in Equation 1.87 for the Γs. This is the FST expression for the m-value of protein denaturation studies, which indicates that the above derivatives often appear to be constant over a range of cosolvent concentration (Greene and Pace 1974).

1.3.9 TECHNICAL ISSUES SURROUNDING THE APPLICATION OF FLUCTUATION SOLUTION THEORY

The expressions provided by FST are exact within the normal assumptions of statistical thermodynamics. This is a major advantage of the theory. There are, however, various technical issues that arise when applying the theory. These basically fall into two categories. The first is related to the analysis of the experimental thermodynamic data, and the second related to obtaining the KBIs from simulation data.

The KB inversion process involves the extraction of KBIs from the available experimental data. The experimental data required for this process—derivatives of the chemical potentials, partial molar volumes, and the isothermal compressibility—are all generally obtained as derivatives of various properties of the solution. Obtaining reliable derivatives can be challenging and will depend on the quality of the source data and the fitting function. Unfortunately, the experimental data often appear without a reliable statistical analysis of the errors involved, and hence the quality of the data is difficult to determine. Matteoli and Lepori have performed a fairly rigorous analysis of a series of binary mixtures and concluded that, for systems under ambient conditions, the quality of the resulting KBIs is primarily determined by the chemical potential data, followed by the partial molar volume data, whereas errors in the compressibility data have essentially no effect on the KBI values (Matteoli and Lepori 1984). Excess chemical potentials are typically obtained from partial pressure data, either isothermal or bubble point determinations, and from osmotic pressure or even electrochemical measurements. The particle number fluctuations can also be extracted directly from light scattering data (Kirkwood and Goldberg 1950; Blanco et al. 2011). Comparison of these types of data indicates that one can observe sizeable differences in KBI values obtained using different activity datasets (Perera et al. 2005), although the major differences are usually restricted to the KBI values between species that are present at low concentrations. The differences are also usually largest for systems that display far from ideal behavior. A database of binary TCFIs and DCFIs for a wide variety of small molecule binaries has been provided (Wooley and O'Connell 1991).

Computer simulations represent one of the most common approaches to determining the local fluctuations. However, there are some technical difficulties, which can arise during the analysis of a typical simulation (see Chapter 6 for a full discussion). Most evaluations of the KBIs have used the integration approach, in contrast to the actual particle number fluctuations. Furthermore, as the vast majority of simulations are performed for closed periodic systems, one is naturally limited to performing the integration out to some cutoff distance from the particle of interest. This seems reasonable given the similarities between the RDFs in open and closed systems (Weerasinghe and Pettitt 1994). Hence, one can define distance-dependent KBIs (and even distance-dependent thermodynamic functions) such that

$$G_{ij}(R) \approx 4\pi \int_0^R \left[g_{ij}^{(2)}(r) - 1 \right] r^2 dr \qquad (1.100)$$

where R is some distance at which the RDFs are essentially unity.

In favorable cases, the integral converges and one observes a limiting constant value for the KBI (Weerasinghe and Smith 2003c; Bentenitis, Cox, and Smith 2009). More typically, there can be significant statistical noise or artifacts that obscure the real limiting behavior, depending on the system size used in the simulation (Perera et al. 2006; Wedberg et al. 2010). Several studies have investigated this problem with a variety of suggested solutions (see also Chapters 6 and 7). The effect of system size on the KBI values has also been studied (Schnell et al. 2011). The problem seems to be particularly acute when determining the isothermal compressibility. The situation appears to be significantly improved when examining the partial molar volumes and chemical potential derivatives, presumably as these involve differences in the KBIs (Nichols, Moore, and Wheeler 2009). However, this might not always be the case as indicated in Chapter 6. Fortunately, the values of the resulting properties can often be tested by other approaches—typically finite difference compressibilities or partial molar volumes obtained from the simulated densities. However, checking the chemical potential derivatives is quite time consuming.

Another issue arises when determining the KBIs or local distributions around infinitely dilute solutes. In particular, special care must be taken when determining the solvent and cosolvent distributions around a single protein solute, properties that are required input for calculation of preferential binding described by Equation 1.87. When using finite systems, it is sometimes necessary to adjust the bulk distribution (m_3) during the analysis. For instance, the ratio of cosolvent to solvent molecules in the bulk region is not equal to the ratio of total molecules used in the simulation when the cosolvent and solvent molecules can exchange with other molecules in the local vicinity of the protein (Kang and Smith 2007). This is especially true for small systems and/or large values of Γ_{23}. In this case the value of the preferential binding should be adjusted so that

$$\Gamma_{23}^{\infty} = \langle N_3 \rangle_2 - m_3 \langle N_1 \rangle_2 = \langle N_3 \rangle_2 - \frac{N_3^{\circ} - \langle N_3 \rangle_2}{N_1^{\circ} - \langle N_1 \rangle_2} \langle N_1 \rangle_2 \qquad (1.101)$$

where the zero superscript indicates the total number of molecules in the system. Hence, as molecules move in or out of the local region the bulk ratio is correspondingly adjusted. This adjustment may appear small but is magnified by the fact that $\langle N_1 \rangle_2$ may be very large for biological solutes.

1.4 CONCLUSIONS

In the previous sections, we have attempted to provide a reasonable summary of FST, including a general derivation, and illustrated a series of known applications of the theory. In our opinion, the use of FST to analyze experimental and simulation data represents both the most appropriate approach and also many times the only approach, for a variety of systems and their properties. The absence of any approximations in the theory provides a solid theoretical foundation for the analysis of solutions. An ability to relate these properties to the underlying molecular distributions also ensures a rigorous link between the atomic and macroscopic picture of solutions.

2 Global and Local Properties of Mixtures

An Expanded Paradigm for the Study of Mixtures

Arieh Ben-Naim

CONTENTS

Abstract: Traditionally, the properties of liquid mixtures were studies in terms of excess thermodynamic functions such as excess Gibbs energy, entropy, enthalpy, and volume. These quantities convey *global* or macroscopic information on the system; global, in the sense that they do not reveal any molecular origin of these quantities. Recently, a complementary view based on the *local* properties of the same system was suggested. These properties are richer and more informative regarding the local properties such as local densities, composition, and the solvation effect. Here, two sets of quantities are calculated for two component mixtures in one dimension. The local properties were calculated using two different methods; one based on the Kirkwood–Buff theory, and the second is based on direct calculation of the pair correlation functions.

2.1   INTRODUCTION

Traditionally, the properties of mixtures were studied by examining the various excess quantities, such as the excess free energy, excess entropy, enthalpy, volume, and so forth (Prigogine with contributions from A. Bellemans and V. Mathot 1957; Rowlinson and Swinton 1982). These quantities convey *global* properties of the mixtures. They are global in the sense that they convey information on the *macroscopic properties* of the mixture. Recently, an alternative approach to the study of the properties of liquid mixtures at a *local* level has been suggested. This view is based on the inversion of the Kirkwood–Buff (KB) theory of solutions (Kirkwood and Buff 1951; Ben-Naim 1977, 1987, 1988, 1989, 1990b, 1992, 2006) on one hand, and on the solvation thermodynamics on the other hand. This approach is more informative, since it provides *microscopic information* on the surroundings of a single molecule in the mixture.

There are essentially three significant quantities that can be derived from the inversion of the KB theory. The first is a measure of the extent of deviation from symmetrical ideal (SI) solution behavior, Δ_{AB}, defined below in the next section. It also provides a necessary and sufficient condition for SI solution. The second is a measure of the extent of *preferential solvation* (PS) around each molecule. In a binary system of A and B, there are only two independent PS quantities; these measure the preference of, say, molecule A to be solvated by either A or B molecules. Deviations from SI solution behavior can be expressed in terms of either the sum or difference of these PS quantities. Finally, the Kirkwood–Buff integrals (KBIs) may be obtained from the inversion of the KB theory. These provide information on the *affinities* between any two species; for instance, $\rho_A G_{AA}$ measures the excess of the average number of A particles around A relative to the average number of A particles in the same region chosen at a random location in the mixture. All these quantities can be obtained from the KB integrals.

The second group consists of the solvation thermodynamic quantities. These quantities also convey local information on the average interaction free energy of a particular molecule with its surroundings, the effect of the solute on the solvent structure, and so forth (Ben-Naim 2009, 2011).

In the next section, the local quantities are defined and a brief discussion of the procedure for calculating these quantities is described. In subsequent sections, we present a complete set of results for a one-dimensional mixture of two components.

2.2   DEFINITIONS OF THE GLOBAL AND LOCAL QUANTITIES

The fundamental *global* quantity is the excess Gibbs energy of the entire mixture. For a two-component system, this is defined by

$$G_m^E = \frac{G^E}{(N_A + N_B)} = x_A(\mu_A - \mu_A^\circ - k_B T \ln x_A) + x_B(\mu_B - \mu_B^\circ - k_B T \ln x_B) \quad (2.1)$$

where μ_A and μ_A° are the chemical potentials of A in the mixture, and in the pure state at some fixed temperature T, and pressure p. The excess is defined with respect to a *symmetric ideal* solution. There are other excess quantities defined with respect to an ideal gas mixture, or with respect to ideal dilute solution of A in B (Ben-Naim 1992, 2006). Each of these excess quantities measures different properties of the solution. In our case, G_m^E measures the deviation from *similarity*, (Ben-Naim 1992, 2006) between A and B.

Once we have the excess Gibbs energy, either from experiments or from theoretical calculations, we can derive all other excess quantities from the relationships provided in Section 1.1.1 in Chapter 1. Other quantities such as excess heat capacity, excess compressibility, and so forth, can also be derived by standard methods (Ben-Naim 1992, 2006).

We next turn to the *local* quantities. The most fundamental quantities to be studied are the KBIs (Kirkwood and Buff 1951; Ben-Naim 2006), $G_{\alpha\beta}$, defined as

$$G_{\alpha\beta} = \int_0^{\infty} [g_{\alpha\beta}(r) - 1]4\pi r^2 dr \qquad (2.2)$$

where $g_{\alpha\beta}$ is the (angular averaged) pair correlation function and the integration is extended over the entire macroscopic volume of the system. It should be stressed that these functions are defined in the open system with respect to both A and B. In a closed system, the normalization of the pair correlation functions is

$$G_{\alpha\beta}^{closed} = -\delta_{\alpha\beta}/\rho_\alpha \qquad (2.3)$$

where ρ_α is the number density of the species α. The significance of $G_{\alpha\beta}$ in Equation 2.2 as a *local property* stems from the following considerations: $\rho_\alpha g_{\alpha\beta}(r)4\pi r^2 dr$ is the average number of α particles in a spherical shell of radius r and width dr, around a β particle, and $\rho_\alpha 4\pi r^2 dr$ is the average number of α particles in the same volume element $4\pi r^2 dr$ chosen at a random point in the system. Therefore, the integral,

$$\rho_\alpha G_{\alpha\beta}(R_{cor}) = \rho_\alpha \int_0^{R_{cor}} [g_{\alpha\beta}(r) - 1]4\pi r^2 dr \qquad (2.4)$$

provides a measure of the change in the average number of α particles in a spherical region of radius R_{cor} brought about by placing a β particle at the center of this region.

The significance of $G_{\alpha\beta}(R_{cor})$ as a measure of the local properties around a β particle follows the meaning of the pair correlation function. For most systems of interest (excluding solid solutions or systems near the critical point), $g_{\alpha\beta}(r)$ approaches unity (that is, no correlation) for r values of the order of a few molecular diameters; hence, R_{cor} in Equation 2.4 can be replaced by infinity, as in Equation 2.2. Thus, the major contribution to $G_{\alpha\beta}$ arises from a small *local* region around the β particle. Positive $\rho_\alpha G_{\alpha\beta}$ values indicates that when a β particle is placed at the center of the correlation volume (i.e., the volume $4\pi R_{cor}^3/3$), an excess of α particles will be attracted around β, compared to the average number of particles in the same volume chosen at a random point in the system. Because of this property and because $G_{\alpha\beta}$

is defined symmetrically with respect to α and β, that is, $G_{\alpha\beta} = G_{\beta\alpha}$, the suggestion was made of referring to $G_{\alpha\beta}$ as a measure of the extent of *affinity* between the α and β species (Ben-Naim 2006).

It should be noted that the existence of a correlation radius (i.e., a R_{cor} distance such that for $r > R_{cor}$, $g_{\alpha\beta}(r)$ is practically unity), is based on what is known both experimentally and theoretically concerning the pair correlation functions. There is no proof of the existence of such a correlation distance (Ben-Naim 2006).

The second local quantity is the PS. In a two-component system, we focus on a small V_{cor} region. We ask how the *composition* in this region changes by placing, say, an A molecule in its center. The PS of the A particles with respect to composition, measured as the mole fraction of A in V_{cor}, is defined as the difference,

$$\delta_{A,A}(V_{cor}) = x_A^L(A) - x_A \tag{2.5}$$

where x_A is the bulk mole fraction of A, and x_A^L is the *local* mole fraction of A around an A particle in the volume V_{cor}. From the definition of the KBIs, it follows (Ben-Naim 2006)

$$\delta_{A,A}(V_{cor}) = \frac{x_A x_B (G_{AA} - G_{AB})}{x_A G_{AA} + x_B G_{AB} + V_{cor}} \tag{2.6}$$

Note that this definition is valid for V_{cor} *larger* than the correlation volume, since $G_{\alpha\beta}$ was taken as defined in Equation 2.2. Application of Equation 2.6 for *any* V_{cor}, requires the use of $G_{\alpha\beta}(R_{cor})$ as defined in Equation 2.4. Similarly, the PS of B with respect to composition (in terms of the mole fraction of A) is defined as

$$\delta_{A,B}(V_{cor}) = x_A^L(B) - x_A = \frac{x_A x_B (G_{AB} - G_{BB})}{x_A G_{AB} + x_B G_{BB} + V_{cor}} \tag{2.7}$$

Clearly, these quantities depends on the choice of the correlation volume V_{cor}; if V_{cor} is very large, then the local composition should approach the bulk composition; hence, the PS should approach zero. Therefore, in order to eliminate the dependence on V_{cor} and obtain an *intrinsic* PS measure of A and B, the suggestion has been made (Ben-Naim 1988, 1990b, 1992, 2006) of introducing the first-order term in the expansion of $\delta_{A,B}(V_{cor})$ in power series around V_{cor}^{-1}. The coefficient of the first-order term in the expansion is (Ben-Naim 2006)

$$\delta_{A,A}^0 = x_A x_B (G_{AA} - G_{AB}) \tag{2.8}$$

$$\delta_{A,B}^0 = x_A x_B (G_{AB} - G_{BB}) \tag{2.9}$$

It should be noted that in expanding the quantities $\delta_{A,A}$ and $\delta_{A,B}$, the $G_{\alpha\beta}$ are not viewed as functions of V_{cor}, since use of the definition in Equation 2.2 implies

$R_{\text{cor}} \to \infty$, rendering all $G_{\alpha\beta}$ independent of V_{cor}. The only dependence on V_{cor} is through the denominator in Equations 2.6 and 2.7. If one wishes to calculate the PS in a small V_{cor} region around A or B, then one needs to take also the finite limit of the KBI as in Equation 2.4.

The third quantity is a measure of the deviation of the mixture from SI solution behavior (see Section 1.3.7 in Chapter 1). This quantity is defined as

$$\Delta_{AB} = G_{AA} + G_{BB} - 2G_{AB} \tag{2.10}$$

and depends on all three KBIs (note that $G_{AB} = G_{BA}$). From the KB theory, one obtains the exact relationship for two-component mixtures of A and B (Kirkwood and Buff 1951; Ben-Naim 2006),

$$\left(\frac{\partial \mu_A}{\partial x_A} \right)_{p,T} = k_B T \left(\frac{1}{x_A} - \frac{x_B \rho \Delta_{AB}}{1 + \rho x_A x_B \Delta_{AB}} \right) \tag{2.11}$$

The SI solution is defined as a solution for which the chemical potential of, say, A is

$$\mu_A = \mu_A^o + k_B T \ln x_A \quad \text{for } 0 \leq x_A \leq 1 \tag{2.12}$$

Equation 2.12, combined with the Gibbs–Duhem relationship leads to

$$\mu_B = \mu_B^o + k_B T \ln x_B \quad \text{for } 0 \leq x_B \leq 1 \tag{2.13}$$

Clearly, $\Delta_{AB} = 0$ is a necessary and sufficient condition for SI behavior (Ben-Naim 2006). Therefore, any finite Δ_{AB} value represents a measure of the extent of deviation from SI behavior. More precisely, in the general case we write

$$\mu_A = \mu_A^o + k_B T \ln x_A + k_B T \ln \gamma_A^{SI} = \mu_A^{SI} + \mu_A^E \tag{2.14}$$

where

$$\mu_A^E = k_B T \ln \gamma_A^{SI} = k_B T \int_0^{x_B} \frac{\rho x_B' \Delta_{AB}}{1 + \rho x_A' x_B' \Delta_{AB}} dx_B' \tag{2.15}$$

with $x_B = 1 - x_A$, Δ_{AB} being regarded as a function of x_B'. Thus, knowing $\Delta_{AB}(x_B)$, one can estimate the deviation from SI, either in terms of an activity coefficient or in terms of an excess chemical potential.

The condition for a solution to be SI, $\Delta_{AB} = 0$, does not necessarily imply anything about the PS, for finite volumes V_{cor}; the reason is that the condition $\Delta_{AB} = 0$ involves use of the infinite limit in the integration, as in Equation 2.2, whereas the PS values use the finite limit R_{cor}. However, if the limiting behavior of the PS at $R_{cor} \to \infty$ is introduced, then the condition $\Delta_{AB} = 0$ is equivalent to

$$\delta^0_{A,A} = \delta^0_{A,B} = -\delta^0_{B,B} \qquad (2.16)$$

Hence, in a symmetric ideal solution, the limiting PS, $\delta^0_{A,A}$ and $\delta^0_{A,B}$, have equal value, but are not necessarily zero. Thus, both Δ_{AB} and the components $\delta^0_{A,A}$ and $\delta^0_{A,B}$ are important in analyzing the sources of nonideality of the mixture.

All the $G_{\alpha\beta}$ quantities defined above may be computed from the inversion of the Kirkwood–Buff theory (Ben-Naim 2006). The relevant relations are provided in Section 1.2.4 in Chapter 1.

The last local quantity of interest is the excess solvation Gibbs energy of A in the mixture relative to that in pure A. This is defined as (Ben-Naim 1977, 1990b, 1992)

$$\Delta\Delta G^*_A = \Delta G^*_A(\text{in the mixture}) - \Delta G^*_A(\text{in pure A}) \qquad (2.17)$$

where ΔG^*_A is the solvation Gibbs energy, as defined for the process of transferring A from a fixed position in an ideal gas phase to a fixed position in the liquid phase (Ben-Naim 1992).

This quantity may also be calculated from experimental data on excess Gibbs energy and excess volume data, that is, (Ben-Naim 1977, 1990b, 1992)

$$\Delta\Delta G^*_A = k_B T \ln\left[\frac{V^E_m + x_A V^o_A + x_B V^o_B}{V^o_A}\right] + G^E_m + x_B\left(\frac{\partial G^E_m}{\partial x_A}\right)_{p,T} \qquad (2.18)$$

2.3 STATISTICAL THERMODYNAMICS OF MIXTURES IN ONE-DIMENSIONAL SYSTEMS

In this section, we present a brief derivation of the partition function (PF) for the mixture in the 1-D system. More details can be found in Ben-Naim (1992). We start with the canonical partition function,

$$Q(T,L,\{N\}) = \frac{1}{\prod_i \Lambda_i^{N_i} N_i!} \int_0^L d\{X\}\exp\left[-\beta U(\{N\})\right] \qquad (2.19)$$

where we used the shorthand notation $d\{X\} = dX_1 \cdots dX_N$, L is the system length, and Λ is the momentum PF in one dimension. The total energy is assumed to be pairwise additive,

$$U(\{N\}) = \sum_{i=1}^N U_{a_i,a_{i+1}}(X_i, X_{i+1}) \qquad (2.20)$$

where $U_{a_i,a_{i+1}}(X_i, X_{i+1})$ is the pair potential corresponding to the pair of species a_i and a_{i+1}, at two consecutive sites X_i and X_{i+1}.

In the one component system, we use an ordering of particles, that is, $0 < X_1 < X_2 < X_3 < \cdots < X_N < L$, to *label* the particles according to their relative locations. But, in our case, since there are c different species, we have to distinguish between a specific ordering of the species (SOOS) and a specific ordering of the particles (SOOP). The configurational integral on the right-hand side of Equation 2.19 can be rewritten by first fixing the order of the species and then summing over all orderings of the species. Thus, we have

$$\int_0^L \cdots \int_0^L = \sum_{\substack{\text{all orderings} \\ \text{of species with} \\ \text{fixed } N_1, \ldots, N_c}} \int_0^L \cdots \int_0^L = \sum_{\substack{\text{all orderings} \\ \text{of species with} \\ \text{fixed } N_1, \ldots, N_c}} \prod_i N_i! \int_{0}^{L} dX_N \cdots \int_{0}^{X_2} dX_1 \qquad (2.21)$$

where in the first step, we have a SOOS, and in the second step, we have a SOOP, and we multiplied by the factor $\prod_i N_i!$

From Equation 2.19 and Equation 2.21, we have,

$$Q(T,L,\{N\}) = \frac{1}{\prod_i \Lambda_i^{N_i}} \sum_{\substack{\text{all orderings} \\ \text{of species with} \\ \text{fixed } N_1, \ldots, N_c}} \int_{0}^{L} dX_N \cdots \int_{0}^{X_2} dX_1 \exp\left[-\beta U(\{N\})\right] \qquad (2.22)$$

Next, we open the system with respect to all particles. The corresponding *grand* PF is,

$$\Xi(T,L,\{\lambda\}) = \sum_{\{N\}} \prod_i \lambda_i^{N_i} \Xi(T,L,\{N\}) = \exp(\beta pL) \qquad (2.23)$$

where $p = (T, L, \{\lambda\})$ is the thermodynamic pressure, given as a function of the variables T, L, $\{\lambda\}$, and λ_i is the absolute activity of the species i. We now take the Laplace transform of Equation 2.23 with respect to the variable L by introducing the new unspecified variable p^*,

$$\Psi(\Xi) = \int_0^\infty \exp\left(-\beta p^* L\right) \Xi\left(-T, L, \{\lambda\}\right) dL \qquad (2.24)$$

Applying the convolution theorem to Equation 2.23, we obtain

$$\Psi(\Xi) = \int_0^\infty \exp(-\beta p^* L) \sum_N (\lambda_i / \Lambda_i)^{N_i}$$

$$\sum_{\substack{\text{all orderings} \\ \text{of species with} \\ \text{fixed } N_1, \ldots, N_c}} \int_{\text{SOOP}} \cdots \int = \sum_{N=0}^{\infty} \sum_S M_{S_1 S_2} M_{S_2 S_3} \cdots M_{S_N S_1} = \sum_{N=0}^{\infty} \mathrm{Tr}(\mathbf{M}^N) \qquad (2.25)$$

where the matrix elements are defined for each pair of species $\alpha\beta$ by

$$M_{\alpha\beta}(p^*) = (\lambda_\alpha/\Lambda_\alpha)^{1/2}(\lambda_\beta/\Lambda_\beta)^{1/2}\Psi_{\alpha\beta}(p^*) \tag{2.26}$$

and

$$\Psi_{\alpha\beta}(p^*) = \int_0^\infty \exp\left[-\beta p^* r - \beta U_{\alpha\beta}(r)\right] dr \tag{2.27}$$

where $\Psi_{\alpha\beta}$ is the Laplace transform of $\exp[-\beta U_{\alpha\beta}(r)]$.

Assuming that the sum in Equation 2.25 converges, we rewrite it as

$$\Psi(\Xi) = \sum_{N=0}^\infty \text{Tr}\mathbf{M}^N = \sum_{N=0}^\infty \sum_{j=1}^c \gamma_j^N = \sum_{j=1}^\infty \left[1 - \gamma_j\left(p^*\right)\right]^{-1} \tag{2.28}$$

where γ_j is the jth eigenvalue of the matrix $\mathbf{M}$. We note the dependence of γ_j on the variable p^*.

From the definition of $\Psi(\Xi)$ in Equation 2.24, and from Equation 2.23, we have

$$\Psi(\Xi) = \int_0^\infty \exp[-\beta L(p^* - p)] dL \tag{2.29}$$

If we choose p^* as the thermodynamic pressure p, then $\Psi(\Xi)$ is referred to as the generalized PF. It is clear from Equation 2.29 that the integral in this case diverges. The reason is that $\sum_{N=0}^\infty \Xi(T, L, \{\lambda\})$ is a function of the single extensive variable L. Transforming L into the thermodynamic intensive variable p gives a partition function that is a function of intensive variables $T, L, \{\lambda\}$ only. However, the Gibbs–Duhem relation (see Equation 1.11) indicates that the intensive variables $T, p, \{\mu\}$ or $T, p, \{\lambda\}$ are not independent. For this reason, we have denoted by p^* the new variable in the Laplace transform taken in Equation 2.24. Suppose that we choose $p^* > p$, then the integral in Equation 2.29 converges, and we have

$$\frac{1}{\beta(p^* - p)} = \sum_{j=1}^{n_c} \left[1 - \gamma_j(p^*)\right]^{-1} \tag{2.30}$$

It is clear that since in the limit $p^* \to p$ the left-hand side of Equation 2.30 diverges, there must be at least one of $\gamma_j(p^*)$ equal to 1. The secular equation of the matrix $\mathbf{M}$ is

$$|\mathbf{M} - \gamma\mathbf{I}| = 0 \tag{2.31}$$

where $\mathbf{I}$ is the unit matrix of the same dimensions as of $\mathbf{M}$.

Since the elements of $\mathbf{M}$ are functions of T, p, and $\{\lambda\}$, we can use the implicit Equation 2.31 to derive all thermodynamic quantities of interest.

We now discuss the special case of a two-component system of A and B for which the secular equation is

$$\begin{vmatrix} M_{AA} - \gamma & M_{AB} \\ M_{AB} & M_{BB} - \gamma \end{vmatrix} = 0 \qquad (2.32)$$

or, equivalently,

$$\gamma_{\pm} = \frac{1}{2} \left[M_{AA} + M_{BB} \pm \sqrt{(M_{AA} - M_{BB})^2 + 4M_{AB}^2} \right] \qquad (2.33)$$

The correct solution can be identified by taking the limit $\lambda_B = 0$, that is, for pure A we must have

$$\gamma = \frac{1}{2}(M_{AA} + M_{AA}) = M_{AA} \qquad (2.34)$$

Therefore, we take the γ_+ solution of Equation 2.34 and equate it to unity to obtain the implicit equation of state,

$$f(T, p, \lambda_A, \lambda_B) = M_{AA} + M_{BB} + \sqrt{(M_{AA} + M_{BB})^2 + 4M_{AB}^2} - 2 = 0 \qquad (2.35)$$

Taking the total differential of f, we have

$$df = \frac{\partial f}{\partial T} dT + \frac{\partial f}{\partial p} dp + \sum \frac{\partial f}{\partial \lambda_i} d\lambda_i = 0 \qquad (2.36)$$

Using the Gibbs-Duhem (GD) expression (see Equation 1.11), we can derive all the thermodynamic quantities of the system from the implicit equation of state Equation 2.35. As an example, the average density of A is

$$\rho_A = \left(\frac{\partial p}{\partial \mu_A} \right)_{T, \mu_B} = -\frac{(\partial f / \partial \mu_A)_{p, T, \mu_B}}{(\partial f / \partial p)_{p, \mu_A, \mu_B}} \qquad (2.37)$$

The entropy of the system is calculated from

$$S = \left(\frac{\partial (pV)}{\partial T} \right)_{V, \mu_A, \mu_B} = V \left(\frac{\partial p}{\partial T} \right)_{\mu_A, \mu_B} = -V \frac{(\partial f / \partial T)_{p, \mu_A, \mu_B}}{(\partial f / \partial p)_{p, \mu_A, \mu_B}} \qquad (2.38)$$

A particular simple case is when

$$\Psi_{AB}^2 = \Psi_{AA} \Psi_{BB} \qquad (2.39)$$

or, equivalently,

$$M_{AB}^2 = M_{AA} M_{BB} \tag{2.40}$$

In this case, the equation of state reduces to

$$f(T, p, \lambda_A, \lambda_B) = 2(M_{AA} + M_{BB}) - 2 = 0 \tag{2.41}$$

The density of A is now

$$\rho_A = -\frac{\beta \lambda_A}{\Lambda_A} \frac{2\Psi_{AA}}{\partial f / \partial p} \tag{2.42}$$

and the mole fraction of A is

$$x_A = \frac{\rho_A}{\rho_A + \rho_B} = \frac{\lambda_A \Psi_{AA}}{\Lambda_A} \tag{2.43}$$

This may be written in a more familiar form as

$$\mu_A = k_B T \ln(\Lambda_A / \Psi_{AA}) + k_B T \ln x_A = \mu_A^\circ + k_B T \ln x_A \tag{2.44}$$

Equation 2.44 is the familiar form of the chemical potential of a symmetric ideal solution. We shall see in Section 2.6 that a mixture of hard rods always forms a symmetric ideal solution.

2.4   CALCULATION OF THE GLOBAL QUANTITIES IN A ONE-DIMENSIONAL SYSTEM

First, we calculate the chemical potential from the equation of state (Equation 2.35). This is calculated using a series of steps. The number densities are calculated from the equation of state (Equation 2.35),

$$\rho_\alpha = -\frac{\partial f / \partial \mu_\alpha}{\partial f / \partial p} \tag{2.45}$$

Since f in Equation 2.35 is given as a function of T, p, λ_A, λ_B, the densities in Equation 2.45 are also functions of these variables. However, these are not independent variables. We can eliminate λ_B from the equation of state to obtain

$$\lambda_B = \frac{\Lambda_B(\Lambda_A - \lambda_A \Psi_{AA})}{\lambda_A \Psi_{AB}^2 + (\Lambda_A - \lambda_A \Psi_{AA})\Psi_{BB}} \tag{2.46}$$

This can be substituted on the right-hand side of Equation 2.45 to obtain the density ρ_A as a function of T, p, λ_A. The latter can be inverted to obtain $\lambda_A = \lambda_A(T, p, \rho_A)$. The last equation can be used to calculate the solvation Gibbs energy, which in our case is (Ben-Naim 2006)

$$\Delta\mu_A^* = \mu_A(T, p, \rho_A) - k_B T \ln \rho_A \Lambda_A \tag{2.47}$$

However, for calculating the excess Gibbs energy, we do not need to calculate the pressure derivative in Equation 2.45. In our case, the mole fraction x_A is calculated from

$$x_A = \frac{\rho_A}{\rho_A + \rho_B} \tag{2.48}$$

having calculated $\rho_A(T, p, \lambda_A)$ and $\rho_B(T, p, \lambda_A)$, we also get $x_A(T, p, \lambda_A)$, which in our case is

$$x_A = \frac{\lambda_A \Lambda_A \Psi_{AB}^2}{\lambda_A (2\Lambda_A - \lambda_A \Psi_{AA})\Psi_{AB}^2 + (\Lambda_A - \lambda_A \Psi_{AA})^2 \Psi_{BB}} \tag{2.49}$$

The last equation can be solved to obtain the absolute activity λ_A as a function of T, p, x_A. The result is

$$\lambda_A = \frac{\Lambda_A \left[(1 - 2x_A)\Psi_{AB}^2 + 2x_A \Psi_{AA}\Psi_{BB} \pm \Psi_{AB}\sqrt{\ } \right]}{2x_A \Psi_{AA}(\Psi_{AA}\Psi_{BB} - \Psi_{AB}^2)} \tag{2.50}$$

where

$$\sqrt{\ } = \sqrt{(1 - x_A)^2 \Psi_{AB}^2 + 4x_A(1 - x_A)\Psi_{AA}\Psi_{BB}} \tag{2.51}$$

To determine the correct solution, we take the two limits; at $x_A \to 0$ and $x_A = 1$, we find that the "+" solution gives the correct limiting quantities,

$$\lambda_A = \frac{\Lambda_A \Psi_{BB} x_A}{\Psi_{AB}^2} + O(x_A)^2 \tag{2.52}$$

and for $x_A = 1$, we get the absolute activity of the pure A,

$$\lambda_A^\circ = \frac{\Lambda_A}{\Psi_{AA}} \tag{2.53}$$

Thus, from the "+" solution in Equation 2.50, we can get the chemical potential of A. Similarly, we get the chemical potential of B and from these we get the excess Gibbs energy per particle as a function of (T, p, x_A),

$$G_m^E = x_A(\mu_A - \mu_A^\circ - k_B T \ln x_A) + (1 - x_A)\left[\mu_B - \mu_B^\circ - k_B T \ln(1 - x_A) \right] \tag{2.54}$$

From the excess Gibbs free energy, we can obtain all the other global excess quantities by taking derivatives with respect to temperature, pressure, and composition.

2.5 CALCULATION OF THE LOCAL QUANTITIES IN A ONE-DIMENSIONAL SYSTEM

In the one-dimensional system, the KBI is defined by

$$G_{\alpha\beta} = \int_{-\infty}^{\infty} \left[g_{\alpha\beta}(R) - 1 \right] dR \tag{2.55}$$

Note that the integration here is over the entire range from $(-\infty)$ to $(+\infty)$. In practice, the integrand is zero for large value of R of the order of a few molecular diameters. Therefore, the integration is over the entire *volume* of the system. The KBIs, the PS, and the measure of deviations from SI isolation are calculated according to the inversion procedure outlined in Section 2.2.

The solvation Gibbs energy of, say, A relative to the solvation Gibbs energy of A in pure A is defined in the same manner as Equation 2.17 and may be determined using Equation 2.18. The solvation Gibbs energy of A in pure A is obtained from the expression for the chemical potential,

$$\Delta G_A^* = \mu_A - k_B T \ln \rho_A \Lambda_A = k_B T \ln \left\langle \exp\left[-\beta B_A\right] \right\rangle \tag{2.56}$$

where B_A is the total interaction energy of a single A with its entire surrounding molecules, and the average is taken over all the configurations of the molecules in the system, except the A molecule we have chosen to place at the center of the coordination system (Ben-Naim 2006). The local character of all the solvation quantities then follows from the short-range nature of the intermolecular interactions. In the next sections we calculate the local quantities for some specific systems.

2.6 MIXTURES OF HARD RODS (HRs) IN ONE-DIMENSIONAL SYSTEMS

We start with the simplest systems consisting of mixtures of hard rods (HRs) in one dimension (Figure 2.1). The HRs are defined in terms of the pair potential shown in Figure 2.2 and given by

$$U(r) = \begin{cases} \infty & \text{for} \quad r \leq \sigma \\ 0 & \text{for} \quad r > \sigma \end{cases} \tag{2.57}$$

where σ is the *diameter*, or the length of the rods. In these systems, the only molecular parameter that can be varied is the ratio of the diameters of the particles.

FIGURE 2.1 A system of hard rods in one dimension.

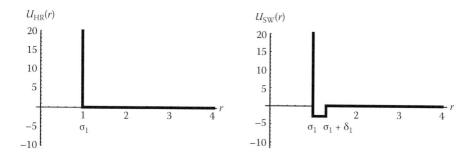

FIGURE 2.2 Hard rod and square-well potentials.

As we have noted in the introduction, the *local* properties provide more detailed information on the mixture than the *global* properties provide. This fact is obviously true for mixtures of HRs, for which all the excess thermodynamic quantities are zero, but the local quantities are not.

For HRs, the functions $\Psi_{\alpha\beta}$ are

$$\Psi_{\alpha\beta}(T,p) = \int_0^\infty \exp\left[-\beta pr - \beta U_{\alpha\beta}(r)\right]dr = \exp\left[-\beta p\sigma_{\alpha\beta}\right]/\beta p \tag{2.58}$$

We assume that

$$\sigma_{\alpha\beta} = \frac{\sigma_{\alpha\alpha} + \sigma_{\beta\beta}}{2} \tag{2.59}$$

In this system, the condition of Equation 2.39 is fulfilled, namely,

$$\Psi_{AB}^2 = \Psi_{AA}\Psi_{BB} \tag{2.60}$$

Hence, a mixture of HRs always forms a SI solution (Ben-Naim 2006). Therefore, for such mixtures, all the excess thermodynamic quantities are zero.

The chemical potentials of A and B are given by,

$$\mu_A(T,p,x_A) = p\sigma_{AA} + k_B T \ln\left(\beta p \Lambda_A x_A\right) \tag{2.61}$$

$$\mu_B(T,p,x_A) = p\sigma_{BB} + k_B T \ln\left[\beta p \Lambda_B (1-x_A)\right] \tag{2.62}$$

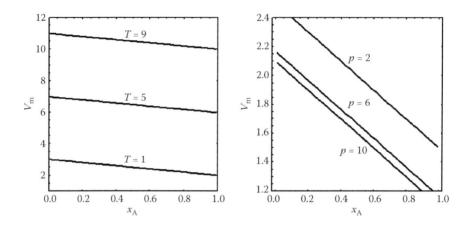

FIGURE 2.3   The volume of the mixture A and B as a function of the mole fraction of A, for hard rods with $\sigma_{AA} = 1$ and $\sigma_{BB} = 2$.

The partial molar volumes of the two components are

$$\overline{V}_A = \sigma_{AA} + (\beta p)^{-1}, \quad \overline{V}_B = \sigma_{BB} + (\beta p)^{-1} \tag{2.63}$$

The volume per particle of the mixture is

$$V_m = x_A\overline{V}_A + (1 - x_A)\overline{V}_B = x_A\sigma_{AA} + (1 - x_B)\sigma_{BB} + (\beta p)^{-1} \tag{2.64}$$

Figure 2.3 shows the volume per particle of a mixture of HRs with diameters $\sigma_{AA} = 1$ and $\sigma_{BB} = 2$. Clearly, as the pressure increases, or as the temperature decreases, the linear curves converge to the limiting linear curve,

$$V_m = x_A\sigma_{AA} + (1 - x_A)\sigma_{BB} \tag{2.65}$$

This is true either for $p \to \infty$ or for $T \to 0$. On the other hand, for $T \to \infty$ or for $p \to 0$, we get the ideal gas behavior of the mixture.

The isothermal compressibility of the mixture, defined in Equation 1.5, is given by

$$\kappa_T = \frac{1}{p\left[1 + \beta p\left(x_A\sigma_{AA} + (1 - x_A)\sigma_{BB}\right)\right]} \tag{2.66}$$

Figure 2.4 shows the isothermal compressibility as a function of x_A, at a fixed temperature ($k_BT = 1$) and various pressures. As expected, the isothermal compressibility is large at low pressures, but at high pressures it converges to zero, for any composition of the system.

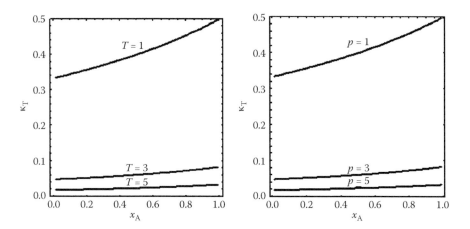

FIGURE 2.4 The isothermal compressibility of the same system as in Figure 2.3.

Next, we calculate the KBI from the inversion of the KB theory. The results are

$$G_{AA} = \rho\langle\sigma^2\rangle - 2\sigma_{AA}$$

$$G_{BB} = \rho\langle\sigma^2\rangle - 2\sigma_{BB} \qquad (2.67)$$

$$G_{AB} = \rho\langle\sigma^2\rangle - \sigma_{AA} - \sigma_{BB}$$

and

$$\langle\sigma^2\rangle = x_A\sigma_{AA}^2 + (1-x_A)\sigma_{BB}^2 \qquad (2.68)$$

In the limit, $\rho_B \to 0$ (or $x_A \to 1$), these quantities reduce to ($\rho \to \rho_A$),

$$G_{AA} = \rho_A\sigma_{AA}^2 - 2\sigma_{AA}$$

$$G_{BB} = \rho_A\sigma_{AA}^2 - 2\sigma_{BB} \qquad (2.69)$$

$$G_{AB} = \rho_A\sigma_{AA}^2 - \sigma_{AA} - \sigma_{BB}$$

and similar equations for the limit $\rho_A \to 0$ (or $x_A \to 0$).

Two other limiting cases are of interest. At very low densities $\rho \to 0$, we have

$$G_{AA} = -2\sigma_{AA}$$

$$G_{BB} = -2\sigma_{BB} \qquad (2.70)$$

$$G_{AB} = -\sigma_{AA} - \sigma_{BB}$$

Clearly, at this limit, the pair correlation functions are given by

$$g_{\alpha\beta}(r) \approx \exp\left[-\beta U_{\alpha\beta}(r)\right] \tag{2.71}$$

Hence, the KBIs in Equation 2.70 are all negative, and are due to the repulsive part of the hard core potential function Equation 2.57.

The other limit is of very high densities. There is a limit on the total density, which can be obtained from the equation of state (Equation 2.64). The limiting density can be obtained either at $p \rightarrow \infty$ or at $T \rightarrow 0$, and is

$$\rho_{\max} \rightarrow \left[x_A \sigma_{AA} + (1 - x_A)\sigma_{BB}\right]^{-1} = \langle\sigma\rangle^{-1} \tag{2.72}$$

At this limit we have

$$G_{AA} = \frac{\langle\sigma^2\rangle}{\langle\sigma\rangle} - 2\sigma_{AA}$$

$$G_{BB} = \frac{\langle\sigma^2\rangle}{\langle\sigma\rangle} - 2\sigma_{BB} \tag{2.73}$$

$$G_{AB} = \frac{\langle\sigma^2\rangle}{\langle\sigma\rangle} - \sigma_{AA} - \sigma_{BB}$$

Figure 2.5 shows G_{AA}, G_{BB}, and G_{AB} as a function of x_A for three total densities (all the $G_{\alpha\beta}$ are measured in units of $\sigma_{AA} = 1$). Note that the maximum density of pure A is $\rho_{\max,A} = 1$, and for B $\rho_{\max,B} = \frac{1}{2}$. Therefore, we plotted the values of $G_{\alpha\beta}$ for densities below the maximal density of the mixture. All the values of the $G_{\alpha\beta}$ are negative. At very low total densities, $G_{\alpha\beta}$ is equal to minus twice the distance of closest approach between α and β.

The limiting coefficients of the preferential solvation are

$$\delta_{A,A}^0 = x_A x_B (G_{AA} - G_{AB}) = (\sigma_B - \sigma_A)x_A x_B$$
$$\delta_{B,B}^0 = x_A x_B (G_{BB} - G_{AB}) = (\sigma_A - \sigma_B)x_A x_B \tag{2.74}$$

and

$$\Delta_{AB} = G_{AA} + G_{BB} - 2G_{AB} = (\sigma_B - \sigma_A) + (\sigma_A - \sigma_B) = 0 \tag{2.75}$$

Clearly, since the system is an SI solution, $\Delta_{AB} = 0$, but from $G_{\alpha\beta}$ we see that the ideality arises from the cancellation of the two terms δ_{AA}^0 and δ_{BB}^0, which in general are nonzero, and in our case $\sigma_B - \sigma_A = 1$ and $\sigma_B - \sigma_A = -1$.

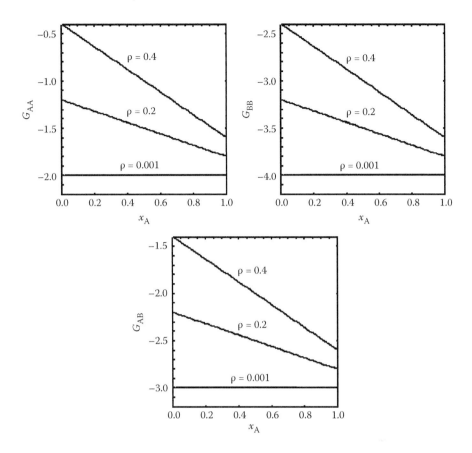

FIGURE 2.5   Values of the KBIs for the same system as in Figure 2.3 at different total densities.

Finally, Figures 2.6 and 2.7 show the solvation Gibbs energies of A and B at various temperatures and at various pressures. For the HR mixtures the solvation Gibbs energies are

$$\Delta G_A^* = p\sigma_{AA} + k_B T \ln\left(1 + \beta p \langle \sigma \rangle\right)$$
$$\Delta G_B^* = p\sigma_{BB} + k_B T \ln\left(1 + \beta p \langle \sigma \rangle\right)$$

(2.76)

Here, the ΔG_α^* are in dimensionless units. In subsequent calculations we chose $\varepsilon_{AA} = -1$, and expressed ΔG_α^* in units of ε_{AA}.

In the system of HR, the solvation Gibbs energy is related to the work of creating a cavity produced by the HR particles, which are *solvated* in the mixture (Ben-Naim 1977; Rowlinson and Swinton 1982). The larger the temperature, or the pressure, the greater the work required to create such a cavity.

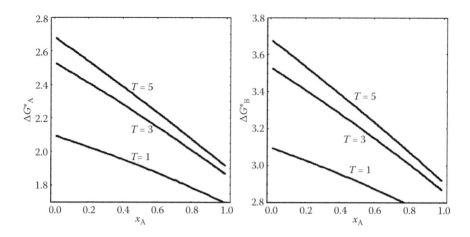

FIGURE 2.6   The solvation Gibbs energy of A and B for the same system as in Figure 2.3 at different temperatures.

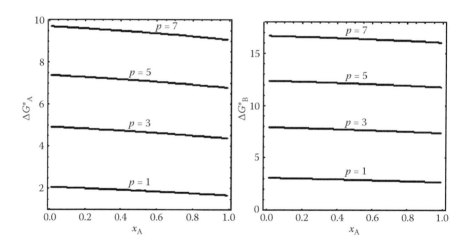

FIGURE 2.7   The solvation Gibbs energy of A and B for the same system as in Figure 2.3 at different pressures.

2.7   MIXTURES OF SQUARE-WELL PARTICLES IN A ONE-DIMENSIONAL SYSTEM

2.7.1   GLOBAL PROPERTIES

In this section, we present some results on the global and local properties of mixtures of particles interacting via square-well potentials. The square-well potential is defined by

$$U(r) = \begin{cases} \infty & \text{for} & r < \sigma \\ \varepsilon & \text{for} & \sigma \le r < \sigma + \delta \\ 0 & \text{for} & r > \sigma + \sigma \end{cases} \qquad (2.77)$$

We have seen that systems of hard rods form SI solutions. Therefore, all the excess thermodynamic quantities are zero. We have already examined the dependence of the local properties on the ratio of diameters in Section 2.6. Therefore, in this section, we choose equal diameters for the particles $\sigma_{AA} = \sigma_{BB} = 1$, and explore the dependence of the thermodynamic properties of the mixture on the ratio of the energy parameter ε. In the succeeding calculation, we choose dimensionless parameters,

$$\sigma_{AA} = \sigma_{AB} = \sigma_{BB} = 1$$

$$\varepsilon_{AA} = -1 \qquad (2.78)$$

$$\varepsilon_{BB} = \varepsilon \qquad \varepsilon_{AB} = \sqrt{\varepsilon_{AA}\varepsilon_{BB}}$$

and choose $T = 1$ and $p = 1$ for the numerical illustrations.

Figure 2.8 shows the volume per particle as a function of the mole fraction x_A for various values of the energy parameter ε. Since we have chosen equal diameters for the particles, the case $\varepsilon = -1$ corresponds to a SI solution. Since T and p are kept constant, the volume of the system increases or decreases according to the corresponding increase or decrease of $|\varepsilon|$.

Figure 2.9 shows the excess Gibbs energy of the mixture for the same set of molecular parameters at $\varepsilon = -1$; we have a SI solution and $g^E = 0$. As $|\varepsilon|$ either increases or decreases, we find positive deviations from SI behavior.

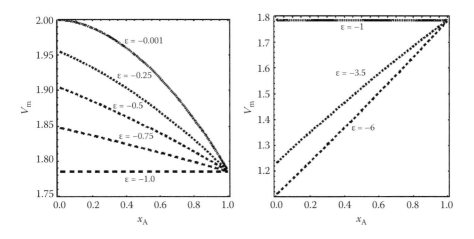

FIGURE 2.8 The volume of a system of A and B particles interacting via a square-well potential (Equation 2.77) with parameters as in Equation 2.78 for different values of ε.

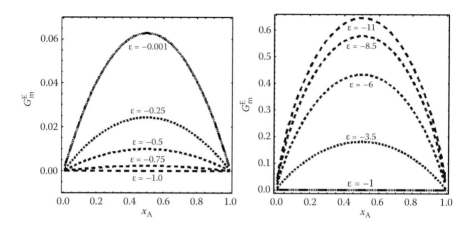

FIGURE 2.9   The excess Gibbs energy for the same system as in Figure 2.8.

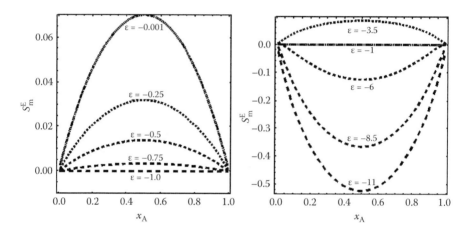

FIGURE 2.10   The excess entropy for the same system as in Figure 2.8.

Figure 2.10 shows the excess entropy of the system. Note that as $|\varepsilon|$ decreases, we get increasingly more positive S_m^E. As $|\varepsilon|$ increases above $|\varepsilon| = 1$, S_m^E initially remains positive, but it becomes negative for $|\varepsilon| \geq 5$.

Figure 2.11 shows the excess enthalpies of the system for the same set of parameters. As can be seen from the figure, all values of the excess enthalpies are positive.

Figure 2.12 shows the excess volume for the same set of parameters. For $|\varepsilon| \leq 1$, V_m^E is positive and decreases with $|\varepsilon|$. This makes sense; by weakening the overall interactions in the system, while keeping T and p constant, the volume expands. For $|\varepsilon| \leq 1$, the smaller the $|\varepsilon|$, the larger the excess volume. On the other hand, for $|\varepsilon| \geq 1$ we find an initial increase in V_m^E, but as $|\varepsilon|$ becomes larger, the strong interactions cause an overall contraction of the system, and the volume becomes negative.

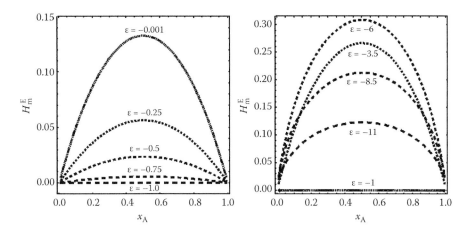

FIGURE 2.11   The excess enthalpy for the same system as in Figure 2.8.

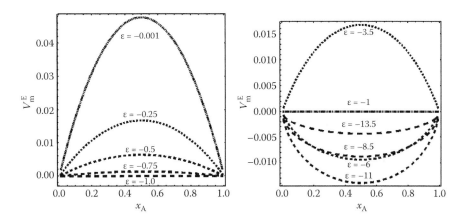

FIGURE 2.12   The excess volume for the same system as in Figure 2.8.

2.7.2   LOCAL PROPERTIES CALCULATED FROM THE INVERSION
OF THE KIRKWOOD–BUFF THEORY

We now turn to a few illustrations of the local quantities for these systems. We use the same set of molecular parameters, and fixed temperature $T = 1$, and pressure $p = 1$, as in the previous illustration.

As a general comment, we note that when $\varepsilon = -1$, the system is a SI solution, and as we have seen, all the excess thermodynamic functions are zero. The local quantities, on the other hand, are not zero. What we observe in Figure 2.13 through Figure 2.15 is that the values of the KBIs do not depend on the composition when $|\varepsilon| = 1$, simply because the particles are identical. However, the values of the KBIs are nonzero, as is the case for real liquid mixtures.

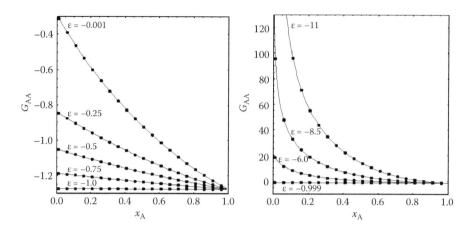

FIGURE 2.13 The KBI G_{AA} for the same system as in Figure 2.8. The dots are the values calculated by the inversion of the KB theory. The continuous curves are calculated as described in Section 2.7.3.

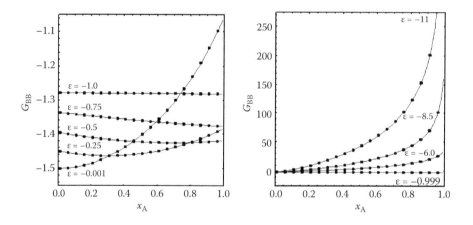

FIGURE 2.14 The KBI G_{BB} for the same system as in Figure 2.8.

The interpretation of the results is sometimes straightforward, sometimes more difficult. In most cases, we can interpret the sign and the magnitude of the KBI by considering the split of the integral as follows (Ben-Naim 2006),

$$G_{\alpha\beta} = \int_{-\infty}^{\infty} \left[g_{\alpha\beta}(r) - 1 \right] dr = -2\sigma_{\alpha\beta} + A_{\alpha\beta} \tag{2.79}$$

where $-2\sigma_{\alpha\beta}$ is a result of the direct repulsion between the two particles α and β, and $A_{\alpha\beta}$ is due to the integration over distances beyond the hard-core diameters of the particles. Clearly, the first term is negative and proportional to the size of the

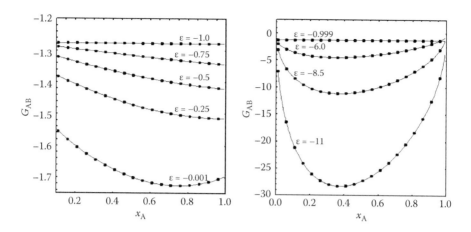

FIGURE 2.15   The KBI G_{AB} for the same system as in Figure 2.8.

particles. The second can be either positive or negative depending on the form of the pair correlation function in the region $\sigma_{\alpha\beta} \le r \le \infty$.

Figure 2.13 shows G_{AA} as a function of the mole fraction of A. For small values of $|\varepsilon| \le 1$, we see that G_{AA} is negative. This is mainly due to the repulsive part of the interaction as we observed for HRs in Figure 2.5. As we increase $|\varepsilon|$, the KBI becomes larger and less dependent on the composition. At $|\varepsilon| = 1$, the values of G_{AA} are independent of the composition and attain the value of G_{AA} for pure A. Note also that all the curves converge to the same value for $x_A = 1$, that is, the value of G_{AA} for pure A, which in this case is about -1.22. Once we increase $|\varepsilon|$ beyond $|\varepsilon| > 1$, the affinities between a pair of A particles increase due to the stronger cohesive forces of the *solvent*, which is the B component when $x_A \approx 0$. The larger $|\varepsilon|$, the larger and more positive are the values of G_{AA}. Note again that all the curves converge to the limiting value of pure A at $x_A \to 1$.

Figure 2.14 shows the values of G_{BB} for the same set of parameters. Here, the behavior is quite different from that of G_{AA}. When $x_A = 0$, that is, for pure B, the values of G_{BB} still change as a function of ε. As we increase $|\varepsilon|$, the affinities between two B particles become less and less negative. Note again that for $|\varepsilon| = 1$, G_{BB} is independent of x_A, but it is not zero. Again, we see that when $|\varepsilon|$ increases beyond one, the values of G_{BB} become large and positive.

It should be noted that the molecular reasons for the large positive values of G_{BB} at $x_A \approx 1$ are similar to the large positive values of G_{AA} at $x_A \approx 0$. In the case of G_{BB} at $x_A \approx 1$, we have B diluted in A. The stronger the BB interaction is, the larger the positive affinities between two B particles are.

Figure 2.15 shows the behavior of G_{AB}, which is quite different from the behavior of both G_{AA} and G_{BB}. The reason is that in neither $x_A = 0$, nor in $x_A \approx 1$, we have a KBI of "pure" AB. Here again we find that for $|\varepsilon| = 1$, G_{AB} is nonzero, but it is independent of x_A. For small values of $|\varepsilon|$ (≤ 1), G_{AB} is negative, but decreases in absolute magnitude when $|\varepsilon|$ increases toward $|\varepsilon| = 1$. On the other hand, for $|\varepsilon| > 1$, we see that G_{AB} becomes large and negative as $|\varepsilon|$ increases. Another phenomenon that is different

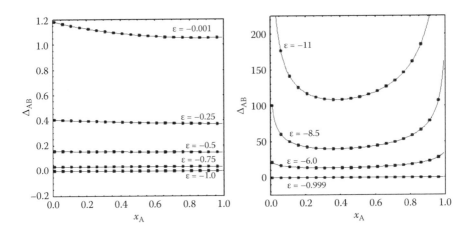

FIGURE 2.16 The measure of the deviation from an SI solution Δ_{AB} for the same system as in Figure 2.8.

from the case of G_{AA} and G_{BB} is that G_{AB} converges to the same value at both ends of the composition range.

Next, we turn to the quantity Δ_{AB} shown in Figure 2.16, which is a measure of the deviations from an SI solution. As expected, for $\varepsilon = -1$, we find that $\Delta_{AB} = 0$. Since the particles are identical, they form an SI solution. For $|\varepsilon| \leq 1$, Δ_{AB} is the larger dissimilarity between the particles, that is, the smaller the value of $|\varepsilon| \leq 1$. However, for $|\varepsilon| \geq 1$, we observe larger positive deviations from SI. The deviations seem to get very large at both ends of the composition range.

In Figure 2.17, we present the limiting coefficient of the PS of A with respect to A and B. These quantities are of interest in their own right. First, we note that for $|\varepsilon| = 1$ values of δ_{AA}^0 are all zero and independent of x_A. This result is due to the fact

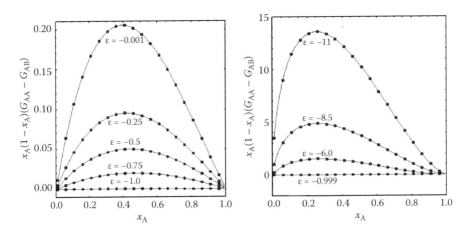

FIGURE 2.17 The limiting coefficient of the preferential solvation of A with respect to A and B for the same system as in Figure 2.8.

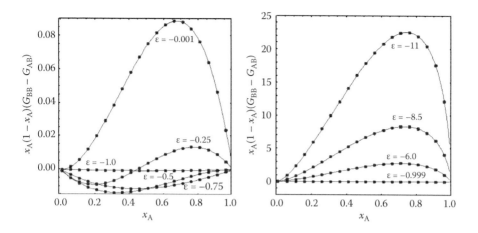

FIGURE 2.18 The limiting coefficient of the preferential solvation of B with respect to A and B for the same system as in Figure 2.8.

that at $|\varepsilon| = 1$, the two components are identical, and therefore there is no preferential solvation. It should be noted that this lack of PS is not a result of the SI behavior but a result of the *identity* of the particles. In general, SI solutions can occur even for different particles, but SI behavior does not imply zero PS. The limiting coefficient of the PS of B with respect to A and B is shown in Figure 2.18. Two interesting features that distinguish these curves from the ones in Figure 2.17 are: first, we find both positive and negative values of the PS; second, the sign of the PS changes when the composition of the mixture changes.

Finally, we present some values of the solvation Gibbs energies of A and B in the entire range of compositions. Figure 2.19 shows ΔG_A^* as a function of composition. Note again that for $|\varepsilon| = 1$, ΔG_A^* is independent of the composition. This is a result of the identity of the particles. Note also that the value of ΔG_A^* is nonzero even at

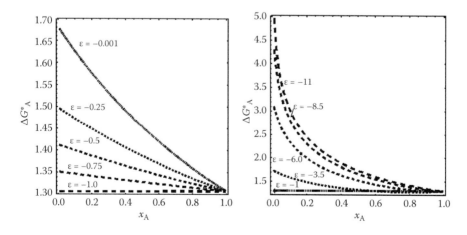

FIGURE 2.19 The solvation Gibbs energy ΔG_A^* in the same system as in Figure 2.8.

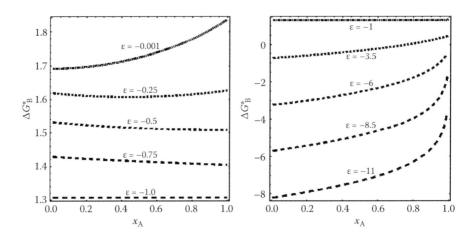

FIGURE 2.20 The solvation Gibbs energy ΔG_B^* in the same system as in Figure 2.8.

$|\varepsilon| = 1$. ΔG_A^* measures the average value of the quantity $\exp[-\beta B_A]$, where B_A is the total interaction energy of a single A with the rest of the system. This quantity is often referred to as the *free energy of interaction* of the species A. As $|\varepsilon| \leq 1$ becomes smaller, we see that the values of ΔG_A^* become larger. On the other hand, for $|\varepsilon| \geq 1$, the reverse trend is true. Also note that in all cases, ΔG_A^* converge to the value of ΔG_A^* for the solvation of A in pure A (i.e., at $x_A \rightarrow 1$). We note also that the behavior of ΔG_A^* as a function of x_A for $|\varepsilon| \geq 1$ is similar to the case of argon in mixtures of argon and xenon (Ben-Naim 1989). The interpretation of this phenomenon is simple. Starting from $x_A = 0$, we have the solvation of A surrounded by pure B. As x_A increases from $x_A = 0$ to $x_A = 1$, the surroundings of A changes from all B to all A particles; hence, the free energy of interaction decreases strongly toward the value of ΔG_A^* for pure A.

A different behavior is exhibited by ΔG_B^* shown in Figure 2.20. Here again, we find that ΔG_B^* is composition independent for $|\varepsilon| = 1$. However, in contrast to the case of ΔG_A^* (Figure 2.19), the values of ΔG_B^* do not converge to a single value for pure B ($x_A = 0$). The reason is simple. In the case of ΔG_A^*, "pure" A (i.e., $x_A = 1$) means a unique system of A particles with fixed molecular parameters $\sigma_{AA} = 1$ and $\varepsilon_{AA} = -1$. On the other hand, when we say pure B (i.e., $x_A = 0$), there are different "pure" systems with $\sigma_{BB} = 1$, but varying values of ε_{BB}.

2.7.3 LOCAL PROPERTIES CALCULATED DIRECTLY FROM THE PAIR CORRELATION FUNCTIONS

In this section we present the results of a recent paper (Ben-Naim and Santos 2009) where we directly recalculated the KBIs for two component mixtures of particles interacting via square-well potential. The theoretical background is lengthy and will not be presented here. Instead, we show a sample of results for mixtures of square-well particles. It is shown that the results are in quantitative agreement with those obtained from the inversion of the Kirkwood–Buff theory of solution. We also

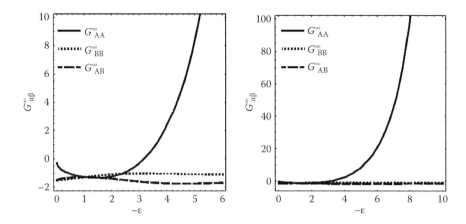

FIGURE 2.21 The limiting values of $G_{\alpha\beta}$ as $x_A \to 0$, as a function of $-\varepsilon$.

calculated the limiting values of the KBIs at $x \to 0$ which we could not have done from the partition function methods.

Another issue that has been examined both numerically and theoretically is the deviation from symmetric ideal solution behavior, and its relation with the stability of the mixtures. It was shown that no miscibility gap can occur in such mixtures (Ben-Naim and Santos 2009). It should be noted that Equation 2.46 through Equation 2.48 in the original study contain some errors, although this did not affect the conclusions (Ben-Naim and Santos 2009). The second term on the right-hand side of Equation 2.46 should contain x_B instead of x_A in the numerator. This also affects the expressions provided in Equation 2.47 and Equation 2.49 of that paper.

In Figure 2.21 we show the KBIs in the limit of $x_A \to 0$ as functions of $-\varepsilon$ for the same system as that of Figure 2.13 through Figure 2.15. We observe that both G_{BB} and G_{AB} are hardly sensitive to the value of ε. In contrast, the solute–solute KBI, G_{AA}, is strongly influenced by the solvent–solvent potential depth, increasing both for small and for large values of $|\varepsilon|$. A careful inspection of the explicit expressions for G_{AB} in the limit $|\varepsilon| \to \infty$ shows that, while G_{BB} and G_{AB} tend to the *same* constant value, G_{AA} diverges as $G_{AA} \sim \exp[(|\varepsilon_{BB}| - 2|\varepsilon_{AB}|)/k_B T)]$. This phenomenon might be relevant to the study of hydrophobic interactions, as discussed in Ben-Naim (2011).

It is interesting to note that the divergence of G_{AA} is due to the increase in the height of the first peak in the radial distribution function $g_{AA}(r)$, see Figure 2.22. On the other hand, both $g_{BB}(r)$ and $g_{AB}(r)$ show long range oscillatory behavior in Figures 2.23 and 2.24.

2.8 CONCLUSIONS

In this chapter, we study very simple systems. Nevertheless, the main message as expressed in the introductory section is quite clear. The *local* properties of the mixtures are richer and more informative than the *global* properties. It is therefore advisable to study both sets of properties of liquid mixtures. This conclusion becomes

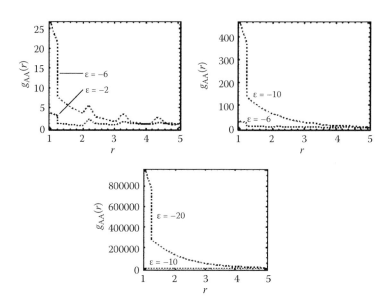

FIGURE 2.22 $g_{AA}(r)$ as $x_A \to 0$ for various values of ε.

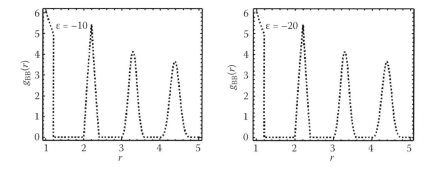

FIGURE 2.23 $g_{BB}(r)$ as $x_A \to 0$ for various values of ε.

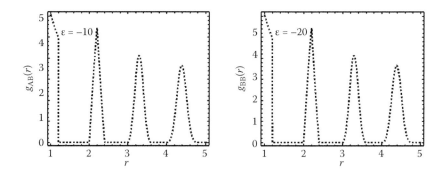

FIGURE 2.24 $g_{AB}(r)$ as $x_A \to 0$ for various values of ε.

a fortiori true for more complex mixtures, particularly aqueous solutions. The agreement between the values of $G_{\alpha\beta}$ obtained by the two methods proves that the inversion of the KB theory provides reliable results on the KBI.

ACKNOWLEDGMENT

I would like to express my thanks to Andres Santos for his help in the calculation of the Kirkwood–Buff integrals from the pair correlation functions in one-dimensional systems.

3 Preferential Solvation in Mixed Solvents

Yizhak Marcus

CONTENTS

Abstract: In a binary mixture of solvents A and B, the surroundings of a molecule of A (or of B) generally differ in terms of the relative amounts of A and B molecules from the bulk composition due to preferential solvation. An eminent method for studying this situation is the use of fluctuation theory, in terms of the Kirkwood–Buff integrals derived from thermodynamic data, provided the latter are sufficiently accurate. The interactions among the components are obtained from this approach when their relative sizes are taken into account. This method has been applied to a large number of binary aqueous-organic solvent mixtures as well as to nonaqueous mixtures and to a small number of ternary solvent mixtures. In some cases, the preferential solvation occurs beyond the first solvation shell, for example, in aqueous mixtures of tetrahydrofuran or acetonitrile, where self-interactions of the water molecules far outweigh those between water and organic cosolvent molecules. The merits and disadvantages of using the Kirkwood–Buff integral approach are briefly compared with those of other approaches, such as the quasi-lattice, quasi-chemical one or the use of solvatochromic probes and other spectroscopic methods.

3.1 INTRODUCTION

In a binary solvent mixture of solvents A and B, the surroundings of a molecule of A (or of B) in terms of the relative amounts of A and B molecules generally differ from the bulk composition due to preferential solvation. In some mixtures, mutual solvation dominates over self-association and in others, the opposite occurs. This

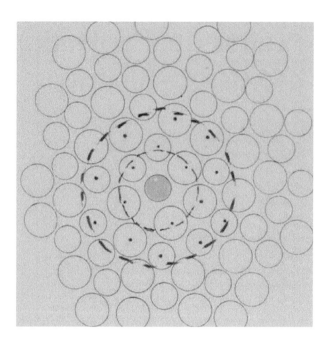

FIGURE 3.1   Schematic (two-dimensional) presentation of preferential solvation in a binary mixture of solvents A (large circles) and B (small circles). In the first solvation shell of the central (grey) molecule, within the inner dashed circle, there are three A and one B molecule, so A preferentially solvates the central molecule. In the outer solvation shell, which includes the inner shell, there are eight A and eight B molecules, the same ratio as in the bulk, so no further preferential solvation takes place.

situation is presented schematically (in two dimensions) in Figure 3.1. In order to ascertain which kind of preferential solvation (mutual or self) takes place, if any, and its quantitative extent, it is necessary to apply some methodology that can be applied to experimental data for the binary mixtures. Fluctuation theory, in the form of the inversed Kirkwood–Buff integrals (Ben-Naim 1977), is an eminent method to deal with this problem, although not the only one.

The preferential solvation in a binary mixture of solvents A and B is expressed most clearly by the local composition x_{AB}^{L} of A molecules around a central B one, where necessarily $x_{AB}^{L} + x_{BB}^{L} = 1$ (see Section 1.3.4 in Chapter 1). The preferential solvation parameter is defined as $\delta x_{AB} = x_{AB}^{L} - x_{A}$. Positive values of δx_{AB} denote preferential mutual solvation of B by A molecules and again necessarily $\delta x_{AB} = -\delta x_{BB}$ (self-solvation of B molecules is disfavored). Conversely, positive values of $\delta x_{AA} = -\delta x_{BA}$ denote preferential self-solvation of A molecules. Thus, only two of the four possible parameters need to be presented. The composition dependence of δx_{AB} and δx_{AA} is of interest, showing at which compositions such preferences, if they do occur, are appreciable. Values of $|\delta x_{AB}| \leq 0.01$ and $|\delta x_{AA}| \leq 0.01$ are not significant, due to the inherent errors in the data, and signify essentially the absence of preferential solvation at compositions where such values are obtained. If $\delta x_{AB} = x_{B}$ this denotes selective solvation, namely $x_{AB}^{L} = 1$, and molecules of B are surrounded by those of

A only, excluding B molecules. Therefore, values of δx_{AB} cannot exceed x_B. Similar considerations are valid for δx_{AA} that they may not exceed x_A.

Another way to represent preferential solvation is by means of the solvation ratio,

$$K_{AB} = \left(x_{AB}^L/x_{BB}^L\right)/\left(x_B/x_A\right) = \left(x_{AB}^L/x_A\right)/\left(x_{BB}^L/x_B\right) \tag{3.1}$$

Preferential solvation takes place if $K_{AB} \neq 1$; however, this presentation loses the details of the $\delta x_{AB}(x_B)$ and $\delta x_{AA}(x_B)$ curves and is less desirable.

Ternary mixtures, in which the third component, for example, S in the A + B mixtures (Ben-Naim 1990b), is a solute present at vanishing concentrations do not pertain to the subject of this chapter and are described elsewhere. However, proper mixtures of three solvents, A + B + C, are briefly dealt with in this chapter.

3.2   CALCULATIONS

The three Kirkwood–Buff integrals (KBIs) for binary mixtures of solvents A and B: G_{AA}, G_{AB}, and G_{BB}, are obtained from proper thermodynamic data for the pure components and the mixtures as described elsewhere (see Section 1.3.2 in Chapter 1). The preferential solvation parameters are related to the KBIs as follows (Ben-Naim 1990b),

$$\delta x_{BB} = x_{BB}^L - x_B = x_B x_A (G_{BB} - G_{AB})/\left[x_B G_{BB} + x_A G_{BA} + V_{cor,B}\right] \tag{3.2}$$

$$\delta x_{BA} = x_{BA}^L - x_B = x_B x_A (G_{BA} - G_{AA})/\left[x_B G_{BA} + x_A G_{AA} + V_{cor,A}\right] \tag{3.3}$$

as also illustrated in Section 1.3.4 in Chapter 1. The quantities $V_{cor,B}$ and $V_{cor,A}$ represent the correlation volumes around the central molecules B and A in which the preferential solvation occurs. These volumes could correspond to only the first solvation sphere (Figure 3.1), but may also include a second or further solvation sphere. It is the determination of the correlation volumes as functions of the bulk composition of the solvent mixtures that leads to definite values of the preferential solvation parameters according to Equations 3.2 and 3.3. The calculation of the correlation volumes requires the determination of the correlation radii, R_{cor} of sequential solvation shells around a given molecule, to which the correlation volumes, $V_{cor} = (4\pi N_A/3)R_{cor}^3$, are related. Such radii of solvation shells are obviously sensitive to the composition of these shells (for molecules of different sizes); hence, an iterative calculation is required (Marcus 1990). The hard sphere diameter $2r_i$ of a solvent molecule was related to its pure molar volume V_i^o as (Kim 1978; Marcus 1990)

$$2(r_i/\text{nm}) = 0.1363\left(V_i^o/\text{cm}^3\text{mol}^{-1}\right)^{1/3} - 0.085 \tag{3.4}$$

The correlation volume, $V_{cor}/\text{cm}^3\text{mol}^{-1}$, can be calculated for m consecutive spherical solvation shells, taking into account partial penetration of molecules from farther shells into nearer ones (Figure 3.1) as well as the preferential solvation in these shells (Marcus 1990). The correlation volumes around A and B molecules are then

$$V_{\text{cor,A}} \qquad\qquad\qquad\qquad\qquad\qquad\qquad\qquad\qquad\qquad (3.5)$$

$$= 2522.7\left\{-0.085m + 0.1363/2\left(V_A^o\right)^{1/3} + 0.1363(m-0.5)\left[x_{AA}^L V_A^o + (1-x_{AA}^L)V_B^o\right]^{1/3}\right\}^3$$

$$V_{\text{cor,B}} \qquad\qquad\qquad\qquad\qquad\qquad\qquad\qquad\qquad\qquad (3.6)$$

$$= 2522.7\left\{-0.085m + 0.1363/2\left(V_B^o\right)^{1/3} + 0.1363(m-0.5)\left[x_{AB}^L V_A^o + \left(1-x_{AB}^L\right)V_B^o\right]^{1/3}\right\}^3$$

The preferential solvation parameters calculated from the KBIs G_{AA}, G_{AB}, and G_{BB} and Equations 3.2 and 3.3, with the aid of Equations 3.5 and 3.6 for the correlation volumes, represent the total values. These are the formally correctly derived values from the inverse Kirkwood–Buff theory applied to thermodynamic data.

However, these preferential solvation parameters do not pertain directly to the molecular *interactions* leading to mutual or self-association. Even for ideal mixtures of solvents A and B, where the interactions of all the molecules are the same, preferential solvation occurs due to the different sizes of the molecules of A and B as pointed out by Matteoli (1997). An example he showed is the *n*-hexane + *n*-hexadecane system, where the ratio of the molar volumes of the components is ~2.5. These mixtures are athermal, signifying that the interactions of all the molecular segments (in the Flory–Huggins sense) are the same. Still, application of the inverse Kirkwood–Buff theory leads to appreciable preferential solvation. This preferential solvation does take place, but is irrelevant if the interactions between the molecules of the components are to be explored. Thus, volume-corrected preferential solvation parameters, $\delta x_{AA}'$ and $\delta x_{AB}'$ should be calculated if such interactions are to be investigated. They result when the KBIs G_{ij} ($i = j$ and $i \neq j$) in Equations 3.2 and 3.3 are replaced by $\Delta G_{ij} = G_{ij} - G_{ij}^{\text{id}}$. In turn, the G_{ij}^{id}, which are the ideal KBIs, are calculated from the thermodynamic data (the molar volumes of the components) on setting to zero the excess quantities $G^E(x_B) = 0$ and $V^E(x_B) = 0$, leading to $\Delta_{AB} = G_{AA} + G_{BB} - 2G_{AB} = 0$ (see Section 1.3.7 in Chapter 1). The G_{ij}^{id} are based solely on the relative molar volumes of A and B. The volume-corrected preferential solvation parameters are then

$$\delta x_{BB}' = x_B x_A (\Delta G_{BB} - \Delta G_{AB}) / [x_B \Delta G_{BB} + x_A \Delta G_{BA} + V_{\text{cor,B}}] \qquad (3.7)$$

$$\delta x_{BA}' = x_B x_A (\Delta G_{BA} - \Delta G_{AA}) / [x_B \Delta G_{BA} + x_A \Delta G_{AA} + V_{\text{cor,A}}] \qquad (3.8)$$

and describe the preferred mutual interactions or self-interactions of the molecules of the two components, with also $\delta x_{AB}' = -\delta x_{BB}'$ and $\delta x_{AA}' = -\delta x_{BA}'$.

Another important point to be mentioned is the requirements for accuracy of the thermodynamic quantities that are to be employed in the calculations. These quantities are the isothermal compressibility κ_T of the mixtures, the partial molar volumes of the two components, $\bar{V}_A$ and $\bar{V}_B$, and the derivative of the chemical potential, activity, or activity coefficient of one of the components, say $(\partial \ln \gamma_B / \partial x_B)_{p,T}$, or alternatively the second derivative of the excess Gibbs energy of mixing of the components, $(\partial^2 G_m^E / \partial x_B^2)_{p,T}$, all as functions of the composition at a given temperature (and pressure).

It turns out that the accuracy requirement from κ_T is numerically minimal, because $k_B T \kappa_T$ is small compared with the other terms contributing to the G_{ij} values. If the values of κ_T are not known for the mixtures to be studied, they may be sufficiently well approximated by $\kappa_T \approx x_A \kappa_{T,A}^o + x_B \kappa_{T,B}^o$. Otherwise, if the adiabatic compressibility, κ_S, obtained from the ultrasound velocities and densities, is known as a function of the composition, the approximation $\kappa_T \approx 1.1 \times \kappa_S$ may be applied.

The need for accurate partial molar volumes of the components as functions of the composition can be met when the excess volume of the mixture, V_m^E, is known. The KBIs are proportional to the partial molar volumes, and, hence, have the same relative accuracy as these quantities. If no partial molar volumes are known, then the approximations $\bar{V}_A \sim V_A^o$, $\bar{V}_B \sim V_B^o$, and $V_m \sim x_A V_A^o + x_B V_B^o$ may be used for the calculation of the G_{ij}.

The crucially important quantity with regard to the needed accuracy is $(\partial \ln \gamma_B / \partial x_B)_{p,T}$ or the corresponding derivative of a_B or μ_B. If these quantities are not available, the less desirable $(\partial^2 G_m^E / \partial x_B^2)_{p,T}$ may serve, this being a second derivative, but then G_m^E may be the only quantity that has been reported. The seriousness of this requirement for accuracy (not of precision) is best shown by the case of aqueous acetonitrile mixtures (Marcus 2008). The quantity D (see Section 1.3.2 in Chapter 1),

$$D = 1 + x_B (1 - x_B) \left[\partial^2 \beta G_m^E / \partial x_B^2 \right]_{T,p} \tag{3.9}$$

(or corresponding quantities if γ_B, a_B, or μ_B are to be used) occurs in the denominators of the expressions used for the calculation of the KBIs, G_{ij} (see Equation 1.72), which ultimately provide the preferential solvation parameters, δx_{ij}. If D is negative, phase separation occurs, but even if it is rather small, though positive, say $D < 1$, it leads to diverging values of the integrals. In the case of water (A) + acetonitrile (B) mixtures, the onset of $D < 1$ as x_B is increased is near $x_B = 0.25$ at all the temperatures studied, but the range where $D < 1$ becomes narrower as T increases, the upper limit being: $x_B \sim 0.65$ at 278 K, 0.60 at 288 K, 0.55 at 298 K, 0.50 at 313 K, and 0.45 at 323 K. This composition range agrees substantially with that corresponding to the occurrence of microheterogeneity derived from various other measurements. The resulting δx_{AA} and δx_{AB} values have positive and negative signs respectively throughout the composition range, suggesting water–water and acetonitrile–acetonitrile clustering. The point to be made, however, is that some of the data used here, for example, (Treiner et al. 1976) for 298.15 K, (French 1987) for 278.15 and 288.15 K, and (Nikolova et al. 2000) for 293.15 K, lead to $D < 0$ over a narrow portion of the composition range, signifying phase separation. Such separation, however, occurs only at lower temperatures, ≤ 271.86 K. Hence, these data are not sufficiently accurate for obtaining the true preferential solvation picture for aqueous acetonitrile mixtures (see below for the detailed presentation of this system).

Another quantity besides the (volume-corrected) preferential solvation parameters that can be extracted from the KBIs is the molar second virial coefficient of a component, say A, in the mixture A + B. The limiting value of the G_{AA} as its concentration tends to zero, G_{AA}^{∞}, is minus twice the molar osmotic pairwise virial coefficient of this component (Matteoli and Lepori 1984). This quantity is related to the self-association of the component A as a solute at infinite dilution in component B as a solvent. When G_{AA}^{∞} is positive then self-association of A is appreciable, in spite of solvation by component B, and is the more pronounced as it becomes larger. Contrarily, if G_{AA}^{∞} is negative then its solvation by component B (the solvent) dominates over its self-association.

3.3   PREFERENTIAL SOLVATION IN SOME BINARY SYSTEMS

Following are descriptions of the preferential solvation in a large variety of binary solvent mixtures, in terms of the preferential solvation parameters. These have been determined mostly by the present author, either from KBIs calculated by him from thermodynamic data or from those calculated by other authors. The sources of the thermodynamic data that have been employed for obtaining them are listed in the papers quoted. In the present context, only the results of the application of the fluctuation theory are presented, so that detailed discussions of the chemistry involved in the preferential solvation that was noted for the mixtures dealt with here should be sought in the publications quoted. The temperatures quoted in the following are rounded to integer values.

3.3.1   AQUEOUS MIXTURES

Matteoli and Lepori were among the first who calculated the KBIs for binary aqueous-organic solvent mixtures from thermodynamic data over the entire composition range (Matteoli and Lepori 1984). They did not, however, report preferential solvation parameters, although they discussed the significance of the KBIs in terms of the interactions that take place. The present author (Marcus 1990) took up these cases, as well as those dealt with by Ben-Naim (1990b), and by adding the explicit expression for the correlation volume (Equations 3.5 and 3.6) was able to present preferential solvation parameters for all these systems. In the following, the component water is described by the subscript $_w$ and the organic solute by the subscript $_s$. For aqueous systems this replaces the A and B notation used above. Fluctuation theory in terms of KBIs has by now been applied to a fairly large number of mixtures and is summarized in a review paper (Marcus 2001) and in a book (Marcus 2002b), which also presents references to the works of other authors. The KBIs of the lower alkanols and some other aqueous systems at 298.15 K that have been presented by other authors (Matteoli and Lepori 1984; Ben-Naim 1990b) are in good agreement with those calculated later (Marcus 2001).

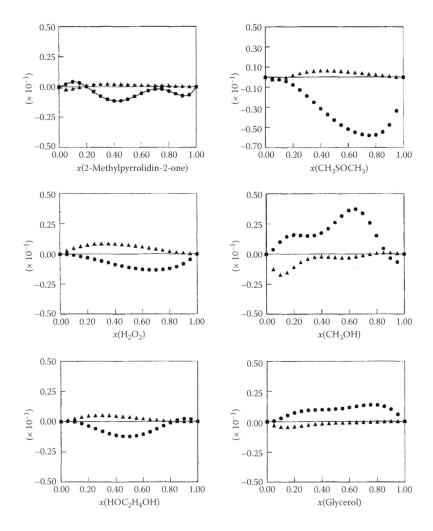

FIGURE 3.2 Volume-corrected preferential solvation parameters $\delta x'_{WW}$ (●) for water–water interactions and $\delta x'_{WS}$ (▲) for water–solute interactions in the first solvation shell of aqueous mixtures with solutes identified in the label of the abscissa. (From Y. Marcus, 2001, Preferential Solvation in Mixed Solvents, Part 10, Completely Miscible Aqueous Co-Solvent Binary Mixtures at 298.15 K, *Monatshefte für Chemie*, 132, 1387, by permission of the publisher, Springer.)

Hydrogen peroxide (Figure 3.2). The $\delta x'_{WW}$ has small negative values (down to −0.013 at $x_S = 0.70$) and $\delta x'_{WS}$ has small positive values (up to 0.009 at $x_S = 0.35$). This system is close to ideal and preferential solvation is hardly significant.

Methanol (Figure 3.2). This system was studied over a wide temperature range, from subambient to near the critical point. The $\delta x'_{WW}$ at 298 K has small positive values (up to 0.037 at $x_S = 0.65$) and $\delta x'_{WS}$ has small negative values (down to -0.017 at $x_S = 0.10$). Nonvolume-corrected preferential solvation parameters for temperatures from 280 K to 523 K have been calculated (Marcus 1999). Double-humped curves of δx_{WW} result at the higher temperatures and their values range up to 0.07, and those of δx_{WS} range down to a deep dip of -0.04 near $x_S = 0.10$, but go to positive values beyond $x_S \sim 0.25$. Correction for the unequal volumes brings $\delta x'_{WS}$ to slight negative values only (Marcus 2001). It is noteworthy that preferential solvation persists in a second solvation shell but practically vanishes in the third shell.

Ethanol, 1-propanol, and t-butanol (Figure 3.3). Self-association of the water in aqueous ethanol is manifested by $\delta x'_{WW}$ at 298 K going up to 0.16 near $x_S = 0.45$ and disfavored mutual association of water and ethanol by $\delta x'_{WS}$ going down to -0.04 near $x_S = 0.35$. KBIs (G_{SS}) at 323 K and 363 K were also shown beside the 298 K one (Matteoli and Lepori 1984) yielding similar trends. For 1-propanol and *t*-butanol the corresponding values for the first solvation shell at 298 K are $\delta x'_{WW}(max) = 0.42$ and 0.40 and $\delta x'_{WS}(min) = -0.28$ and -0.24, respectively. Preferential solvation was also calculated for the second solvation shell: $\delta x'_{WW}(max) = 0.22$ and 0.28 and $\delta x'_{WS}(min) = -0.09$ and -0.08 respectively for 1-propanol and *t*-butanol (second solvation shell calculations were not carried out for aqueous ethanol).

2-Propanol and 1-butanol. The preferential solvation of 2-propanol is comparable with that of ethanol, being considerably smaller than for 1-propanol and *t*-butanol. The values of $\delta x'_{WW}(max) = 0.165$ and $\delta x'_{WS}(min) = -0.031$ were found at 298 K. Aqueous 1-butanol has a phase separation region and the KBIs were shown mainly for the butanol-rich region (Matteoli and Lepori 1984). Interestingly, Ben-Naim (1990b) ignored the phase separation and calculated derived quantities (local mole fractions) as if it did not occur, but perhaps *t*-butanol was meant instead.

1,2-Ethanediol (ethylene glycol) (Figure 3.2). This aqueous system is, as expected, nearly ideal, resembling the hydrogen peroxide one but being more symmetrical. The extreme values are $\delta x'_{WW}(min) = -0.012$ and $\delta x'_{WS}(max) = 0.005$ at 298 K, the latter being hardly significant at all.

1,2-Propanediol (propylene glycol) and polyethyleneglycols. The preferential solvation curves become less symmetric as the sizes of the glycols increase. In the series ethylene glycol, diethylene glycol, triethyleneglycol, tetraethylene glycol, PEG 300 (polyethylene glycol with an average relative molar mass of 300), and PEG 400 at the equimolar composition, $x_S = 0.5$, the water self-preference, $\delta x'_{WW}$, decreases from $= -0.005$ to -0.035, whereas at $x_S = 0.2$ it first decreases from -0.012 to a minimum of -0.025 for tetraethylene glycol but increases again for the PEGs and is positive, $\delta x'_{WW} = 0.025$, for PEG 400. The mutual preferential solvation parameters, $\delta x'_{WS}$, remain insignificant, ≤ 0.010, along the series. For propylene glycol, however, slightly positive $\delta x'_{WW}$ and slightly negative $\delta x'_{WS}$ values resulted at all compositions, remaining below the significance value, ± 0.010 (Marcus 2003) (see also Matteoli 1997).

Glycerol (Figure 3.2). This aqueous system was dealt with in detail (Marcus 2000) and the system is close to ideal, but $\delta x'_{WW} > 0$ and $\delta x'_{WS} < 0$, contrary to aqueous ethylene glycol, though both parameters are very small: $\delta x'_{WW}(max) = 0.014$ and $\delta x'_{WS}(min) = -0.005$ (hardly significant) at 298 K.

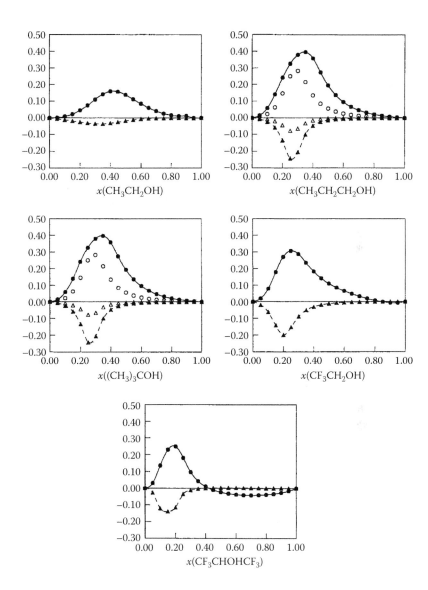

FIGURE 3.3 Volume-corrected preferential solvation parameters $\delta x'_{WW}$ (●) for water–water interactions and $\delta x'_{WS}$ (▲) for water–solute interactions in the first solvation shell of aqueous mixtures with solutes identified in the label of the abscissa. The empty circles pertain to the second solvation shell. (From Y. Marcus, 2001, Preferential Solvation in Mixed Solvents, Part 10, Completely Miscible Aqueous Co-Solvent Binary Mixtures at 298.15 K, *Monatshefte für Chemie*, 132, 1387, by permission of the publisher, Springer.)

Trifluoroethanol, hexafluoro-2-propanol (Figure 3.3). These fluoroalkanols have rather unsymmetric preferential solvation parameter curves with extrema near x_S = 0.2 to 0.25. The $\delta x'_{WW}(max)$ are 0.31 and 0.28 and the $\delta x'_{WS}(min)$ are −0.20 and −0.13, respectively, for the trifluoro- and hexafluoro-alkanol. In both cases water self-associates more for the former, having an exposed hydrophobic methylene group.

2-Methoxy-, 2-ethoxy-, and 2-butoxyethanol (Figure 3.4). At 313 K the preferential solvation curve for aqueous 2-methoxyethanol is symmetric with $\delta x'_{WW}(max)$ = 0.10 and $\delta x'_{WS}(min)$ = −0.008, similar to the values for 1,2-ethanediol at 298 K (see above, the temperature effect appears to be small, since the 343 K data are similar to the 313 K ones). As the alkyl substituent becomes longer, the curves become skewed, with the extrema near x_S = 0.2 for $\delta x'_{WW}(max)$ = 0.18 and 0.075 and near x_S = 0.15 for $\delta x'_{WS}(min)$ = −0.06 and −0.019 for the ethoxy and butoxy components. The parameters for the latter, being smaller than those for the former, are unexpected.

Ethanolamines (Figure 3.4). Data for 2-ethanolamine, *N*-methyl-2-ethanolamine, *N*,*N*-dimethyl-2-ethanolamine, diethanolamine, and triethanolamine have been dealt with in detail (Marcus 1995), and the results are summarized here. KBIs for 2-ethanolamine have also been reported previously (Matteoli and Lepori 1984) in substantial agreement with the later ones. For this aqueous system mutual association dominates over the self-association: $\delta x'_{WW}(min)$ = −0.046 and $\delta x'_{WS}(max)$ = 0.007, the latter (or − $\delta x'_{SS}$ for self-association of the ethanolamine) being insignificant. The results for *N*-methyl-2-ethanolamine resemble closely those for the free amine group. When another methyl group is substituted on the amine the system becomes even more ideal, the preferential solvation parameters do not exceed 0.02 in absolute value, but in water-rich mixtures $\delta x'_{WW}$ is slightly positive. For di- and triethanolamines, data are available only for 273 K as are data for monoethanolamine, that is, 2-aminoethanol. For all three systems the (nonvolume-corrected) preferential solvation curves are skewed, $\delta x_{WW}(min)$ being near x_S = 0.75 and $\delta x_{WS}(max)$ being near x_S = 0.25, the former extremum is ≥ −0.02, the latter is ≤ 0.04. The more ethanol groups are substituted on the amine, the more ideal the mixtures become, see also Matteoli (1997).

Tetrahydrofuran, 1,4-dioxane (Figure 3.5) *and 1,3-dioxolane.* Here appreciable preferential solvation occurs—for the former mixtures with $\delta x'_{WW}(max)$ = 0.33 near x_S = 30 and $\delta x'_{WS}(min)$ = −0.20 near x_S = 0.25 in the first solvation shell at 298 K. The water–water preference and disfavored water–tetrahydrofuran interactions persist to the second solvation shell. For the aqueous dioxane mixtures with two ether linkages and the same number (four) of hydrophobic methylene groups, the self-association of the water is even larger with $\delta x'_{WW}(max)$ = 0.57 near x_S = 0.65 in the first solvation shell at 298 K and persists with $\delta x'_{WW}(max)$ = 0.20 in the second shell. However, the disfavored mutual interaction is much smaller than for tetrahydrofuran, being only $\delta x'_{WS}(min)$ = −0.052 in the first shell and hardly significant in the second. For aqueous 1,3-dioxolane (Marcus 2002a), with only three methylene groups, the preferential solvation curves show less preferences than for aqueous tetrahydrofuran and 1,4-dioxane (data are available at 323 K): $\delta x'_{WW}(max)$ = 0.17 near x_S = 0.30 and $\delta x'_{WS}(min)$ = −0.075 near x_S = 0.20.

Acetone (Figure 3.5). Here, again, the KBIs were reported previously (Matteoli and Lepori 1990). The self-association of the water, $\delta x'_{WW}(max)$ = 0.31 near x_S = 0.50

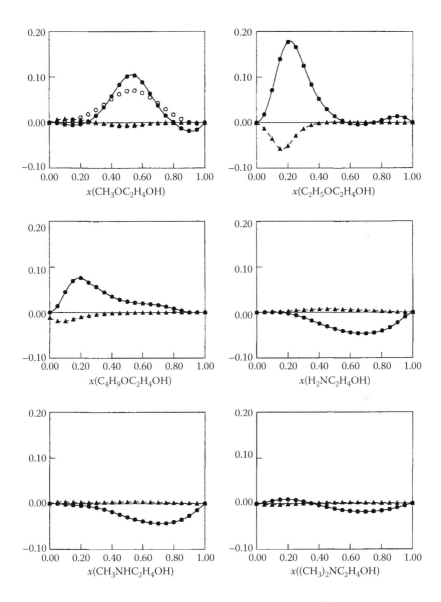

FIGURE 3.4 Volume-corrected preferential solvation parameters $\delta x'_{WW}$ (●) for water–water interactions and $\delta x'_{WS}$ (▲) for water–solute interactions in the first solvation shell of aqueous mixtures with solutes identified in the label of the abscissa. (From Y. Marcus, 2001, Preferential Solvation in Mixed Solvents, Part 10, Completely Miscible Aqueous Co-Solvent Binary Mixtures at 298.15 K, *Monatshefte für Chemie*, 132, 1387, by permission of the publisher, Springer.)

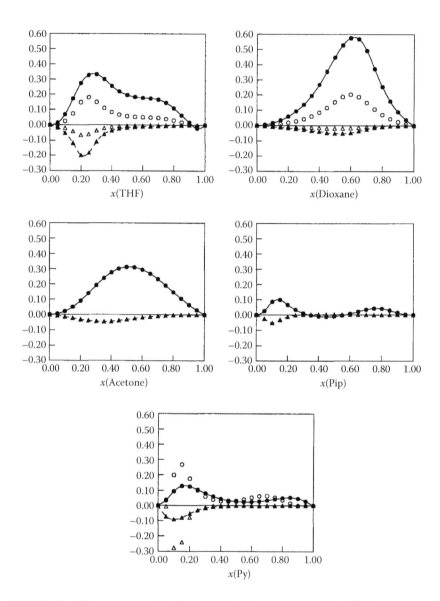

FIGURE 3.5 Volume-corrected preferential solvation parameters $\delta x'_{WW}$ (●) for water–water interactions and $\delta x'_{WS}$ (▲) for water–solute interactions in the first solvation shell of aqueous mixtures with solutes identified in the label of the abscissa (THF = tetrahydrofuran, Pip = piperidine, Py = pyridine). Empty symbols for tetrahydrofuran and dioxane pertain to second solvation shell, for pyridine to data from a second source. (From Y. Marcus, 2001, Preferential Solvation in Mixed Solvents, Part 10, Completely Miscible Aqueous Co-Solvent Binary Mixtures at 298.15 K, *Monatshefte für Chemie*, 132, 1387, by permission of the publisher, Springer.)

is less than for dioxane, but the disfavored mutual interactions, $\delta x'_{WS}(\min) = -0.045$ near $x_S = 0.35$, are about the same as for this ether.

Formic, acetic, and propanoic acids (Figure 3.6). The KBIs are inaccurate for the formic acid system outside the range $0.1 \leq x_S \leq 0.9$ due to the quality of the data. Within this range, mutual association is favored, $\delta x'_{WS}(\max) = 0.033$ near $x_S = 0.15$,

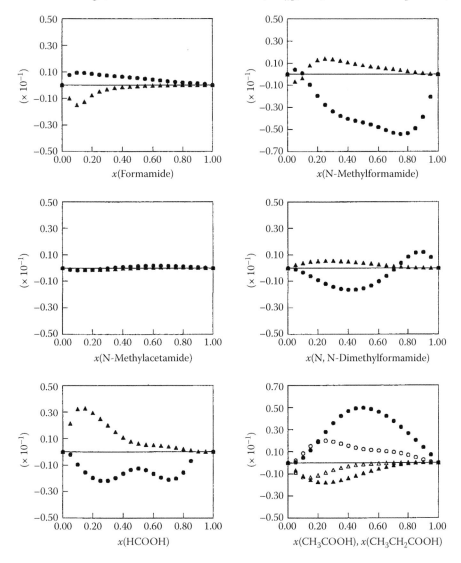

FIGURE 3.6 Volume-corrected preferential solvation parameters $\delta x'_{WW}$ ($\bullet$) for water–water interactions and $\delta x'_{WS}$ ($\blacktriangle$) for water–solute interactions in the first solvation shell of aqueous mixtures with solutes identified in the label of the abscissa. (From Y. Marcus, 2001, Preferential Solvation in Mixed Solvents, Part 10, Completely Miscible Aqueous Co-Solvent Binary Mixtures at 298.15 K, *Monatshefte für Chemie*, 132, 1387, by permission of the publisher, Springer.)

and water–water association is disfavored, $\delta x'_{WW}$ having two minima of –0.022, near $x_S = 0.3$ and 0.7. The system is nearly ideal, in view of the small values of the extrema, for this very hydrophilic solute. For the aqueous acetic acid system, water self-association is favored, $\delta x'_{WW}(max) = 0.050$ near $x_S = 0.50$ and for aqueous propanoic acid, $\delta x'_{WW}(max) = 0.020$ near $x_S = 0.25$, is smaller than for acetic acid, contrary to expectations (the reverse ratio is found for the nonvolume-corrected δx_{WW} values [Marcus 1990]). The mutual interactions are disfavored: for acetic acid $\delta x'_{WS}(min)$ = –0.018 near $x_S = 0.25$ and for aqueous propanoic acid, $\delta x'_{WS}(min) = -0.013$ near $x_S = 0.15$.

Propylene carbonate. This system has a miscibility gap at 298 K, so only the organic solvent-rich region was dealt with (Marcus 1990). The preferential solvation of water around propylene carbonate, $\delta x_{WS} \sim 0$ for the homogeneous region $x_S > 0.68$, but the water–water preference, δx_{WW} has a maximum of 0.36 near $x_S = 0.76$ and this preference extends even to the third solvation shell, where $\delta x_{WW}(max) = 0.040$ still. This preference is expected, in view of the phase separation that takes place.

γ-Butyrolactone. This system was dealt with (Marcus 2002a) at 343 K, for which data were available (higher temperature data showed unreasonable KBIs at $x_S \leq$ 0.35). Only small disfavored mutual association and moderate self-association of the water were observed for these mixtures: $\delta x'_{WS}(min) = -0.015$ near $x_S = 0.3$ and $\delta x'_{WW}(max) = 0.055$ near $x_S = 0.75$.

Di- and triethylamine. For aqueous diethylamine, data were available for 313 K (Marcus 2002a) and yielded preferred water–water association with a double-humped curve: $\delta x'_{WW}(max) = 0.14$ near $x_S = 0.20$ and $\delta x'_{WW}(max) = 0.08$ near $x_S = 0.80$. The disfavored mutual association has a single extremum: $\delta x'_{WS}(min) = -0.06$ near $x_S =$ 0.15, whereas at $x_S \geq 0.5$, $\delta x'_{WS}$ is only very slightly negative. Triethylamine has a lower critical consolute temperature of 292 K, and even at 283 K, the excess Gibbs energy is not sufficiently accurate (see above) at $x_S \leq 0.4$ to obtain meaningful KBIs (Marcus 1990). At this temperature, $\delta x_{WW}(min) = -0.060$ at $x_S = 0.77$ and $\delta x_{WS}(max)$ = 0.028 at $x_S = 0.47$. No volume-corrected preferential solvation parameters were calculated for this system, so it is still an open question why the interactions appear to be of opposite sign for the di- and triethylamines.

1,2-Diaminoethane (ethylene diamine). Data at 336 K appear to be unreliable (Marcus 2002a) and only those at 373 K could be used in a reasonable manner. These lead to nearly ideal behavior of aqueous ethylene diamine at this high temperature: $\delta x'_{WS}$ is essentially zero over the entire composition range and $\delta x'_{WW}$ is slightly negative with $\delta x'_{WW}(min) = -0.02$ at $x_S = 0.65$.

Pyrrolidine, piperidine, morpholine, and pyridine. The aqueous pyrrolidine system was studied at 333 K (Marcus 2002a) and showed a double-humped $\delta x'_{WW}$ curve and a single trough $\delta x'_{WS}$ curve approaching zero, similar to the diethylamine case treated above (having the same number of carbon atoms but two ethyl chains rather than the tetramethylene chain of pyrrolidine). The extrema are: $\delta x'_{WW}(max) = 0.055$ near $x_S = 0.15$ and $\delta x'_{WW}(max) = 0.008$ near $x_S = 0.80$ and $\delta x'_{WS}(min) = -0.04$ near $x_S = 0.10$, much smaller for pyrrolidine than for diethylamine. It is interesting that for the former of these two systems, $\delta x'_{WW}$ turns slightly negative for $0.40 \leq x_S \leq$ 0.65, where $\delta x'_{WS}$ is already very nearly zero. Aqueous piperidine (Marcus 2001),

Figure 3.5, was studied at 298 K and has preferential solvation parameters qualitatively similar to those of pyrrolidine but with somewhat larger values: $\delta x'_{WW}(max) =$ 0.010 near $x_S = 0.15$ and $\delta x'_{WW}(max) = 0.030$ near $x_S = 0.80$ and $\delta x'_{WS}(min) = -0.050$ near $x_S = 0.10$. Morpholine, being intermediate between 1,4-dioxane with two –O– bridges and piperidine with two –NH– bridges between two ethylene chains by having one of each kind, was studied at 348 K (Marcus 2002a). Its aqueous mixtures have positive values for the mutual association $\delta x'_{WS}(max) = 0.008$ near $x_S = 0.15$ declining to very near zero values at $x_S \geq 0.55$, but an S-shaped curve for self-association of the water: it has $\delta x'_{WW}(min) = 0.017$ near $x_S = 0.25$ rising to $\delta x'_{WW}(max) = 0.040$ at $x_S = 0.75$. The aqueous pyridine system (Marcus 2001), Figure 3.5, at 298 K resembles the piperidine system with double-humped ($\delta x'_{WW}$) and single trough ($\delta x'_{WS}$) curves of approximately the same magnitudes as those reported above.

Acetonitrile (Figure 3.7). As mentioned above, most of the data available for this system at ambient temperatures for the calculation of the variable D of Equation 3.9

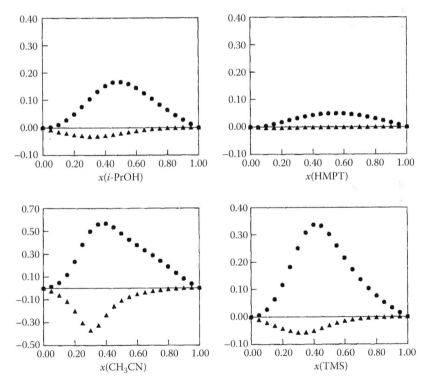

FIGURE 3.7 Volume-corrected preferential solvation parameters $\delta x'_{WW}$ (●) for water–water interactions and $\delta x'_{WS}$ (▲) for water–solute interactions in the first solvation shell of aqueous mixtures with solutes identified in the label of the abscissa (HMPT = hexamethyl phosphoric triamide, TMS = tetramethylene sulfone [sulfolane]). (From Y. Marcus, 2001, Preferential Solvation in Mixed Solvents, Part 10, Completely Miscible Aqueous Co-Solvent Binary Mixtures at 298.15 K, *Monatshefte für Chemie*, 132, 1387, by permission of the publisher, Springer.)

were not sufficiently accurate (Marcus 2008). Only those of Nikolova et al. (2000), and of these only the partial pressure data of acetonitrile disregarding the partial pressure data for water, could yield usable values of D not smaller than 1.0. The resulting preferential solvation parameter curves for 298 K are shown in Figure 3.7. A very large $\delta x'_{WW}(max) = 0.56$ at $x_S = 0.35$ and a large negative $\delta x'_{WS}(min) = -0.37$ at $x_S = 0.30$ is observed. Aqueous acetonitrile mixtures were studied over a wide temperature range, 283 to 323 K (Marcus and Migron 1991) and, although the upper critical consolute temperature is 272 K, microheterogeneity still occurs at 298 K. This leads to domains where water–water interactions dominate and water–acetonitrile interactions take place only at their fringes. Indeed, positive δx_{WW} values persist up to seven solvation layers, although the values decline with increasing temperatures.

Formamide, N-methylformamide N,N-dimethylformamide, and N-methylacetamide (Figure 3.6). The preferential solvation curves for these aqueous systems at 298 K show only small deviations from ideality (absolute values ≤ 0.05). Formamide has small positive $\delta x'_{WW}$ values with $\delta x'_{WW}(max) = 0.010$ at $x_S = 0.12$ and corresponding small negative $\delta x'_{WS}$ values with $\delta x'_{WS}(min) = -0.012$ at the same composition. Aqueous *N*-methylformamide in water-rich mixtures ($x_S < 0.1$) shows preferences similar to (but even smaller than) formamide. At larger solute contents, the opposite preferences are manifested: $\delta x'_{WS}(min) = -0.054$ at $x_S = 0.75$ and $\delta x'_{WW}(max) = 0.014$ at $x_S = 0.25$. Small mutual association preference is also shown by aqueous *N,N*-dimethylformamide: $\delta x'_{WS}(max) = 0.006$ at $x_S = 0.30$, but the water–water association curve is S-shaped: it has $\delta x'_{WW}(min) = -0.016$ at $x_S = 0.40$ but rises to $\delta x'_{WW}(max) = 0.012$ at $x_S = 0.90$; see also Matteoli (1997). The aqueous *N*-methylacetamide mixtures are essentially ideal, in that no preferential solvation can be discerned at all in the $\delta x'_{WS}$ and $\delta x'_{WW}$ curves.

N-Methylpyrrolidin-2-one (Figure 3.2). Data are available for 303 K (Marcus 2001) and the mixtures are nearly ideal. Above $x_S = 0.2$, mutual association of the components is preferred and water–water interactions are disfavored. These aqueous mixtures resemble in this respect the ethanolamine mixtures.

Tetramethylurea. The data are available for 298 K (Marcus 2002a) and the preferential solvation curves resemble those of the cyclic amines described above (Figure 3.5), in that a double humped $\delta x'_{WW}$ curve and a $\delta x'_{WS}$ curve with a single trough are found. However, contrary to the cyclic amines, the organic solvent-rich $\delta x'_{WW}(max) = 0.014$ at $x_S = 0.15$, rather than the other way round. The disfavored mutual association $\delta x_{WS}(min) = -0.008$ at $x_S = 0.10$ is hardly significant and very near zero values of δx_{WS} dominate over the composition range at $x_S \geq 0.4$.

Hexamethyl phosphoric triamide (Figure 3.7). The data at 298 K show some water–water preferential association but only to a minor extent. The value of $\delta x'_{WW}(max) = 0.050$ at $x_S = 0.55$ indicates essentially zero preferential mutual solvation.

Dimethylsulfoxide (Figure 3.2) and *tetramethylene sulfone (sulfolane)* (Figure 3.7). The aqueous dimethylsulfoxide data for 298 K show moderately strong disfavored self-association of the water, $\delta x'_{WW}(min) = -0.058$ at $x_S = 0.75$ and a rather small preference for mutual association, $\delta x'_{WS}(max) = 0.006$ at $x_S = 0.45$. On the contrary, for aqueous sulfolane mixtures at 298 K, the self-association of the water is strongly favored, $\delta x'_{WW}(max) = 0.34$ at $x_S = 0.40$ and the mutual association is disfavored,

$\delta x'_{WS}(min) = -0.06$ at $x_S = 0.30$. The pronounced positive $\delta x'_{WW}$ persist over several solvation shells as in the cases of aqueous acetone, tetrahydrofuran, and acetonitrile.

3.3.2 Summary of Binary Aqueous Mixtures of Solutes

Table 3.1 summarizes the results for the aqueous mixtures in terms of whether $\delta x'_{WW}$ and $\delta x'_{WS}$ are positive or negative and the relative sizes of the extrema. It is interesting to compare the water self-association results in the aqueous mixtures, $\delta x'_{WW}$, with results from other kinds of studies. The mixtures marked with ++ in Table 3.1 for this

TABLE 3.1

Summary of the Preferential Solvation in Binary Mixtures of Water with a Series of Solutes at Ambient Conditions, or as near as the Availability of Data Permit

Solute	$\delta x'_{WW}$	$\delta x'_{WS}$	Solute	$\delta x'_{WW}$	$\delta x'_{WS}$
Hydrogen peroxide	–	0	Formic acid	–	+
Methanol	+	–	Acetic acid	+	–
Ethanol	++	–	Propanoic acid	+	–
1-Propanol	++	– –	Propylene carbonate[a]	++	–
2-Propanol	++	–	γ-Butyrolactone	+	–
t-Butanol	++	– –	Diethylamine	++	–
1,2-Ethanediol	–	0	Triethylamine[a]	–	+
Diethylene glycol	0	–	1,2-Diaminoethane	–	0
1,2-Propanediol	0	0	Pyrrolidine	+	–
Glycerol	+	0	Piperidine	+	–
2,2,2-Trifluoroethanol	++	– –	Morpholine	– +	+
Hexafluoro-2-propanol	++	– –	Pyridine	+	–
2-Methoxyethanol	+	0	Acetonitrile	++	– –
2-Ethoxyethanol	+	–	Formamide	+	–
2-Butoxyethanol	+	–	N-Methylformamide	–	+
2-Aminoethanol	–	0	N,N-Dimethylformamide	– +	0
N-Methyl-2-aminoethanol	–	0	N-Methylacetamide	0	0
Diethanolamine	–	+	N-Methylpyrrolidin-2-one	0	0
Triethanolamine	–	+	Tetramethylurea	+	0
Tetrahydrofuran	++	– –	Hexamethyl phosphoramide	+	0
1,3-Dioxolane	++	–	Dimethylsulfoxide	–	0
1,4-Dioxane	++	–	Tetramethylene sulfone	++	–
Acetone	++	–			

Note: The extrema in the curves are marked as ++ or – – if larger than 0.1, as + or – if larger than 0.01, all in the absolute sense, or as zero otherwise, or as – + if they change sign as the concentration varies.

[a] No volume-corrected preferential solvation parameters are available, only the δx_{WW} and δx_{WS}.

TABLE 3.2

The Self-Association of Various Solutes at Infinite Dilution in Water[a]

Solute	$G_{SS}^{\infty}/cm^3 \cdot mol^{-1}$	Solute	$G_{SS}^{\infty}/cm^3 \cdot mol^{-1}$
Hydrogen peroxide	−53	Tetrahydrofuran	+28[b]
Methanol	−45	1,4-Dioxane	−127
Ethanol	−50	Acetone	−76
1-Propanol	+102	Formic acid	−140
2-Propanol	+40	Acetic acid	−8
1-Butanol	+160[b]	Propanoic acid	+50
t-Butanol	+58	Piperidine	+238[b]
1,2-Ethanediol	−98	Pyridine	+380[b]
Glycerol	−95	Acetonitrile	+155
Trifluoroethanol	−61	Formamide	>600
Hexafluoro-2-propanol	−89	N-Methylformamide	−129
2-Methoxyethanol	−230	N,N-Dimethylformamide	−162
2-Ethoxyethanol	−107	N-Methylacetamide	−132
2-Butoxyethanol	< −500	Hexamethyl phosphoramide	−370
2-Aminoethanol	−116	Dimethylsulfoxide	−76
N,N-Dimethyl-2-aminoethanol	−180[b]	Tetramethylene sulfone	+5

[a] From Y. Marcus, 2001, Preferential Solvation in Mixed Solvents, Part 10, Completely Miscible Aqueous Co-Solvent Binary Mixtures at 298.15 K, *Monatshefte für Chemie*, 132, 1387, unless otherwise noted.

[b] From E. Matteoli and L. Lepori, 1984, Solute–Solute Interactions in Water, 2, An Analysis through the Kirkwood–Buff Integrals for 14 Organic Solutes, *Journal of Chemical Physics*, 80, 2856.

quantity correlate well with results obtained for the enhancement of the water structure in water-rich mixtures, related to the change from compact domains to bulky ones (also from low to high heat capacity domains) (Marcus 2011, 2012). Where no such enhancement was observed in these studies, $\delta x'_{WW}$ is marked by either 0 or – in Table 3.1.

The self-association of the solutes at infinite dilution in water is represented as the values of G_{SS}^{∞} and, as mentioned above, are the molar osmotic pairwise virial coefficients of this component. The values are shown in Table 3.2 but the extrapolation to infinite dilution of the KBIs is beset by errors not smaller than ±10%. Hydrophilic solutes with good electron pair donation abilities have large negative values of G_{SS}^{∞} and conversely, relatively hydrophobic solutes that are still miscible with water and are poor electron pair donors, have large positive values of this quantity (Matteoli and Lepori 1984).

3.3.3 Nonaqueous Binary Solvent Mixtures

A large number of nonaqueous binary solvent mixtures have been studied by the present author and others, including some fluids in the liquid state that cannot be termed *solvents*. KBIs have been obtained, for instance, for mixtures of argon and krypton (Matteoli 1997), from which slight self-preference was deduced for both components, but this is not further elaborated here.

In the following, mixtures of methanol, ethanol, acetone, and triethylamine, marked as *solvent A*, each with many cosolvents marked as *B* (Marcus 1991) are described. Mixtures of 1,4-dioxane (A) (Marcus 2006b) and of tetrahydrofuran (A) (Marcus 2006a) with cosolvents (B) are also presented as well as a few studies by others. It is, of course, arbitrary which solvent is marked as *A* and which as *B*. The numerical values of the preferential solvation parameter extrema quoted below were read from figures and have an uncertainty of ±10%.

Methanol and cosolvents. Preferential solvation curves δx_{AA} for the self-association of methanol, A, and δx_{AB} for preferential solvation of the cosolvent B by methanol (neither being volume-corrected) were obtained (Marcus 1991) from thermodynamic data. For B = benzene and toluene, $\delta x_{AA}(max) \sim 0.50$ in the first solvation shell (for toluene, even in the second shell $\delta x_{AA}(max) = 0.35$), for B = nitromethane, $\delta x_{AA}(max) = 0.40$, and smaller values were obtained for B = sulfolane, $\delta x_{AA}(max) = 0.20$, for B = chloroform, $\delta x_{AA}(max) = 0.16$, and for B = acetonitrile, $\delta x_{AA}(max) = 0.08$. Correspondingly, the solvation of the cosolvent by methanol is disfavored. For B = benzene, $\delta x_{AB}(min) = -0.25$, for B = toluene and nitromethane, $\delta x_{AB}(min) = -0.30$, for B = sulfolane, $\delta x_{AB}(min) = -0.12$, and for B = chloroform and acetonitrile, $\delta x_{AB}(min) = -0.09$. Smaller values are obtained for B = tetrahydrofuran: $\delta x_{AB}(min) = -0.06$ whereas for the self-association of methanol, $\delta x_{AA}(max) = 0.01$ only. For B = acetone, the δx_{AA} curve is S-shaped but with very small values, >0 in methanol-rich mixtures and <0 in acetone-rich ones, and $\delta x_{AB}(min) = -0.042$. For some other systems, results were reported qualitatively, in that mixtures of methanol with B = chlorobenzene resemble those with B = acetonitrile, those with B = pyridine do not have appreciable preferential solvation, those with B = formamide resemble those with tetrahydrofuran mentioned above, whereas those with B = *N,N*-dimethyl- and *N,N*-diethylformamide have positive mutual solvation, $\delta x_{AB} > 0$ and disfavored self-association, $\delta x_{AA} < 0$, see also Matteoli (1997). Mixtures of methanol and B = 1-decanol (Matteoli 1997) showed the expected effect of the large difference in the molar volumes of the components (their ratio is ~5). The ΔG_{BB} integrals are positive, contrary to the negative G_{BB} integrals, hence some self-association of the decanol in the presence of the methanol, due to dispersion forces between the alkyl chains, is manifested.

Ethanol and cosolvents. Preferential solvation curves δx_{AA} and δx_{AB} (not volume-corrected) for A = ethanol and several cosolvents, B, were obtained (Marcus 1991). For all the mixtures, self-association of the ethanol is favored whereas mutual solvation is disfavored. For instance, when B = *n*-heptane, $\delta x_{AA}(max) = 0.40$; for B = sulfolane, $\delta x_{AA}(max) = 0.32$; for B = nitromethane, $\delta x_{AA}(max) = 0.22$; for B = benzene, $\delta x_{AA}(max) = 0.16$; for B = acetonitrile, $\delta x_{AA}(max) = 0.10$; and, for B = chloroform, $\delta x_{AA}(max) = 0.06$, in descending order. The mutual solvation parameter extrema are: for B = *n*-heptane $\delta x_{AM}(min) = -0.12$; for B = sulfolane $\delta x_{AM}(min) = <-0.30$; for B = nitromethane $\delta x_{AB}(min) = -0.28$; for B = benzene, $\delta x_{AB}(min) = -0.25$; for B = acetonitrile, $\delta x_{AB}(min) = -0.12$; and, for B = chloroform, $\delta x_{AB}(min) = -0.08$. For some other systems, results were reported qualitatively, in that mixtures of ethanol with B = toluene and chlorobenzene resemble those with B = benzene, those with B = n-hexane resemble those for B = n-heptane, those for B = c-hexane resemble those with B = acetonitrile, those with B = triethylamine and B = acetone resemble those with A = methanol and B = acetone, and those with B = 1,4-dioxane resemble those with A = methanol and B = acetonitrile described above. Mixtures of A = ethanol and B = pyridine do not have any appreciable preferential solvation. For the two alcohols as solvents A, there exists an approximately linear correlation of G_{BB}^{∞} for solutes B with their Kamlet–Taft electron pair donation parameters β_B.

Acetone and cosolvents. Preferential solvation curves δx_{AA} and δx_{AB} (not volume-corrected) for A = acetone and several cosolvents, B, were obtained (Marcus 1991). The preferences vary among the systems, some showing preferential mutual solvation, others preferential self-association of the acetone. For B = chloroform, δx_{AB}(max) = 0.048 and δx_{AA}(min) = –0.01, that is, small preferences are manifested. For B = methanol, the δx_{AA} curve is S-shaped but with very small values, <0 in methanol-rich mixtures and >0 in acetone-rich ones, and the mutual δx_{AB}(min) – –0.042. For B = toluene, curves qualitatively similar to those for B = methanol were obtained. For B = n-heptane, δx_{AA}(max) = 0.23 and δx_{AB}(min) = –0.15 and qualitatively similar curves were obtained for B = n-hexane, B = c-hexane, and B = 1,2 = ethanediol. For B = benzene, δx_{AB} is small and positive, δx_{AA} is small and negative; for B = chlorobenzene, δx_{AB} is small and positive, δx_{AA} is near zero. For B = pyridine and B = N,N-diethylformamide, both parameters are near zero, and for B = formamide both parameters are negative, δx_{AB} moderately and δx_{AA} less so.

Triethylamine and cosolvents. Preferential solvation curves δx_{AA} and δx_{AB} (not volume-corrected) for A = triethylamine and two cosolvents, B, were shown by Marcus (1991), but qualitative results for other cosolvents were also mentioned there. For B = chloroform, δx_{AB}(max) = 0.020 and δx_{AA}(min) = –0.25, that is, mutual solvation is somewhat preferred. For B = methanol, δx_{AB}(min) = –0.05 and δx_{AA}(min) = –0.02, meaning that self-association of the methanol δx_{BB} = $-\delta x_{AB}$(min) = 0.05 is preferred. Qualitative results are that for B = n-heptane no appreciable preferential solvation occurs; for B = benzene and chlorobenzene, both δx_{AA} and δx_{AB} are negative and small; for B = 1-butanol, δx_{AB} is positive and small, but δx_{AA} is near zero; for B = 1-propanol and B = t-butanol, δx_{AB} is small and negative and δx_{AA} is near zero. Values of both the infinite dilution KBIs, G_{AA}^{∞} and G_{BB}^{∞} of these systems have been calculated (Marcus 1991).

Tetrahydrofuran and cosolvents. A detailed study of the preferential solvation in mixtures of tetrahydrofuran, A, and cosolvents, B, was reported by Marcus (2006a). For B = n-hexane and c-hexane at 303 K, n-heptane and i-octane at 298 K, the self-association of tetrahydrofuran is preferred: $\delta x_{AA}'$(max) = 0.025, 0.030, 0.025, and 0.034, respectively. The mutual interaction is small: the extrema in $\delta x_{AB}'$ are: –0.010, 0, 0.004, and 0.004, that is, hardly significant. For B = benzene, toluene, ethylbenzene, dichloromethane, chloroform, tetrachloromethane, and 1-chlorobutane at 303 K, there is essentially no preferential solvation, both $\delta x_{AA}'$ and $\delta x_{AB}'$ being less than 0.01. The smallness of the preferential solvation parameters for mixtures of tetrahydrofuran and tetrachloromethane has already been noted by Ben-Naim (1990b) who concluded that these systems are symmetric ideal mixtures. For B = hexafluorobenzene at 303 K $\delta x'_{AA}(max) = 0.015$, but $\delta x'_{AB}(max) = 0.005$, and although positive, these are hardly significant. Small preferential solvation was also noted for B = pyrrolidine at 313 K and for B = triethylamine at 323 K: $\delta x'_{AA}(max) = 0.007$ and 0.020 and $\delta x'_{AB}(min) = -0.008$ and -0.006, these not being significant. More significant interactions were found with the alkanols at 298 K: $\delta x'_{AA}(max) = 0.014$ for B = ethanol, 0.009 for B = 1-propanol, and only 0.003 for 2-propanol, but the disfavored $\delta x'_{AB} < 0$ corresponds to favored self-interactions of the alkanols, $\delta x'_{BB}(max) = 0.031$, 0.008, and 0.024, respectively. For B = dibutylether at 298 K, both $\delta x'_{AA}$ and $\delta x'_{AB}$ are positive, 0.010 and 0.017, whereas for B = acetic acid at 298 K, $\delta x'_{AA}(min) = -0.015$, signifying preferred mutual interaction, $\delta x'_{BA}(max) \approx \delta x'_{AB}(max) = 0.015$, obvious for the Lewis acid-base interaction of tetrahydrofuran and acetic acid. Small preferential solvation is also obtained for interaction of the basic tetrahydrofuran with the basic B = acetone: $\delta x'_{AA}(max) = 0.009$ and $\delta x'_{AB}(min) = -0.012$. More significant are those with B = acetonitrile: $\delta x'_{AA}(max) = 0.016$ and $\delta x'_{AB}(min) = -0.038$. (It should be noted that in the published paper by Marcus [2006a], the preferential solvation parameter curves have erroneously been replaced by the KBI curves.) The largest preferential solvation was obtained for the two sulfur containing cosolvents, dimethylsulfoxide at 298 K and sulfolane at 353 K: $\delta x'_{AA}(max) = 0.044$ and 0.049 and $\delta x'_{AB}(min) = -0.101$ and -0.079, respectively, the latter two values signifying appreciable self-association ($\delta x'_{BB} > 0$) of the highly basic and dipolar cosolvents in the presence of the basic tetrahydrofuran.

1,4-Dioxane and cosolvents. A detailed study of the preferential solvation in mixtures of 1,4-dioxane, A, and cosolvents, B, was reported by Marcus (2006b). Significant self-association of dioxane in mixtures with B = *c*-hexane at 308 K, $\delta x'_{AA}(max) = 0.24$, with B = *n*-heptane at 303 K, $\delta x'_{AA}(max) = 0.19$, with B = *n*-hexane at 308 K, and B = methyl-*c*-hexane at 303 K, $\delta x'_{AA}(max) = 0.10$ for both, and correspondingly disfavored mutual interaction, $\delta x'_{AB}(min) = -0.15$, -0.065, -0.060, and -0.060, respectively, for these four cosolvents. For B = *c*-hexane, the preferential solvation was demonstrated to extend to the second and third solvation shells. Considerably smaller preferential solvation was observed for mixtures with B = benzene and toluene at 293 K and with ethylbenzene at 303 K: $\delta x'_{AA}(max) = 0.004$, 0.018, and 0.018, respectively, and $\delta x'_{AB}(min) = -0.002$, -0.003, and -0.011, respectively, with the values approaching zero even more closely in cosolvent rich mixtures. More significant preferential solvation was observed with alkanols as the cosolvents: for B = methanol, ethanol, and 1-propanol at 333 K and for B = 2-propanol at 323 K, $\delta x'_{AA}(max) = 0.015$, 0.023, 0.023, and 0.041, respectively, and $\delta x'_{AB}(min) = -0.023$, -0.044, -0.068, and -0.068, respectively. The values for *c*-pentanol at 298 K resemble those for methanol. With B = acetic acid at 298 K $\delta x'_{AB}(min) = -0.006$ only, but $\delta x'_{AA}(min) = -0.017$, which corresponds to $\delta x'_{BA}(max) = 0.017$ in cosolvent rich mixtures, signifying Lewis acid-base interactions in this system. Appreciable self-association of dioxane in mixtures with 1,2-ethanediol at 303 K take place, $\delta x'_{AA}(max) = 0.11$, but the mutual solvation is almost completely absent, $\delta x'_{AB}$ approaching the limit $-x_B$ up to $x_B = 0.35$, with $\delta x'_{AB}(min) = -0.30$ at this composition. The diol thus almost selectively prefers to associate with itself rather than to solvate the dioxane. For the chloromethanes, B = dichloromethane at 298 K, chloroform at 323 K, and tetrachloromethane at 293 K, only moderate preferential solvation was observed: $\delta x'_{AA}(min) = -0.022, -0.021$, but $\delta x'_{AA}(max) = 0.010$ and $\delta x'_{AB}(max) = 0.012, 0.013$, but $\delta x'_{AB}(min) = -0.004$, respectively. Tetrachloromethane, which is nonpolar, behaves differently in these mixtures than the dipolar dichloromethane and chloroform (but such a difference was not observed for tetrahydrofuran instead of the dioxane). The smallness of the preferential solvation parameters for mixtures of 1,4-dioxane and tetrachloromethane has already been noted by Ben-Naim (1990b) who concluded that these systems are symmetric ideal mixtures. Similar results, that is, only moderate preferential solvation, was found for mixtures with haloethanes. For B = 1,2-dichloroethane at 298 K, 1,2-dibromoethane at 293 K, trichloroethene and tetrachloroethene at 303 K, and halothane (CHBrClCF$_3$) at 293 K, the extrema in $\delta x'_{AA}$ are 0.013, 0.016, −0.009, 0.032, and −0.026 and those in $\delta x'_{AB}$ are −0.008, 0.008, −0.009, −0.018, and 0.018, respectively. For B = hexafluorobenzene at 303 K $\delta x'_{AA}(max) = 0.010$ and $\delta x'_{AB}(min) = -0.006$, but $\delta x'_{AB}$ becomes slightly positive at $x_B = 0.3$. Moderate selective solvation was observed with B = triethylamine at 323 K and with piperidine at 298 K: $\delta x'_{AA}(max) = 0.048$ and 0.025, and $\delta x'_{AB}(min) = -0.033$ and −0.014, whereas with B = acetonitrile at 313 K $\delta x'_{AA}(min) = -0.004$ and $\delta x'_{AB}(min) = -0.033$ (signifying self-association of the acetonitrile $\delta x'_{BB} > 0$, as for the system with A = tetrahydrofuran). For B = formamide at 313 K, $\delta x'_{AA}(min) = -0.10$, but becomes slightly positive at $x_B \geq 0.8$, and $\delta x'_{AB}(min) = -0.032$. That is, dioxane prefers to have some formamide in its solvation shell but not vice versa, while formamide prefers to self-associate. For B = DMF at 333 K, $\delta x'_{AA}(max) = 0.008$ and $\delta x'_{AB}(min) = -0.024$. With the sulfur containing cosolvents, B = dimethylsulfoxide at 293 K and sulfolane at 353 K, $\delta x'_{AA}(max) = 0.015$ and 0.042 and $\delta x'_{AB}(min) = -0.060$ and −0.014, respectively.

Other mixtures. Mixtures of various other nonaqueous solvents have also been studied (Matteoli 1997). Mixtures of benzene (A) and *c*-hexane (B) showed that the dispersion interactions lead to nearly ideal mixing, but the small molar volume difference (the ratio is 1.30) does cause a small deviation from ideality in the KBIs. Mixtures of *n*-hexane (A) and *n*-hexadecane (B), being athermal, again do show the effect of the discrepancy in the molar volumes (their ratio is ~2.5). Contrary to the positive G_{AA} curve, the ΔG_{AA} curve is negative, showing that *n*-hexane does not self-aggregate in these mixtures. Mixtures of the globular tetrachloromethane (A) and octamethyl-*c*-tetrasiloxane (B) behave similarly to the latter mixtures.

Table 3.3 summarizes the self-association at infinite dilution of the two components of the nonaqueous binary mixtures, G_{AA}^{∞} and G_{BB}^{∞}, as reported in Marcus (1991, 2006a, 2006b). The values are rounded to the nearest 10 cm^3·mol^{-1}, but it should be realized that the extrapolation to infinite dilution cannot yield values better than ±10%. Positive values of G_{ii}^{∞} mean that component *i* tends to self-associate at infinite dilution in the other component and negative values mean that component *i* is well solvated by the other component and does not have *i* molecules in its solvation shell. For example, each of ethanol and *n*-heptane extensively self-associate at infinite dilution in the other component, but each of chloroform and acetone at infinite dilution are solvated exclusively by the other component.

TABLE 3.3

The Self-Association at Infinite Dilution of the Two Components of Nonaqueous Binary Mixtures A + B, G_{AA}^∞/cm³·mol⁻¹, and G_{BB}^∞/cm³·mol⁻¹

Solvent B	Solvent A									
	Methanol		Ethanol		Acetone		Triethylamine		Tetrahydrofuran	1,4-Dioxane
	G_{AA}^∞	G_{BB}^∞	G_{AA}^∞	G_{BB}^∞	G_{AA}^∞	G_{BB}^∞	G_{AA}^∞	G_{BB}^∞	G_{BB}^∞	G_{BB}^∞
n-Hexane			500	1250	70	870	-170	-120		210
n-Heptane			560	810	230	680			-150	-30
i-Octane									-210	
c-Hexane									-40	160
Methyl-c-hexane									-130	160
Benzene	280	450	680	280	30	20	-40	-160	-170	-80
Toluene			660	530	20	120			-240	-10
Ethylbenzene									-210	-70
Methanol	-80	-20	-20	-80	120	-50			160	30
Ethanol			-90	-40					40	140
1-Propanol							50	0	-20	30
2-Propanol									110	240
1-Butanol	-110	30	-120	0			-160	0		
c-Pentanol										80
1,2-Ethanediol			30	80						450

continued

TABLE 3.3 (continued)

The Self-Association at Infinite Dilution of the Two Components of Nonaqueous Binary Mixtures A + B, G_{AA}^{∞}/cm³·mol⁻¹, and G_{BB}^{∞}/cm³·mol⁻¹

Solvent B	Methanol		Ethanol		Acetone		Triethylamine		Tetrahydrofuran	1,4-Dioxane
	G_{AA}^{∞}	G_{BB}^{∞}	G_{AA}^{∞}	G_{BB}^{∞}	G_{AA}^{∞}	G_{BB}^{∞}	G_{AA}^{∞}	G_{BB}^{∞}	G_{BB}^{∞}	G_{BB}^{∞}
Dibutyl ether									-300	
Acetone	-50	120							-20	
Acetic acid									-190	-10
Dichloromethane									-160	-40
Chloroform	190	-40	350	-140	-220	-60	-20	-210	-280	-90
Tetrachloromethane									-150	-70
1,2-Dichloroethane									-190	-140
1,2-Dibromoethane										-200
1,1,1-Trichloroethene									50	
Trichloroethene										-60
Tetrachloroethene									0	40

Chlorobenzene	−280	−270			10	40	−330	−90	−150	−20
Hexafluorobenzene									−20	
Pyrrolidine										−50
Piperidine										160
Triethylamine	−120	90	−70	0					−140	160
Pyridine	−90	−20	−70	−30	−15					240
Acetonitrile	30	100	240	220					40	
Nitromethane	100	140	350	570	0	−110				
Formamide										70
N,N-Dimethylformamide										120
Dimethylsulfoxide	340	450	450						340	240
Sulfolane	340	450	450	630					250	120

Source: Data reported in Y. Marcus, 1991, Preferential Solvation in Mixed Solvents, Part 6, Binary Mixtures Containing Methanol, Ethanol, Acetone or Triethylamine and Another Organic Solvent, *Journal of the Chemical Society, Faraday Transactions*, 87, 1843; Y. Marcus, 2006a, Preferential Solvation in Mixed Solvents, Part 13, Mixtures of Tetrahydrofuran with Organic Solvents: Kirkwood–Buff, *Journal of Solution Chemistry*, 35, 251; Y. Marcus, 2006b, Preferential Solvation in Mixed Solvents, Part 14, Mixtures of 1,4-Dioxane with Organic Solvents: Kirkwood–Buff Integrals and Volume-Corrected Preferential Solvation Parameters. *Journal of Molecular Liquids*, 128, 115.

3.4 PREFERENTIAL SOLVATION IN TERNARY SOLVENT MIXTURES

Ternary solvent systems were dealt with by Matteoli and Lepori (1995) following an earlier paper in which a binary solvent system (water + 1-propanol) containing urea at substantial concentrations was treated (Matteoli and Lepori 1990). The general expression for relating the KBIs of the A + B + C ternary solvent system, where i and j are any two among the three solvents, to thermodynamic data is (see Equation 1.62)

$$G_{ij} = \frac{\mathbf{A}^{ij}}{c_i c_j |\mathbf{A}|} - \frac{\delta_{ij}}{c_j} \tag{3.10}$$

This equation is treated in more detail in Section 1.2.4 in Chapter 1 and Section 4.3.1 in Chapter 4. There was no way for the estimation of the correlation volume for the application of expressions such as Equation 3.2, so that the preferential solvation coefficients, for example, for solvation of component C by the components A and B is simply defined as

$$\mathrm{psc}_{(A,B)C} = \Delta G_{AC} - \Delta G_{BC} \tag{3.11}$$

where the $\Delta G_{ij} = G_{ij} - G_{ij}^{\mathrm{id}}$ are the volume corrected KBIs as explained in Section 3.2. For a more detailed treatment of preferential solvation in ternary systems see Sections 4.3.2 and 4.3.3 in Chapter 4.

There are six KBIs for each ternary system that could be presented in a three-dimensional triangular prism. It was expedient, however, to show the results in terms of sections parallel to a side of the triangular base. In such sections, for example, the mole fraction x_C is kept constant and x_A is varied between 0 and $1 - x_C$ while $x_B = 1 - x_C - x_A$.

These equations were applied to four ternary solvent systems: water + 1-propanol + N,N-dimethylformamide at 313 K, and water + 1,2-ethanediol + ethanol, water + 1,2-ethanediol + acetonitrile, and ethanol + 1,4-dioxane + chloroform at 323 K. Addition of component C = N,N-dimethylformamide to the water (A) + 1-propanol (B) system decreases the self-association of each of these components drastically, until it practically disappears at $x_C = 0.2$. The $\mathrm{psc}_{(A,B)C}$ in this system changes sign when $x_C = 0.1$, being positive but turning negative in 1-propanol-rich mixtures, where water loses its structure and individual water molecules are preferentially solvating the N,N-dimethylformamide by its better hydrogen bonding ability and smaller size. Acetonitrile (C) is strongly self-associated in both binary mixtures with water (A) and with 1,2-ethanediol (B), and addition of B to A + C mixtures or of A to B + C mixtures has little effect on this. On the other hand, a definite preference for water–1,2-ethanediol association in the presence of acetonitrile is manifested. If ethanol replaces acetonitrile as component C in the mixtures, the same trends are observed, although to a smaller degree. Thus, mixtures of A + B in the last two examples act as if they were a single component, irrespective of their composition, with regard to solvation of the component C.

The fourth example dealt with has a protic component C = ethanol and two aprotic ones, of which one is (mildly) dipolar, A = chloroform, and the other not,

B = 1,4-dioxane. Association of A and B by Lewis acid-base interactions are evident but in the presence of small concentrations of C (that self-associates at high contents) its free molecules compete successfully with A for solvation of B.

Other studies of the preferential solvation for which information can be derived from KBIs in ternary systems have also been made. The system n-heptane + ethanol + 1-propanol at 313 K (Zielkiewicz 1995a) showed that ethanol and 1-propanol mix in a random manner in the presence of n-heptane with no preferential solvation between these two solvents. The same author studied the solvation of N,N-dimethylformamide (C) in mixtures of water (A) and each of methanol, ethanol, and 1-propanol (B) at 313 K (Zielkiewicz 1995b). At $x_C > 0.8$ this component was solvated equally by A and B, but at $x_C < 0.15$ it was preferentially hydrated, that is, solvated by A, except when $x_A > 0.8$, where the solvation of C by A and B was random. N,N-dimethylformamide (C) featured also in the studies (Ruckenstein and Shulgin 2001a) of it in aqueous (A) methanol (B). The KBIs in the system n-hexane + 1-hexanol + methyl benzoate were studied at 298 K (Aparicio et al. 2005). They calculated the excess (or deficit) number of molecules of, say, A, around molecules of B in pseudobinary systems at constant mole fraction of C from

$$n_{AB} = c_A \left(G_{AB} - G_{AB}^{id} \right) \tag{3.12}$$

as suggested in Ruckenstein and Shulgin (2001a). However, although deduction of G_{ij}^{id} could be justified regarding the interactions involved, this is questionable when the composition (number of neighboring molecules of each kind) is to be calculated.

3.5   COMPARISON WITH OTHER APPROACHES TO PREFERENTIAL SOLVATION IN SOLVENT MIXTURES

The Kirkwood–Buff integrals are obtained rigorously from the thermodynamic data, but the calculation of the interactions between the components of the solvent mixtures as volume-corrected preferential solvation parameters depends on the validity of the notion that subtraction of G_{ij}^{id} from G_{ij} cancels out the volume size discrepancies between the components (Matteoli 1997). Although endorsed by the present author and others, this notion has been controversial (Ben-Naim 2007a; Matteoli and Lepori 2007; Ben-Naim 2008; Matteoli and Lepori 2008; Shulgin and Ruckenstein 2008b) (see also Section 4.5 in Chapter 4). The *composition* of the environment of a component is indeed dependent on G_{ij}, but the *interactions* leading to preferences depend on $\Delta G_{ij} = G_{ij} - G_{ij}^{id}$. The numerical values for the (volume-corrected) preferential solvation parameters presented in this chapter depend on the estimation of the correlation volumes by the iterative method proposed (Marcus 1990) that could possibly be improved on. However, the main difficulty with the KBI approach to the preferential solvation is the high accuracy demanded for the derivative of the chemical potential of a component (or for the second derivative of the excess Gibbs energy) that cannot always be met.

One way around this difficulty is to depend on a model, which although not rigorous does lead to reasonable quantitative results regarding the preferential solvation in binary solvent mixtures. Such a model is the quasi-lattice, quasi-chemical (QLQC) model proposed by Marcus (1983, 1989). The quasi-lattice model assumes the number of nearest neighbors a molecule has (the lattice parameter Z) is the weighted mean of the lattice parameters of the pure components. It also assumes that the interaction energy of a molecule of component A with one of a neighboring molecule of component B is independent of the natures of the other neighbors. The model also assumes that ideal volumes and entropies of mixing take place. The quasi-chemical part of the model specifies the number of unlike neighbors in terms of those of like neighbors and the interaction energies. The calculation then leads to the local mole fractions, $x^L_{A(B)}$, and to the preferential solvation parameters $\delta x_{A(B)} = x^L_{A(B)} - x_A$. Values of $\delta x_{A(B)}$ have been presented (Marcus 1989) for A = water and B = methanol, ethanol, 1-propanol, 2-propanol, 1,4-dioxane, tetrahydrofuran, acetone, dimethylsulfoxide, sulfolane, triethylamine, pyridine, acetonitrile, N,N-dimethylformamide, and hexamethyl phosphoric triamide. They were also presented for nonaqueous mixtures: A = acetone and B = nitromethane, n-hexane, chloroform, and ethanol, and for A = ethanol and B = chlorobenzene and acetonitrile, for A = 1-propanol and B = methanol, for A = 1-butanol and B = n-hexane, and for A = 1,4-dioxane and B = c-hexane. Some further systems were studied by means of the QLQC approach involving A = methanol, ethanol, acetone, and triethylamine and a variety of solvents B (Marcus 1991) and aqueous acetonitrile (Marcus and Migron 1991). Qualitative agreement between the results obtained from the KBI and QLQC approaches was demonstrated (Marcus 1991), although the details (magnitudes, not signs) of the preferential solvation parameters differ. This is not surprising, in view of the different premises involved in the two approaches.

The advantage of the QLQC method is that it is much less demanding on the accuracy of the input data, because no derivative functions are required (molar volumes rather than partial molar volumes and in particular excess Gibbs energies rather than derivatives of the chemical potential, activity, or activity coefficient or the second derivative of the excess Gibbs energy). On the other hand, the QLQC approach deals only with the nearest neighbors, but the KBI approach, by means of the calculation of the correlation volume and the parameter m denoting the number of solvation shells involved in Equations 3.7 and 3.8, is applicable to more than the first solvation shell.

Many studies have been made on the behavior of probe solutes in order to characterize the selective solvation in solvent mixtures, in particular depending on solvatochromism. However, what is determined is the selective solvation of the probe, not of the binary solvent components with each other. It is conceivable that a spectroscopic investigation of a binary solvent mixture in the absence of a probe could be carried out if each component had a clear signal (absorption or fluorescence) at a characteristic exclusive wavelength. Such mixtures could be formamide + acetonitrile and formamide + dimethylsulfoxide, in which the C=O bond stretching wave number of the hydrogen bond donor differs from the C≡N and the S=O bond stretching wave number of the acceptors (Alves and Santos 2007).

4 Kirkwood–Buff Integrals in Fully Miscible Ternary Systems

Thermodynamic Data, Calculation, Representation, and Interpretation

Enrico Matteoli, Paolo Gianni, and Luciano Lepori

CONTENTS

Abstract: A complete procedure, starting from the experimental techniques to obtain accurate thermodynamic G^E and V^E data for mixing, is accurately described for the calculation of Kirkwood–Buff integrals (KBIs) and local compositions in the whole mole fraction domain of a ternary mixture of organic compounds. Proper equations to represent G^E and V^E are suggested, and these are used in a computer program, which calculates the values of all the intermediate quantities necessary to obtain the KBIs. The computer routines allow an easy and fast calculation of the KBIs, and of the local compositions, in correspondence with a lattice of mole fraction values that is then used to plot the KBIs as continuous surfaces. This treatment is applied to three ternary mixtures containing components of different polarity, purposely chosen to examine the roles played by different types of interactions (H-bond, dipole–dipole, dipole-induced dipole, dispersion) in the determination of the local composition around each solute species. The mixture of ethanol (Et) + tetrahydrofuran (THF) + cyclohexane (cyH) (system I) is investigated and the capability of THF to perturb the strong Et self-association characteristics of the alcohol-hydrocarbon binary mixtures is examined. In system II, 2-methoxyethanol (MEt) + THF + cyH, the KBIs and local composition behaviors are compared with system I to examine the effects of the additional ether oxygen of MEt on the molecular interactions among the components. Finally, the mixture of trichloromethane (TCM) + THF + cyH (III) is studied to indicate the degree to which the strength of the TCM-THF interaction competes with the TCM and THF respective self-associative interactions shown in their binary mixtures with cyH.

4.1 INTRODUCTION

Among the various applications of the Kirkwood–Buff (KB) theory exploited so far, its use in solvation studies was the first one to be considered by investigators of chemical and physical processes in solution. The KB theory was devised to offer the possibility of predicting, from the knowledge of the g_{ij}s and through their integrals, the KBIs, the excess thermodynamic properties of mixtures (Kirkwood and Buff 1951). Due to the difficulties of obtaining g_{ij} for actual systems, the theory was practically ignored for many years, until Ben-Naim developed a procedure for its inversion, in this way providing the equations for obtaining the KBIs from excess thermodynamic properties (Ben-Naim 1977).

Knowledge of the KBIs for a mixture allows one to calculate the excess or deficiency of the number of moles of component j (solvatant) in the neighborhood of the central molecule (solvaton) i, and therefore, under certain approximations, information on the local composition in the solvation shell of i can be obtained. Despite the numerous developments that this approach offered concerning solvation studies, 7 years passed before a systematic and accurate study of the KBIs for a number of mixtures appeared (Matteoli and Lepori 1984). In the following years, other papers by different authors were published on this subject and a variety of aqueous and non-aqueous mixtures of nonelectrolytes and of electrolytes was considered. Chapter 3 in this book provides a wide variety of information concerning authors and the specific mixtures they studied.

It should be pointed out, however, that only a very small number of studies are devoted to ternary mixtures of compounds miscible in the whole composition range, where the KBIs have been calculated over the whole domain. Yet, ternary systems are more challenging and appealing than binaries from the point of view of the interpretation of the KBIs and the local composition in terms of the prevailing inter-molecular interactions and the microscopic structure of the mixtures.

We believe there are at least two reasons that explain this omission. The main one is the scarcity of the experimental thermodynamic data necessary for the KBI calculation. In fact, it is generally time consuming and not trivial to experimentally determine precise activity coefficient, γ, and V^E data in sufficient quantities to allow a densely detailed coverage of the whole triangular domain, both for the three binary sides and within the internal ternary area. Moreover, a high degree of accurate data, especially of G^E or $\ln \gamma$, is a must because the KBI calculation procedure requires multiple differentiations of the functions representing the above properties. In addi-tion, the absence of data cannot be circumvented by using group contribution models to obtain G^E. These can provide useful predictions for engineering applications, but there is no certainty that the accuracy is enough to warrant reliable values of the required derivatives.

The other reason is the lengthy calculation procedure, which involves two dif-ficult steps. The former is an accurate fitting of the experimental data for $\ln \gamma$ (or G^E) and for V^E by means of thermodynamically consistent equations. The latter is the complexity and awkwardness of the mathematical procedure to derive the final operative equations, which differ in their dependence on the type and analytical form of the equations chosen for fitting G^E and V^E data. All this makes it desirable to provide a software application, which taking as input the analytical forms and the parameters of the fitting equations, *whichever* they are, is able to provide the KBIs and related quantities, such as local composition and preferential solvation, at *any* given composition point in the triangular domain.

In previous papers (Matteoli and Lepori 1990, 1995), the results for six ternary systems were presented. However, these results were limited to the calculation of KBIs and preferential solvation coefficients at a small set of values for the ternary composition ensemble. The same restriction is also found in papers by other authors (see, for instance, Zielkiewicz 1995b; Ruckenstein and Shulgin 2001a; García et al. 2003; Aparicio et al. 2008). A representation of the KBIs and local composition in such a way as to provide a full view in the whole triangular domain would be more interesting and useful for understanding and discussing the structure of mixtures, since their complete behavior would appear thereby allowing comparisons between many different composition areas. This is the approach we have followed in this chapter to report and discuss unpublished KBI results of three ternary mixtures of compounds with different polarities and functional groups: ethanol (Et, 1) + tetra-hydrofuran (THF, 2) + cyclohexane (cyH, 3) (system I), 2-methoxyethanol (MEt, 1) + THF + cyH (system II), trichloromethane (TCM, 1) + THF + cyH (system III). They have in common two components, cyH, which can be considered practically inert in this context, and THF, which is weakly polar and can participate in H-bond formation as a proton acceptor. Components 1 are strongly polar molecules, Et and MEt being able to form H-bonds either as donor or acceptor of protons, while TCM can act only as donor. The experimental data for ln γ and for V^E were all determined in our laboratory (Lepori and Matteoli 1997; Lepori and Matteoli 1998; Lepori et al. 1998; Conti et al. 2003; Gianni, Lepori, and Matteoli 2010; Matteoli, Gianni, and Lepori 2010). The experiments were planned to cover the whole ternary domain in detail, and sufficient care in their treatment was exercised to assure high accuracy during the fitting.

We start this chapter with a description of the experimental techniques used and the equations that we have found to best fit the data. We then outline the mathematical derivation, from the basic equations of KB theory, to the operative equations to calculate the KBIs and the local composition. The algorithms for translating the whole calculation procedure into a software application are also described. Finally, we examine and discuss the information that the KBI and the local composition results convey on the role played by the different interactions (H-bond, dipole–dipole, dipole-induced dipole, dispersion) in determining the local composition in the neighborhood of a molecule of each species. Issues that will be discussed are: (1) the capability of THF to perturb the strong self-association feature of the alcohol-hydrocarbon binary mixtures; (2) the influence of the additional ether oxygen of MEt on the alcohol self-association due to the possibility of MEt forming an intramolecular H-bond or enhancing intermolecular H-bonding; and (3) the interplay of the TCM-THF hetero-association, shown in the binary mixture, and the TCM and THF respective self-associations evident in their binary mixtures with cyH.

4.2 EXPERIMENTAL TECHNIQUES

4.2.1 G^E DETERMINATION

As hinted in the Introduction, and explained in detail in Chapter 3 and in a previous paper (Matteoli and Lepori 1984), in order to obtain reliable values of the KBIs, very precise data for ln γ (or G^E) are necessary. At the same time, an empirical or model equation that accurately describes or fits these experimental data is also necessary in order to be able to calculate significant values of all the derivatives with respect to concentrations of the chemical potentials required by the operative equations. In an effort to fulfill these requirements, during the last decades we have developed, tested, and applied a technique for determining ln γ based on the measurement, via gas chromatography, of the composition of the vapor in equilibrium with a liquid mixture of known composition (Gianni, Lepori, and Matteoli 2010; Matteoli, Gianni, and Lepori 2010, and references therein). To be able to carry out a large number of experiments in a reasonable time, a device, which provides consecutive additions of small aliquots of a pure component to another component, or to a known stock mixture, was implemented. Those areas of the composition domain where ln γ displayed a strong dependence on concentration were covered with a higher density of experimental points. Each experimental point in this technique is a vector of five numbers: x_1, x_2, A_1, A_2, A_3, the liquid composition, and the areas of the three gas-chromatographic peaks of the vapor components. Each vector allows one to calculate the experimental mole fraction of each component in the vapor phase,

$$y_i = 1 \Big/ \sum_{j=1}^{3} r_{ij} \left(A_j / A_i \right) \tag{4.1}$$

where r_{ij} are calibration constants that relate the peak area to the number of moles. Correspondingly, y_i can be calculated as

$$y_i^{\text{calc}} = \left(\gamma_i^{\text{calc}} x_i p_i^{\circ} / \Phi_i \right) \Big/ \sum_{j=1}^{3} \gamma_j^{\text{calc}} x_j p_j^{\circ} / \Phi_j \tag{4.2}$$

where p° is the vapor pressure of the pure compound and Φ is a factor that takes into account vapor nonideality (Van Ness 1995). γ^{calc} is obtained from a parametric expression of $\ln \gamma$ (or, if not available, from G^{E} by differentiation).

By means of a least square routine which minimizes the objective function (O.F.),

$$\text{O.F.} = \sum_{i=1}^{m} \sum_{k} \left(\ln y_{ik}^{\text{exp}} - \ln y_{ik}^{\text{calc}} \right)^2 \tag{4.3}$$

the values of the parameters of the $\ln \gamma$ or G^{E} expression are obtained. The procedure is applied first to each single binary system ($m = 2$), then to the ternary points, in which case the binary parameters are kept fixed.

4.2.2 V^{E} and κ_{T} Determination

For V^{E}, the same degree of accuracy as G^{E} is not necessary. Excess volumes have been determined by measuring the density of mixtures with the commonly used vibrating tube technique. Experimental V^{E} data are fitted to parametric equations, similar to those used for G^{E}, by minimizing the O.F.,

$$\text{O.F.} = \sum_{k} \left(V_k^{E,\text{exp}} - V_k^{E,\text{calc}} \right) \tag{4.4}$$

As for the G^{E} data, the binary systems are studied first, and their parameters are then used as constants in the treatment of the ternary data. Usually, for a ternary mixture of strongly interacting components, the number of experimental points, k, is of the order of one hundred both for G^{E} and V^{E}. The numbers of parameters required is around 8 for G^{E} and around 14 for V^{E}.

Isothermal compressibility values of mixtures are very scarce in the literature, and their determination requires instrumentation not commonly available in laboratories. Fortunately, they play a very modest role in determining the KBI values, so it is enough to have available values of the pure components (not necessarily very accurate and can even be estimated), and to assume ideal behavior for the calculation of κ_T of a mixture,

$$\kappa_T = \sum_{i=1}^{3} \phi_i \kappa_{T,i}^{\circ} \tag{4.5}$$

where ϕ_i is the volume fraction of component i.

4.2.3 Data Treatment

Many equations, either empirical or derived from models, are available to represent excess thermodynamic properties as a function of the liquid mole fraction and of a number of adjustable parameters for use in Equations 4.3 and 4.4 (Prausnitz, Lichtenthaler, and Gomes de Azevedo 1999). In the treatment of our experimental data, we have examined the capability of many of these equations to fit data with the smallest number of parameters and with the least standard deviation of the O.F. (Lepori et al. 1998). For the majority of systems, either binary or ternary, examined by us in the course of about two decades, the rational form (Myers and Scott 1963) of the Redlich–Kister (RK) expression (Redlich and Kister 1948), and the Wilson (Wilson 1964) equation in the extended form (Novák, Matouš, and Pick 1987), have resulted in the most appropriate representations of G^E and V^E. For ternary systems, the excess functions can be expressed as the sum of a contribution (subscript B), which depends only on the parameters of the three binary systems, and a ternary contribution (subscript T), which involves additional parameters,

$$Y^E = Y_B^E + Y_T^E \tag{4.6}$$

where g or V, and $g = G/RT$.

In the case of the Wilson equation, g_B^E and V_B^E are formally different,

$$g_B^E = -\sum_{i=1}^{3} x_i \ln\left(\sum_{j=1}^{3} \Lambda_{ij} x_j\right) + \sum_{i=1}^{2}\sum_{j=i+1}^{3} p_{ij} x_i x_j \quad (\Lambda_{ii}=1) \tag{4.7}$$

$$V_B^E = \sum_{i=1}^{3} x_i \left(\sum_{j=1}^{3} x_j \Lambda_{ij}\lambda_{ij} \Big/ \sum_{j=1}^{3} x_j \Lambda_{ij}\right) \quad (\Lambda_{ii}=1, \lambda_{ii}=0) \tag{4.8}$$

If the RK model is assumed, the same formal equation can be used to express g_B^E and V_B^E,

$$Y_B^E = \sum_{i=1}^{2} \sum_{j=i+1}^{3} Y_{ij}^E \tag{4.9}$$

where Y_{ij}^E refers to the binary $i + j$ mixture,

$$Y_{ij}^E = x_i x_j \sum_{m=0}^{4} a_{m,ij} (x_i - x_j)^m \left/ \left[1 + d_{ij}(x_i - x_j) \right] \right. \tag{4.10}$$

For the ternary contribution, the rational form (Nagata and Miyazaki 1992) of the Van Ness expansion (Morris et al. 1975) is assumed, both for g^E and V^E,

$$Y_T^E = \left(x_1 x_2 x_3 \sum_{m=0}^{3} \sum_{n=0}^{m} b_{mn} x_1^{m-n} x_2^n \right) \left/ \sum_{i=1}^{2} \sum_{j=i+1}^{3} \left(1 + d_{ij}^T(x_i - x_j) \right) \right. \tag{4.11}$$

For the three systems studied here, the extended Wilson equation has been chosen for g^E, and the RK equation for V^E. The parameters are collected in Table 4.1 for g^E and in Table 4.2 for V^E.

TABLE 4.1

Parameters of Equation 4.7 and Equation 4.11 for g^E of Systems I, II, and III at 298.15 K[a]

	Λ_{12}	Λ_{21}	Λ_{13}	Λ_{31}	Λ_{23}	Λ_{32}	p_{13}	b_{00}
I	0.5684	0.8706	0.0276	0.2481	0.7295	0.7517	−0.1233	0
II	0.4989	0.8485	0.0257	0.2474	0.7295	0.7517	0	0.1989
III	1.8441	1.8546	0.8606	0.7056	0.7295	0.7517	0	−0.0887

Sources: From L. Lepori and E. Matteoli, 1997, Excess Gibbs Energies of the Ternary System Ethanol + Tetrahydrofuran + Cyclohexane at 298.15 K, *Fluid Phase Equilibria,* 134, 113; L. Lepori, E. Matteoli, G. Conti, and P. Gianni, 1998, Excess Gibbs Energies of the Ternary System Ethanol + N,N-Dimethylformamide + Cyclohexane at 298.15 K, *Fluid Phase Equilibria.* 153, 293; P. Gianni, L. Lepori, and E. Matteoli, 2010, Excess Gibbs Energies and Volumes of the Ternary System Chloroform + Tetrahydrofuran + Cyclohexane at 298.15 K, *Fluid Phase Equilibria*, 297, 52; E. Matteoli, P. Gianni, and L. Lepori, 2010, Excess Gibbs Energies of the Ternary System 2-Methoxyethanol + Tetrahydrofuran + Cyclohexane and Other Relevant Binaries at 298.15 K, *Journal of Chemical and Engineering Data*, 55, 5441.

[a] Values of parameters not reported in the table are zero. All parameters are dimensionless.

TABLE 4.2

Parameters of Equations 4.10 and 4.11 for V_m^E of Systems I, II, and III at 298.15 K[a]

	a_0	a_1	a_2	a_3	a_4	d
I, Et+THF	−0.0844	−0.2544	0	0	0	0
I, Et+cyH	2.2329	2.2858	0.8137	0.4951	0	0.9697
I,II,III, THF+cyH	2.1217	−0.2426	0.1538	−0.0748	0	0
II, MEt+THF	0.4948	−0.0954	0.0562	0	0	0
II, MEt+cyH	3.5685	3.6412	1.7023	0	−0.8728	0.9348
III, TCM+THF	−1.4061	−0.2420	0.3276	0.3793	0	0
III, TCM+cyH	2.2357	−0.1737	0.0526	0	0	0

	b_{00}	b_{10}	b_{11}	b_{22}	b_{33}	d^T
I	0.24770	1.5576	−1.4841	3.1869	−2.0987	0.99593
II	4.1144	−3.7071	−4.1203	0	0	0.469216
III	−7.7156	4.7686	5.1233	0	0	0

Sources: From L. Lepori and E. Matteoli, 1998, Excess Volumes of the Ternary System Ethanol + Tetrahydrofuran + Cyclohexane at 298.15 K, *Fluid Phase Equilibria*, 145, 69; G. Conti, P. Gianni, L. Lepori, and E. Matteoli, 2003, Volumetric Study of (2-Methoxyethanol + Tetrahydrofuran + Cyclohexane) at 298.15 K, *Journal of Chemical Thermodynamics*, 35, 503; P. Gianni, L. Lepori, and E. Matteoli, 2010, Excess Gibbs Energies and Volumes of the Ternary System Chloroform + Tetrahydrofuran + Cyclohexane at 298.15 K, *Fluid Phase Equilibria*, 297, 52.

[a] Values of parameters not reported in the table are zero. Units are cm³mol⁻¹ for a and b; d are dimensionless. The following values of pure components have been used: V° (cm³ mol⁻¹), 58.74, 79.18, 83.10, 81.74, 108.75 for Et, MEt, TCM, THF, cyH, respectively; κ_T° (MPa⁻¹), 0.00114 for cyH, 0.001 for Et, MEt, TCM, THF.

4.3 FUNDAMENTAL AND OPERATIVE EQUATIONS

4.3.1 KIRKWOOD–BUFF INTEGRALS

In this section, we illustrate the procedure to obtain KBI having available analytical parametric expressions of G^E (or $\ln \gamma_i$) and V^E as functions of the mole fractions x_1, x_2, x_3. The fundamental equation, which relates G_{ij} to thermodynamic properties in the isothermal–isobaric ensemble (in mole units) can be expressed as

$$G_{ij} = \frac{\mathbf{A}^{ij}}{c_i c_j |\mathbf{A}|} - \frac{\delta_{ij}}{c_j} \tag{4.12}$$

where $\mathbf{A}$ is a matrix whose elements are defined by (see Section 1.2.4 in Chapter 1)

$$A_{ij} = \frac{\kappa_T \mu'_{ij} V + \bar{V}_i \bar{V}_j}{RT\kappa_T} \tag{4.13}$$

The chemical potential derivative in Equation 4.13 is defined as

$$\mu'_{ij} = \left(\frac{\partial \mu_i}{\partial N_j} \right)_{T,p,\{N\}'} \tag{4.14}$$

The equations that are commonly used to represent experimental data of Z^E ($Z^E =$ G^E, V^E) and μ_i are expressed as a function of x_i, whereas in Equation 4.14 derivatives with respect to N_j are required. We need therefore to express them in a function of x_i. Taking into account the definition of the excess partial molar quantity, Z_i^E, as a function of N_i, the relationship between x_i and N_i, the differentials of $Z^E = f(x_1, x_2)$ with respect to x_i and of x_i with respect to N_i, and applying the treatment to one mole of mixture, after some substitutions and rearrangements, the diagonal elements μ'_{ii} can be expressed in a function of x_i and of four derivatives of the chemical potential of components 1 and 2,

$$\mu'_{11} = (1 - x_1) \left(\frac{\partial \mu_1}{\partial x_1} \right)_{T,p,x_2} - x_2 \left(\frac{\partial \mu_1}{\partial x_2} \right)_{T,p,x_1} \tag{4.15}$$

$$\mu'_{22} = (1 - x_2) \left(\frac{\partial \mu_2}{\partial x_2} \right)_{T,p,x_1} - x_1 \left(\frac{\partial \mu_2}{\partial x_1} \right)_{T,p,x_2} \tag{4.16}$$

$$\mu'_{33} = \frac{1}{x_3} \left\{ x_1^2 \left(\frac{\partial \mu_1}{\partial x_1} \right)_{T,p,x_2} + x_1 x_2 \left[\left(\frac{\partial \mu_1}{\partial x_2} \right)_{T,p,x_1} + \left(\frac{\partial \mu_2}{\partial x_1} \right)_{T,p,x_2} \right] + x_2^2 \left(\frac{\partial \mu_2}{\partial x_2} \right)_{T,p,x_1} \right\} \tag{4.17}$$

The remaining three off-diagonal μ'_{ij} ($\mathbf{\mu}$ is a symmetrical matrix) can be expressed in terms of the μ_{ii} thanks to the Gibbs–Duhem relationship,

$$\mu'_{ij} = \frac{x_k^2 \mu'_{kk} - x_i^2 \mu'_{ii} - x_j^2 \mu'_{jj}}{2 x_j x_i} \tag{4.18}$$

The derivatives in Equation 4.15 through Equation 4.17 are easily obtained from the expression of the chemical potential on the rational scale,

$$\mu_i = \mu_i^\circ + RT \ln \gamma_i x_i \tag{4.19}$$

$$\left(\frac{\partial \beta \mu_i}{\partial x_j} \right)_{T,p,\{x\}'} = \frac{\delta_{ij}}{x_j} + \left(\frac{\partial \ln \gamma_i}{\partial x_j} \right)_{T,p,\{x\}'} \tag{4.20}$$

Likewise, the expression for the excess partial molar quantity, Z_i^E, as a function of x_i is obtained as

$$Z_i^E = Z^E + \left(x_j + x_k\right)\left(\frac{\partial Z^E}{\partial x_i}\right)_{T,p,x_j} - x_j\left(\frac{\partial Z^E}{\partial x_j}\right)_{T,p,x_i} \qquad (j < k) \qquad (4.21)$$

where Z_i^E represents $RT \ln \gamma_i$ or V_i^E.

4.3.2 LOCAL COMPOSITION

As it is well known, the KB theory allows one to calculate N_{ij}, the excess (or deficit) number (or excess coordination number) of moles of j in the whole space around a central molecule i,

$$N_{ij} = c_j G_{ij} \qquad (4.22)$$

Usually, for systems not very near an instability, the range of influence of the central particle on the surrounding molecules does not exceed a distance, R_{cor}, which is on the order of a few molecular diameters. This means that $g_{ij}(r) = 1$ for $r > R_{cor}$ and that correlations between the central molecule and the surrounding ones are active only at $r \leq R_{cor}$. Denoting V_{cor} as the volume around the central molecule i where correlations are effective, and with V_m the volume of one mole of mixture, the total number of moles n_t in the interior of V_{cor} is given by

$$n_t = \left(\sum_{m=1}^{3} x_m G_{mi} + V_{cor}\right)\bigg/ V_m \qquad (4.23)$$

The local (that is, inside V_{cor}) mole fraction of component j around i is given by

$$x_{ji}^L = \left(x_j G_{ji} + x_j V_{cor}\right)\bigg/ \left(\sum_{m=1}^{3} x_m G_{mi} + V_{cor}\right) \qquad (4.24)$$

By subtracting from x_{ji}^L the bulk mole fraction x_j and rearranging, the following final expression is obtained for the excess local mole fraction δx_{ji},

$$\delta x_{ji} = \left[x_j x_i (G_{ji} - G_{ii}) + x_j x_k (G_{ji} - G_{ki})\right]\bigg/ \left(\sum_{m=1}^{3} x_m G_{mi} + V_{cor}\right) \qquad (4.25)$$

For the general expression of Equation 4.25, see Section 1.3.4 in Chapter 1.

4.3.3 PREFERENTIAL SOLVATION

The availability of the local compositions allows one to calculate a quantity called *preferential solvation*. There is no generally established definition of preferential solvation (see Section 3.1 in Chapter 3). For a binary mixture, the excess local molar fraction has been usually considered synonymous with preferential solvation: the more positive δx_{ji}, the higher the preference of i to be solvated by j than by the other species. However, in a ternary system, where the central particle can be solvated by three different species, this might be insufficient; δx_{ji} may be positive, but less positive than say δx_{ki}: in this case, the preference of i is $k > j > i$. On this basis, as suggested in a previous work (Matteoli and Lepori 1995), a quantitative measure of the preferential solvation, $ps_i^{j,k}$, of solvaton i with respect to solvatant species j and k can be assumed to be the difference of the two corresponding excess local mole fractions,

$$ps_i^{j,k} = \delta x_{ji} - \delta x_{ki} \qquad (4.26)$$

The more positive the value of $ps_i^{j,k}$ is, the larger the preference by i to be solvated by j than by k. To apply Equations 4.25 and 4.26, a procedure to calculate V_{cor} must be devised. This is illustrated in the next section. A procedure to obtain preferential solvation in the form of a "limiting coefficient of preferential solvation" without the need of calculating V_{cor} is illustrated in Section 1.3.4 in Chapter 1. Although we have followed this procedure in previous work, in this chapter we have preferred to calculate the excess mole fractions to show in a more realistic way what happens to the microscopic structure of the mixtures considered here.

4.4 CALCULATION ALGORITHM

4.4.1 KIRKWOOD–BUFF INTEGRALS

It is evident that even with the simplest expressions for G^E and V^E, like the Redlich–Kister equation with only one parameter p for each binary and no ternary parameters, it is not a simple process to derive explicit analytic expressions for each $G_{ij} = f(x_i, p_i)$. While it is possible to obtain the expressions for $\ln \gamma_i$ from G^E using Equation 4.21, the subsequent steps, which concern the derivatives in Equations 4.20 and their combinations in Equation 4.15 through Equation 4.18 and again in Equations 4.13 and 4.12, are so as to discourage following this route. In addition, the final equations would depend on the type of the G^E and V^E expressions used to represent the experimental data. For these reasons, we have developed a software application which, taking as input the analytical forms and the parameters of the equations for G^E and V^E, calculates all six G_{ij} at any given values of x_1, x_2. The trick used to avoid algebraic treatment of the many derivatives required is to approximate their exact value at each x_1, x_2 pair with the corresponding incremental ratio, $\partial f/\partial x \cong [f(x + \delta x) - f(x)]/\delta x$. Using a particularly small value of the increment δx, the error in the final G_{ij} values is absolutely negligible. For the benefit of the reader who would like to write his/her own program, the main steps of this application are now explained.

The fundamental step is the calculation of the four derivatives appearing in Equation 4.15 through Equation 4.17. As an example, we will show how to calculate one of these, $(\partial \mu_1/\partial x_1)_{x_2}$, using Equation 4.19 through Equation 4.21. To calculate the derivative of $\ln \gamma_1$, we first need the value of $\ln \gamma_1$ at the given composition by means of Equation 4.21, which involves the calculation of two derivatives of G^E. After calculating G^E at the given composition, the increment δx is separately applied to x_1 and x_2 together with a decrement by the same entity applied to the mole fraction that is not required to be kept constant, in this case x_3. The G^E values calculated at these new (incremented) compositions, together with those obtained before, allow the calculation, by the incremental ratio, of the two derivatives in Equation 4.21 and therefore of $\ln \gamma_1$ at x_1, x_2. The applied increment is $\delta x = 10^{-7}$, which due to the use of double precision variables (16 significant figures) generates very small but significant G^E variations. To proceed with Equation 4.20, we need now to calculate $\ln \gamma_1$ at $x_1 + 10\delta x$ (and, being x_2 to be kept constant, $x_3 - 10\delta x$); this is again achieved through Equations 4.21 with the procedure just described. At this point we have two values of $\ln \gamma_1$ at very similar compositions, x_1 and $x_1 + 10\delta x$, which allows one to obtain $(\partial \ln \gamma_1/\partial x_1)_{x_2}$ and therefore $(\partial \mu_1/\partial x_1)_{x_2}$. To calculate the incremented value of $\ln \gamma_1$, a ten times larger δx increment has been used to avoid interference between the incremental ratios of G^E and $\ln \gamma_1$, and to ensure enough significant figures in the difference between the two $\ln \gamma_1$ values. With an analogous procedure, the remaining derivatives, $(\partial \mu_1/\partial x_2)_{x_1}$, $(\partial \mu_2/\partial x_1)_{x_2}$, and $(\partial \mu_2/\partial x_2)_{x_1}$, are then calculated.

The subsequent step is the calculation of the quantities that appear in the matrix elements A_{ij}. By means of Equation 4.15 through Equation 4.18, all chemical potential derivatives are calculated. The partial molar volumes are obtained by adding to the molar volume of the pure components the excess partial molar volumes, obtained through Equation 4.21 with $Z^E = V^E$, by applying the same procedure used to calculate $\ln \gamma$. The mixture molar volume is then obtained as $V_m = x_1 V_1 + x_2 V_2 + x_3 V_3$ and κ_T from Equation 4.5. The values of G_{ij} are finally obtained from Equation 4.12, for which the calculations of the concentrations, the determinant, and the cofactors are all straightforward.

It should be observed that the calculation cannot be performed for mole fractions corresponding to the three sides of the composition triangle (that is, a binary mixture and a pure compound, respectively) when one composition is taken to be exactly zero, because the incremental ratios cause some x values to be negative or to exceed one. This drawback is overcome by approximating the zero values with a small value, such as 0.00001. The error on G_{ij} due to this approximation is less than 0.01 cm^3 mol^{-1}.

4.4.2 Local Composition

In our model, V_{cor} is assumed to be the volume of a mole of spheres of radius $r_c + n_l t_l$ minus the volume of a mole of central molecules,

$$V_{cor} = N_A 4\pi \left[(r_c + n_l t_l)^3 - r_c^3 \right]/3 \qquad (4.27)$$

where r_c is the radius of the central molecule assumed as spherical, t_l is the thickness of one solvation layer, and n_l is the number of solvation layers. The radii of the central species and of the other components, r_i are calculated from the respective van der Waals molecular volumes taken from Bondi (1964); t_l is calculated from the composition of the solvation shell according to

$$t_l = 2\sum_{i=1}^{3} \phi_{i,l} r_i \qquad (4.28)$$

where $\phi_{i,l}$ is the volume fraction of i in the solvation shell. Since the composition of the shell is initially not known, at the start of the calculations a shell composed by one layer and with bulk composition is assumed. Then an iterative routine is applied. The routine also checks if any of the local mole fractions exceeds one, and if so, the number of shells is increased by one. Convergence is reached after three or four iterations. This procedure of calculating V_{cor} is very similar to that proposed by Marcus in his works (Marcus 2001) and Chapter 2; the r_i values used by Marcus, obtained through an empirical formula using V_i, are in good agreement with those obtained here from van der Waals volumes.

The software, written in Visual Basic, finally calculates and stores KBI, δx_{ji}, and $ps_i^{j,k}$, as well as the corresponding volume (or reference state, see next section) corrected quantities, for a lattice of composition points which can be read by the freeware GNUPLOT program to plot them as surfaces over the triangular domain.

4.5   ON THE USE OF KIRKWOOD–BUFF INTEGRALS FOR THE STUDY OF THE LOCAL STRUCTURE OF MIXTURES

An exhaustive insight into the local or microscopic structure of a liquid mixture would be provided by the knowledge of the g_{ij} for all couples of components. Unfortunately, these are not known even for mixtures of weakly interacting simple substances. The KBIs, being the integrals of the g_{ij} over the whole space, are a sort of average of the spatial distribution of molecules around the central one, and their value conveys qualitative information if the species i and j have a tendency to attraction (KBI $\gg$ 0) or to repulsion. As described in the previous sections, this indication can be transformed into microscopic information on the local structure around the central molecule by assuming that the correlation between the central molecules and the surrounding ones does not exert at distances larger than a few molecular diameters.

If we examine the literature works that use the KBI and the local composition or preferential solvation to obtain information on the type and strength of the interactions in mixtures, we realize that in most of these works the KBIs have not been used directly, but rather after subtraction of a quantity called the *ideal contribution*, or *volume correction*, or *excluded volume contribution* (Matteoli and Lepori 1995; Matteoli 1997; Marcus 2001; Ruckenstein and Shulgin 2001a; Vergara et al. 2002; Aparicio et al. 2008). The last definition comes from the always-negative contribution to the KBI from zero to the distance of maximum approach. See also Section 2.7.2

in Chapter 2, and Section 3.2 and Section 3.5 in Chapter 3. The word correction was not used (and we do not do so here) to mean that the KBIs as calculated according to the original KB theory were in some way wrong, but simply to indicate the convenience of subtracting from the original value a contribution that does not originate from molecular interactions, but from the intrinsic volume properties of the components, and that therefore is not only useless, but misleading in the interpretation of the KBI in terms of local compositions and interactions.

The *anomalous* features due to this contribution that suggested the convenience and need to correct the KBI can be summarized as follows: (1) For a pure compound i, the KBI has a nonzero (negative) value, therefore implying a nonzero value for the deficit number of molecules of i around i; (2) In the surroundings of a solute at infinite dilution, an unrealistic deficit number of molecules of solvent is calculated when the solute molar volume is much larger than the solvent; (3) For symmetrical ideal mixtures of components with *different molar volumes* the local composition is found to be different from the bulk, the larger the difference, the larger the size mismatch. In particular, for binary mixtures, it is found that in the surroundings of *both i* and *j* components, there is a depletion of the same species, the one with the larger size, and a corresponding enrichment of the other; that is to say, *both i* and *j* are preferentially solvated by the *same* species.

Clear and explanatory actual examples of the unreasonable and meaningless information that is obtained if a correction is not applied were also reported by Vergara et al. (2002) and Shulgin and Ruckenstein (2006a, 2008a) for aqueous solutions of large molecules such as PEG and proteins, respectively.

At least three different correction procedures have been proposed by investigators to remove the consequences of the above drawbacks. They do not differ significantly as to the extent of the correction values (Matteoli 1997; Ruckenstein and Shulgin 2001a; Shulgin and Ruckenstein 2006a). Zielkiewicz (1995b) also realized the importance of the volume contribution and proposed a way of calculating it, but did not use it to correct the KBIs.

A few years ago, a debate occurred in the literature on the correctness of applying such a correction. No agreement has been reached by the supporters of the opposite theses and each has kept to their own opinions (Ben-Naim 2007a, 2007b, 2008; Matteoli and Lepori 2007, 2008; Shulgin and Ruckenstein 2008b). As the first authors to suggest and apply a correction (Matteoli and Lepori 1995; Matteoli 1997), we have been addressed the strongest criticism (Ben-Naim 2007a). However, we believe we have properly refuted all critical arguments, and therefore we are convinced that the correction procedure used by us is well founded and capable of eliminating the source and the consequences of the above described unreasonable features, completely and homogeneously, in the whole domain of binary and ternary mixtures.

To obtain the corrected values of the KBIs, ΔG_{ij}, the values, G_{ij}^{id}, of an ideal system ($G^{\mathrm{E}} = V^{\mathrm{E}} = 0$ at all compositions) whose components have the same molar volumes as the actual system are subtracted from G_{ij},

$$\Delta G_{ij} = G_{ij} - G_{ij}^{\mathrm{id}} \tag{4.29}$$

Since the G_{ij}^{id} have magnitudes on the order of the molar volumes, the correction is important for systems showing a behavior near to ideality. However, it is negligible for large parts of the composition domain if the system presents strong deviations from ideality.

In a similar way, the corrected local excess composition, $\delta x'_{ji}$, is calculated by subtracting from the local composition that of the corresponding ideal system,

$$\delta x'_{ji} = \delta x_{ji} - \delta x_{ji}^{\text{id}} = x_{ji}^{\text{L}} - x_{ji}^{\text{L,id}} \qquad (4.30)$$

where $x_{ji}^{\text{L,id}}$ is obtained with the same procedure as x_{ji}^{L} (Equation 4.24) but assuming $G^{\text{E}} = V^{\text{E}} = 0$ for the system under examination. Also, the preferential solvation can be corrected for the volume contribution by using $\delta x'_{ji}$ instead of δx_{ji} in Equation 4.26,

$$\Delta \text{ps}_i^{j,k} = \delta x'_{ji} - \delta x'_{ki} \qquad (4.31)$$

4.6 PRESENTATION AND DISCUSSION OF THE RESULTS

All ΔG_{ij} (Equation 4.29) values for the three ternary mixtures are shown in Figure 4.1 through Figure 4.3. Note that the units are dm^{3}mol^{-1} for systems I and II, and cm^{3}mol^{-1} for system III. The surfaces of ΔG_{ij} have been depicted with almost the

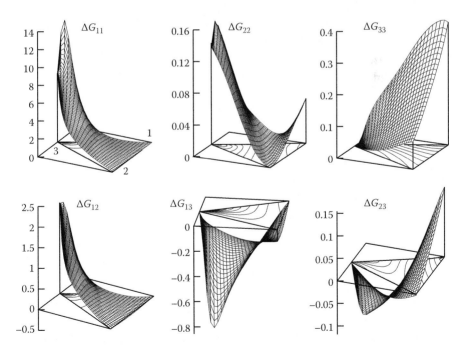

FIGURE 4.1 Corrected KBIs (ΔG_{ij}, Equation 4.29) for the ternary system ethanol(1) + tetrahydrofuran(2) + cyclohexane(3) at 298.15 K. Units are dm^{3}mol^{-1}. For all pictures, components correspond to the three corners of the composition triangle as illustrated in the first panel.

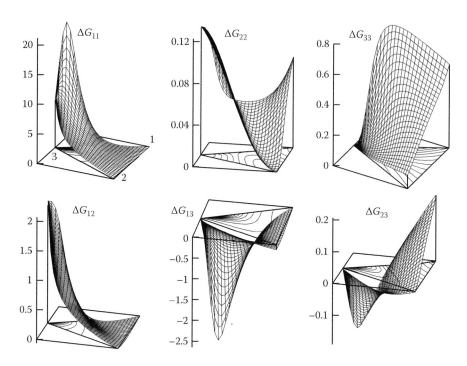

FIGURE 4.2 Corrected KBIs (ΔG_{ij}, Equation 4.29) for the ternary system 2-methoxyethanol(1) + tetrahydrofuran(2) + cyclohexane(3) at 298.15 K. Units are dm^3mol^{-1}. For all pictures, components correspond to the three corners of the composition triangle as illustrated in the first panel.

same orientation in order to facilitate comparison among them, but with different scales for the z-axis. Because of the huge difference in the magnitudes of some ΔG_{ij}, the use of the same scale would have compressed the images with small ΔG_{ij} values with the consequence of practically hiding their features.

Figure 4.4 through Figure 4.6 show the plots of all $\delta x'_{ji}$ values (Equation 4.30) for systems I–III. The surfaces in a row show the excess composition of the solvation shell of the same central species; their comparison indicates which compounds the solvaton prefers to be solvated by. Alternatively, by comparing the pictures in a column, information can be inferred on which solute a component prefers to solvate.

It should be noted that for system I (Figure 4.4) and system II (Figure 4.5), the model for calculating V_{cor} finds a solvation shell composed of two layers, but only one layer for system III (Figure 4.6). It must also be mentioned that this model is somewhat empirical because it was founded on reasonable, but not rigorous, assumptions and other different, more scientifically valid, methods to calculate V_{cor} could be put forward. Therefore, the values of the excess local composition so obtained should not be given a quantitative meaning. However, they can be safely used for comparison since, as the form of Equation 4.25 suggests, the V_{cor} value cannot change the relative magnitude of δx_{ji}.

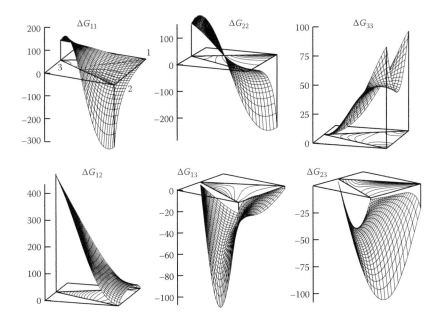

FIGURE 4.3 Corrected KBIs (ΔG_{ij}, Equation 4.29) for the ternary system trichloromethane(1) + tetrahydrofuran(2) + cyclohexane(3) at 298.15 K. Units are cm^3 mol^{-1}. For all pictures, components correspond to the three corners of the composition triangle as illustrated in the first panel.

4.6.1 KIRKWOOD–BUFF INTEGRALS

The main features that first appear after looking at Figure 4.1 are the very high positive values of ΔG_{11}, which reach a maximum of 14 dm^3mol^{-1} in the cyH-rich area of the binary Et+cyH, at $x_1 \approx 0.1$. These values greatly surpass the also high values of ΔG_{12}, which almost parallel ΔG_{11}, and the small but always positive values of ΔG_{22}. ΔG_{33} is also positive in the whole domain, with the largest values in the Et-rich area. ΔG_{13} is fairly strongly negative with a minimum in the cyH-rich region of the Et+cyH binary. ΔG_{23} shows either positive or negative values, but is always very small in magnitude like ΔG_{22}.

These trends suggest that the interaction that dominates is the associative attraction Et-Et, which is much stronger than the hetero-association Et-THF. The relatively high values of ΔG_{12} indicate that this association is fairly effective in spite of the strong competition by the Et-Et H-bond association. It should be noticed that the cyH-cyH affinity is generally stronger than the THF-THF attraction in spite of the lack in cyH of polar functional groups; evidently, this tendency to association, as well as the strong Et-cyH nonaffinity, are the passive result of the Et-Et and Et-THF associations.

The system MEt+THF+cyH behaves qualitatively as system I, as can be seen by comparing the corresponding ΔG_{ij} between the two systems. However, ΔG_{11} is about 80% larger, and correspondingly ΔG_{33} and ΔG_{13} are also much larger in magnitude.

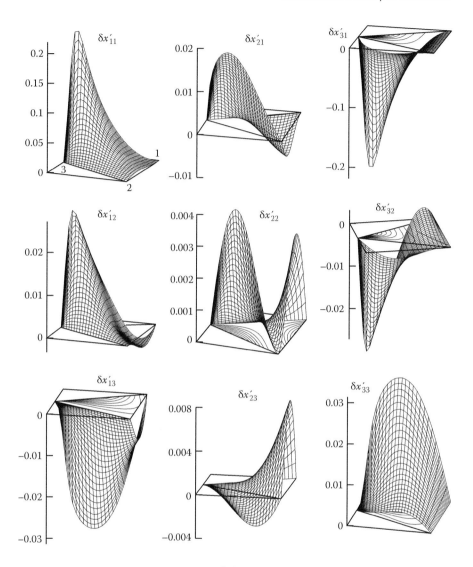

FIGURE 4.4 Excess local compositions ($\delta x'_{ji}$, Equation 4.30) for the ternary system ethanol(1) + tetrahydrofuran(2) + cyclohexane(3) at 298.15 K. For all pictures, components correspond to the three corners of the composition triangle as illustrated in the first panel.

On the contrary, ΔG_{22} and ΔG_{12} are practically identical with system I. These observations indicate that MEt-MEt association is much stronger than Et-Et, an effect which is brought about by the ether oxygen of MEt in two ways: an increased polarity of MEt and an enhancement of intermolecular H-bonding through its lone pairs. Instead, the formation of an intramolecular H-bond in MEt can be ruled out. This H-bond would diminish the number of sites capable of forming intermolecular H-bonds with a consequent decrease of ΔG_{11} with respect to Et-Et and a weakening of the features of the other ΔG_{ij} as described previously. The observed behavior

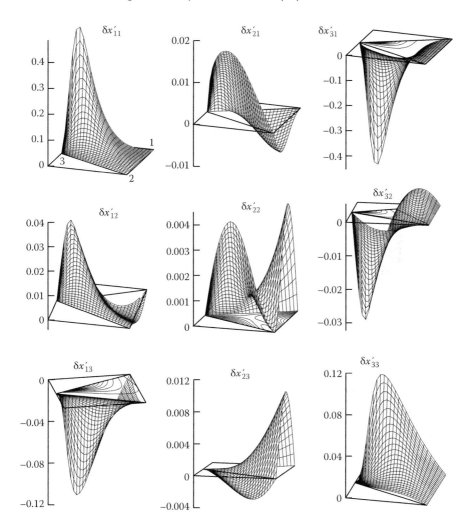

FIGURE 4.5 Excess local compositions ($\delta x'_{ji}$, Equation 4.30) for the ternary system 2-methoxyethanol(1) + tetrahydrofuran(2) + cyclohexane(3) at 298.15 K. For all pictures, components correspond to the three corners of the composition triangle as illustrated in the first panel.

suggests instead that the ether oxygen is engaged in the formation of intermolecular H-bonds. The close similarity of ΔG_{12} and ΔG_{22} between systems I and II indicates that 1-2 and 2-2 affinities are independent of the strength of 1-1 association and that the interaction 1-2 is not due to a specific H-bond formation.

The substitution in systems I and II of the component capable of forming H-bonds either as a donor or as an acceptor of protons with TCM, a substance which can form H-bonds only as a donor and also possesses a fairly strong dipole moment, dramatically changes the aspects of almost all the ΔG_{ij} surfaces. Figure 4.3 shows first of all that the very large magnitudes of some of the ΔG_{ij}

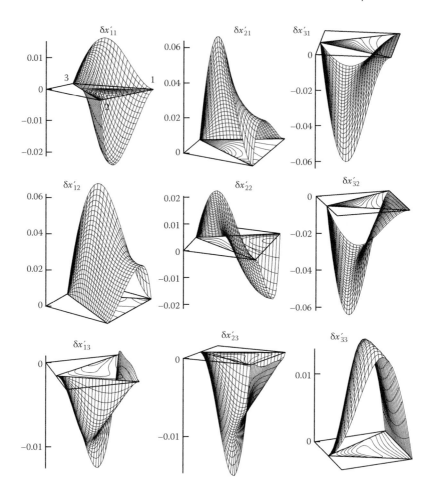

FIGURE 4.6 Excess local compositions ($\delta x'_{ji}$, Equation 4.30) for the ternary system trichloromethane(1) + tetrahydrofuran(2) + cyclohexane(3) at 298.15 K. For all pictures, components correspond to the three corners of the composition triangle as illustrated in the first panel.

values shown by the other systems have disappeared, all the ΔG_{ij} values being small and of the same order of magnitude. The largest values are found for ΔG_{12}, which are positive over the whole domain, while ΔG_{22} values are positive at cyH-rich compositions and negative elsewhere; ΔG_{23} appears negative for all compositions. These features indicate that it is the specific interaction by TCM and THF, probably a CH–O bond, which prevails and is responsible for the behavior of all ΔG_{ij}. Also, in this system ΔG_{33} is found to be positive for the whole domain, evidently for the same reason. Being an inert compound, when the other components undergo association driven by strong interactions, cyH is in a way compelled to self-associate because of the repulsion exercised on it by the other two molecules, as the negative ΔG_{13} and ΔG_{23} suggest. At compositions rich in cyH, some affinity for TCM-TCM and THF-THF association is evident, but much weaker than the TCM-THF affinity.

The MEt-MEt, Et-Et, TCM-THF associations can all be ascribed to H-bonds, but the range of the corresponding ΔG_{ij} values, ΔG_{11} for Et-Et and MEt-MEt, and ΔG_{12} for TCM-TGF, is extremely different: ΔG_{11} maximum $\gg 10^4$ cm^3 mol^{-1}, ΔG_{12} maximum ≈ 450 cm^3 mol^{-1}. This should not be considered a surprise, since in the first two cases the formation of a dimer leaves other sites free for further association to form trimers, tetramers, and so on, until, at certain temperature and pressure values, the clusters become so large as to initiate a phase separation. In the other case, once a couple TCM-THF is formed, further aggregation could at most produce the trimer TCM-THF-TCM thanks to the other available lone pair of the ether oxygen, but this is highly improbable, both for spatial and for electrostatic reasons. In fact, hetero-association cannot bring about very high values of ΔG_{ij} as is evident from the operative equations to calculate G_{ij} in binary systems. These equations (see Equation 1.72 in Chapter 1) contain the quantity D, which has the characteristics of approaching zero when positive G^E approaches the limiting value of a phase separation, with the consequence of generating G_{ij} values that tend to infinity. When G^E is negative, as is the case of the TCM+THF binary mixture, this cannot happen.

4.6.2 LOCAL COMPOSITION

The analysis of the KBI behavior of a given system and the comparison among systems allows one to understand the role played by the various types of molecular interactions in determining associations among molecules of the same or different species. The possibility offered by the KB theory to calculate the composition of the solvation shell of a given solute species, though under the approximations discussed above, allows one to obtain something more: information on the microscopic or molecular structure of the mixtures.

Looking at Figure 4.4, we can see that the two-layered solvation shell of Et is the one that undergoes the largest alterations with respect to bulk composition. It is crowded by Et in the whole domain and in the strongest way at compositions poor with THF reaching a maximum of about 0.22 in the Et-cyH binary mixture. This enrichment occurs at the expense of cyH only, whose concentration practically reduces by the same quantity (compare $\delta x'_{11}$ and $\delta x'_{31}$). $\delta x'_{21}$ shows in fact relatively very low values positive near the 2-3 side and negative in proximity of the 1-2 side, indicating the low interaction-induced participation of THF in the solvation of Et.

The second row of Figure 4.4 represents the excess composition in the solvation shell of THF, which even if very small, appears to be enriched by Et and to a smaller extent by THF and poorly by cyH. The surface of $\delta x'_{22}$ looks like a saddle, showing a zone of composition near the bisector from vertex 2, where the local concentration of THF is the same as in the ideal mixture. In other words, when $x_1 \approx x_3$, THF as a solute does not exhibit any interaction with THF as solvatant agent, but interacts attractively with Et at expense of cyH. The three pictures of the central column reveal the behavior of THF as a solvatant of the three species: all three $\delta x'_{2j}$ are much lower in magnitude than those in the other two columns, suggesting a reluctance of THF to solvate, that is, to enter the solvation shells of Et, THF, and cyH. Regarding cyH as solvaton, we see that its solvation shell is strongly depleted with Et and enriched with cyH, but almost unchanged as to THF.

The comparison of Figures 4.4 and 4.5 reveals strong similarities, as expected on the basis of the G_{ij} analogies. Practically, all surfaces are qualitatively identical, but not for all values. Looking at the pictures in columns 1 and 3, which show the behavior and capability of Et, MEt, and cyH as solvatants, we can see that the values in system II are much larger in magnitude, by about 100%, than in system I. On the contrary, the behavior of THF as solvatant (central column) shows not only the same surface shapes as system I, but also the same values, as if THF could not distinguish the different strength of the MEt-MEt and Et-Et associations, which is instead felt by cyH. Moreover, the same values of $\delta x'_{21}$ in systems I and II are again an indication that THF does not interact via H-bond with MEt and Et.

Let us now examine the nine pictures in Figure 4.6 for system III. First, the range of all values are in the same order of magnitude, and the maximum effects are shown in the solvation shells of TCM and THF. They are characterized by positive values of $\delta x'_{21}$ and $\delta x'_{12}$, respectively, and correspondingly negative values of $\delta x'_{31}$ and $\delta x'_{32}$, which demonstrate the preference of each of these components to be solvated by the other in the whole domain except, of course, in the binary systems where the solvatant is not available. In fact, in the area near the binary 1+3, $\delta x'_{11}$ is positive, and similarly, near the binary 2+3, $\delta x'_{22}$ is positive. In the surrounding of the solute cyH (third row of pictures), the values of the excess local compositions are smaller than those around solvatons TCM and THF, and are positive for $\delta x'_{33}$ and negative for the other two.

The fact that $\delta x'_{33}$ is positive over the whole domain for all three systems suggests that cyH is always preferentially solvated by other cyH molecules. However, the absence of polar functional groups in its structure indicates that is the responsibility of those interactions that are effective among the other components. A small (in system III) or large (in systems I and II) fraction of the number of molecules of species 1 or 2 that would be present in the solvation shell of cyH if it had the bulk composition, due to the interactions between their functional group are attracted in the solvation shells of 1 or 2 and replaced by other cyH molecules.

4.6.3 PREFERENTIAL SOLVATION

Since our definition of preferential solvation is simply a difference of excess local compositions whose behavior has been shown in previous figures, we have not reported figures with the pictures of all $\Delta ps_i^{j,k}$ surfaces. From Figure 4.4 through Figure 4.6, the necessary information can be obtained. Moreover, a large degree of the insight into the local structure that can be inferred from the discussion of $\Delta ps_i^{j,k}$ has already been obtained from the analysis of the behavior of $\delta x'_{ji}$. For this reason, we will limit the discussion of $\Delta ps_i^{j,k}$ to the most salient features.

In system I, the most evident characteristics is the behavior of $\Delta ps_1^{1,3}$. As can be realized by examining the first and third surface in the upper row of Figure 4.4,

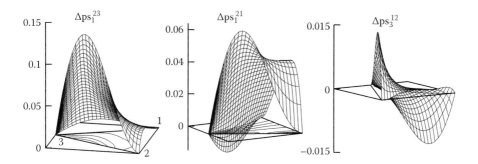

FIGURE 4.7   Preferential solvation ($\Delta ps_i^{j,k}$, Equation 4.31) in the system trichloromethane(1) + tetrahydrofuran(2) + cyclohexane(3) at 298.15 K.

$\Delta ps_1^{1,3}$ is positive over the whole domain, and shows very large values in a wide region near binary 1+3 with a maximum at $x_1 \approx 0.2$, $x_3 \approx 0.8$. This preference of Et to be solvated by Et molecules is also evident with respect to THF, except in a small composition area in and near binary 2+3, due to the high dilution of Et. THF as solvatant is instead preferred by Et, compared with cyH, except in the binary 1+2 mixture. To determine the preferences of THF as solvaton, the three surfaces of the second row should be compared. The first and third frames indicate a preference for Et with respect to cyH ($\Delta ps_2^{1,3} \gg 0$), except in the area near the binary Et+cyH, where $\Delta ps_2^{1,3} \leq 0$. The same preference is shown with respect to THF, but not in and near binary 1+2 and part of 2+3. It is only when the rival solvatant is cyH that THF prefers to be solvated by the same species. In fact, $\Delta ps_2^{2,3} \gg 0$ almost in the whole domain. By comparing the three surfaces in the bottom row, we can determine the solvation preferances of cyH: $\Delta ps_3^{3,i}$ ($i = 1,2$) $\gg 0$ practically at any composition indicates that cyH prefers to be solvated by cyH with respect to Et and THF. When the solvatant capabilities of THF and Et are compared, it is the less polar, THF, that is preferred ($\Delta ps_3^{1,2} \ll 0$).

System II behaves exactly like system I, but with more marked features, as we have found for the KBI and local composition, suggesting again a stronger alcohol-alcohol association than in system I.

System III is characterized by the large preferential solvation of TCM and THF for THF and TCM, respectively, with regard to cyH: $\Delta ps_1^{2,3}$, $\Delta ps_2^{1,3} \gg 0$ in the whole domain. Less pronounced are the values of $\Delta ps_1^{2,1}$ and $\Delta ps_2^{1,2}$, which are positive except in small areas near the binaries 1+3 and 2+3, respectively, and those of $\Delta ps_1^{1,3}$ and $\Delta ps_2^{2,3}$. These results suggest the following order of preferential solvation by THF: TCM>THF>cyH and by TCM: THF>TCM>cyH. This latter behavior is clearly evident from the left and central panels of Figure 4.7. As regards to cyH as a solvaton, both $\Delta ps_3^{1,3}$ and $\Delta ps_3^{2,3}$ are negative for almost the whole domain, whereas $\Delta ps_3^{1,2}$ (third picture in Figure 4.7) is practically zero except in and very near the binaries 1+3 and 2+3, where its values are anyway very small in magnitude. This indicates that cyH prefers, or we would say is compelled, to be solvated by cyH and that is not capable of distinguishing the solvation properties of TCM and THF.

4.7   CONCLUSIONS

The application of the KB theory, in an inverted form, to three ternary mixtures and relevant binaries has provided information on the local (microscopic) structure using simple thermodynamic (macroscopic) properties of the mixture. The procedure illustrated here is useful whenever insights into the chemical composition of a solvation shell or the solvation preferences of a given solute are required to understand the role played by the molecular interactions among the mixture components on many physical and chemical processes in solution.

5 Accurate Force Fields for Molecular Simulation

Elizabeth A. Ploetz, Samantha Weerasinghe, Myungshim Kang, and Paul E. Smith

CONTENTS

Abstract: The analysis of thermodynamic data concerning solutions using Fluctuation Solution Theory (FST) has provided a deeper understanding of local composition in solution mixtures. Over the past decade, we have used this data to help improve the description of intermolecular interactions implemented in common force fields for molecular simulation, with an emphasis on solutes of biological interest. A comparison between experimental and simulation data for small solute–solvent mixtures provides a clear indication of the quality of the force field, provides access to solute (and solvent) activities, and thereby helps to ensure that the correct balance between solute–solute, solute–solvent, and solvent–solvent distributions is attained. Here we discuss the advantages and disadvantages of using such an approach for force field design and parameterization, and provide simulation results for a series of representative solution mixtures obtained with a variety of currently available biomolecular force fields.

5.1   INTRODUCTION

Computer simulation is used for a wide range of purposes, from exploring experimentally inaccessible phenomena, to providing an alternative when the corresponding experiment would be prohibitively costly, time-consuming, difficult, dangerous, or controversial. From a biomolecular perspective, molecular modeling techniques are commonly used for rational drug design, for protein structure prediction, and in a host of other scenarios where the atomic level resolution they provide can lend valuable insights into the systems of interest. Using computer simulation, one can even study processes that are impossible or unphysical in real life (e.g., computational alchemy). Thus, the potential of biomolecular computer simulation is undeniable.

Despite the many exciting uses of molecular simulation, there is no formal guarantee that the results will reflect reality. The quality of a molecular dynamics (MD) or Monte Carlo (MC) simulation depends on the degree of sampling achieved during the simulation, which we shall not consider here, together with a satisfactory description of the intra- and intermolecular interactions in the system, that is, an accurate force field (FF). Decades have passed since the first MD simulations were performed, but the effort to continuously improve the quality of the underlying FFs continues to this day (Cheatham and Brooks 1998; Wang et al. 2001; Mu, Kosov, and Stock 2003; Mackerell 2004; Allison et al. 2011; Schmid et al. 2011; Zhu, Lopes, and Mackerell 2012). It has been expressed that both the general philosophy and the parameters for biomolecular FFs are probably converging (Wang et al. 2001). However, we argue and demonstrate here that there is still significant room for improvement, and that a markedly different parameterization philosophy is probably necessary.

5.2   CURRENTLY IDENTIFIED FORCE FIELD (FF) CHALLENGES

Two of the main biomolecular FF challenges are: (i) The generation of accurate torsional potentials for peptides and proteins, and (ii) improved parameters for the description of nonbonded interactions. A commonly cited example of the former is Simmerling's (and others) exposure of a bias toward the α-helical secondary structure in the AMBER ff94 and ff99 FFs (Okur et al. 2003). Simmerling was able to achieve an improved balance between the possible secondary structure propensities through modification of the torsional potentials, resulting in the AMBER99sb FF (Hornak et al. 2006). While incorrect torsional potentials may have contributed to other reported artifacts (Mu, Kosov, and Stock 2003; Allison et al. 2011), they are not addressed here because FST theory will not help with this part of the parameterization. Our focus will only be on the use of FST/KB (Kirkwood–Buff) theory to help improve intermolecular nonbonded interactions for simple, effective charge, nonpolarizable FFs.

One strict test of the quality of the nonbonded parameters is the ability to reproduce appropriate protein–ligand association interactions and binding free energies. In FF-based computational drug design, it is notoriously difficult to predict binding free energies quantitatively, or to rank ligands based upon their binding affinities, even in cases where the correct ligand-receptor pose has been predicted (Lazaridis 2002; Bonnet and Bryce 2004). Consequently, it has been suggested that the nonbonded parameters may be deficient (Lazaridis, Masunov, and Gandolfo 2002). This issue extends further to question the quantitative features concerning assembly and aggregation equilibria.

Furthermore, because many proteins are only marginally stable (Dill, Ghosh, and Schmit 2011), a correct description of nonbonded interactions is necessary for attaining quantitative conformational equilibria, melting temperatures, and other thermodynamic properties associated with protein folding. For example, it was reported that a 28-residue miniprotein that adopts a $\beta\beta\alpha$ motif had a simulated melting temperature

84 K higher than experiment using OPLS-AA (Feng, Kao, and Marshall 2009), while the native structure of the Trp-cage miniprotein has been shown to be less stable than the unfolded state when using the OPLS-AA FF (Juraszek and Bolhuis 2008). The melting temperature of the Trp-cage protein was also overestimated by 130 K using AMBERff94 (Paschek, Hempel, and Garcia 2008), but was in reasonable agreement with experiment when using the AMBER99sb FF due to an apparent cancellation of errors in which the enthalpy and entropy contributions were both a factor of two or more smaller than experiment (Day, Paschek, and Garcia 2010; Zhang and Ma 2010; Paschek, Day, and Garcia 2011). The Trpzip2 miniprotein simulated with AMBER99sb has also been reported to have an enthalpy and entropy of unfolding about half the magnitude of the experimental values, and a heat capacity change about 1/10th the experimental value (Nymeyer 2009; Zhang and Ma 2010). Interestingly, we have not found a simulated folding equilibrium where both the melting temperature and the underlying entropy and enthalpy contributions agree with experiment. Considering that protein folding is currently a popular area of biomolecular simulation (Bowman, Voelz, and Pande 2011), it seems prudent to place the nonbonded interactions under further scrutiny.

5.3 IMPROVING BIOMOLECULAR FORCE FIELDS

Our interest in the quality of FFs grew naturally from a long-standing effort to understand the effects of cosolvents on biomolecules. Examples from the 1990s include Smith and Pettitt's observations of interesting salt–biomolecule interactions in light of the Hofmeister series (Smith and Pettitt 1991, 1992; Smith, Marlow, and Pettitt 1993). Unfortunately, it was unclear if some of the results described in these early studies represented real physical behavior or were simply artifacts of the FF (Smith, Marlow, and Pettitt 1993).

Consequently, in the early 2000s, the Smith group analyzed several solution mixtures from a KB point of view seeking to decipher if the mainstream biomolecular FFs could be used to quantify the preferential interactions occurring in ternary systems of a solvent, biomolecule, and cosolvent (see Section 1.3.5 in Chapter 1). Specifically, Chitra and Smith simulated aqueous mixtures of a series of cosolvents—including 2,2,2-trifluoroethanol, urea, guanidinium chloride, sodium chloride, and ammonium sulfate—with several commonly available FFs. These studies indicated a general inability of the available FFs to reproduce the experimental Kirkwood–Buff integrals (KBIs) (and hence the chemical potential derivatives), resulting from a tendency for the effective solute–solute and solvent–solvent interactions to be too favorable (Chitra and Smith 2001a, 2001b, 2001c). These studies also implied that established small molecule FFs would not be able to model the correct distributions of common cosolvents around biomolecules (Kang and Smith 2007). The general conclusion from these studies was that, while it was common for different FFs to provide similar accuracy for properties such as densities, diffusion constants, dielectric properties, and so forth, the resulting KBIs usually varied greatly (Chitra and Smith 2000, 2001c).

Others have confirmed the general inability of biomolecular FFs to consistently reproduce the experimental KBIs. For example, Perera and coworkers studied acetone–water mixtures with both OPLS and the acetone FF of Klein (Ferrario et al.

1990) and found that, while both FFs were able to acceptably reproduce the enthalpies, densities, and diffusion coefficients of the mixtures, neither of the models could reproduce the solution KBIs. Indeed, both models phase separated in the acetone mole fraction range of ~0.3–0.7 (Perera and Sokolic 2004). Furthermore, alternative force fields for urea have been observed to exhibit different quantitative behavior concerning their interactions with small peptides (Horinek and Netz 2011).

One of the main advantages of studying solutions using KB theory is the access to chemical potentials, in the form of their composition derivatives. We note that instead of using KB theory, thermodynamic integration or particle insertion methods could be used to calculate the chemical potentials (Kokubo et al. 2007; Joung and Cheatham 2008, 2009). However, these methods often do not have the precision required to detect the typically small changes in the chemical potential with composition, which can often be less than a kJ/mol, without a significant computational investment. Additionally, we emphasize that the chemical potential derivatives with respect to composition, not the chemical potentials themselves, are directly related to the preferential interactions generally used to quantify cosolvent effects on biomolecular systems (Record, Zhang, and Anderson 1998; Timasheff 1998a). Finally, while infinite dilution free energies or enthalpies of solvation probe solute–solvent interactions, they do not provide a check of solute–solute interactions. Thus, we prefer to use KB theory as a guide, because it provides a framework to determine the chemical potential changes for solutes of any size throughout a range of compositions.

Current FFs typically have very similar bonded terms. The vdW parameters are usually developed to reproduce the density and enthalpy of vaporization of pure liquids and the crystal structure dimensions. Most of the variability and uncertainty lies in the Coulomb term. Usually, gas phase charges from *ab initio* calculations of the geometries and minimum interaction energies between dimers are scaled to mimic condensed phase charges, and are then tested to ensure that they reproduce the experimental data for pure liquids. Alternatively, condensed phase partial charge distributions are simply adopted from gas phase values that were calculated using a basis set that created erroneously large charge distributions. We have argued that the primary reason that many current force fields struggle to reproduce the experimental KBIs lies in the approximate nature of the effective charge distributions used to describe solute polarity, that is, the unknown degree of polarization of a solute in a polar solvent (Weerasinghe and Smith 2003a). Consequently, instead of seeking effective pair potentials that reproduce the KBIs, one could pose the legitimate argument that explicitly polarizable FFs should naturally reproduce the KBIs better than any nonpolarizable FF. We agree with this argument in theory. In practice, the increased computational demand of explicit polarization may prohibit those investigators seeking to reach long simulation timescales (a significant subset of the community) from adopting polarizable FFs (Freddolino et al. 2010). This would be especially true for those interested in implicit solvent simulations where explicit polarization is more difficult to include. Polarizable FFs are currently available and examples include Ponder's Atomic Multipole Optimized Energetics for Biomolecular Applications (AMOEBA) FF, which is currently only available for organic molecules (Ren, Wu, and Ponder 2011), and the Chemistry at HARvard Molecular Mechanics (CHARMM) polarizable FF for proteins, nucleic acids, and lipids, which is nearly completed (Zhu, Lopes, and Mackerell 2012). We have not investigated the properties of these FFs here.

5.4   SIMULATED BINARY MIXTURE KIRKWOOD–BUFF INTEGRALS FOR POPULAR BIOMOLECULAR FORCE FIELDS: A CASE STUDY

Our previous studies have investigated a range of FFs for simple cosolvents that often interact with peptides and proteins. Here, we ask the question: how well do current biomolecular FFs reproduce the binary solution properties for small solutes representative of amino acid sidechains? Biomolecular force fields are designed such that a protein is simply a sum of its parts, that is, small solutes can be studied to develop the parameters required for description of functional groups commonly found in proteins. Hence, parameters are assumed to be additive and transferable. To this end, we present the results of a case study showing the KBIs for four binary mixtures that were each simulated using some of the most commonly used biomolecular FFs—AMBER99sb (Hornak et al. 2006), CHARMM27 (Mackerell, Feig, and Brooks 2004; Bjelkmar et al. 2010), OPLS-AA (Jorgensen and Tirado-Rives 1988, 2005; Kaminski et al. 2001), and GROMOS54a7 (Schmid et al. 2011). Thus, for a given FF we will interpret the results from these mixtures as a general indicator of how successful the FF may be at reproducing solute–solute, solute–solvent, and solvent–solvent distributions, which presumably relates to the quality of the underlying interaction potentials.

It is worth noting when the most recent nonbonded updates were published for each of these FFs. The AMBER99sb release only involved updates to the torsional parameters of AMBER99ff (Hornak et al. 2006). Among other changes, AMBER99ff (Wang, Cieplak, and Kollman 2000), based upon AMBER94ff (Cornell et al. 1995), did include updates to the partial charges and the addition of a few new atom types. CHARMM's nonbonded parameters were most recently updated with the release of the CHARMM22 protein FF (Mackerell et al. 1998). GROMOS modified their N–H, C=O repulsion with the release of GROMOS 54A7 (Schmid et al. 2011) and, prior to that, refined their nonbonded parameters for oxygen-containing functional groups with the release of the GROMOS $53A6_{OXY}$ parameter set (Horta et al. 2011). Following the extensive nonbonded OPLS-AA parameterization for all amino acids (Jorgensen and Tirado-Rives 1988), the cysteine and methionine nonbonded parameters were updated in the latest OPLS-AA protein FF (Kaminski et al. 2001).

The systems studied here were methanol (MOH) + water (HOH) at $x_{MOH} = 0.5$ at 300 K, benzene (Ben) + MOH at $x_{Ben} = 0.5$ and $x_{Ben} = 0.75$ at 308 K, N-methylacetamide (NMA) + HOH at $x_{NMA} = 0.1$ and $x_{NMA} = 0.2$ at 313 K, and 1 m and 3 m aqueous zwitterionic glycine (Gly) at 300 K. The aqueous simulations were performed with the appropriate water models: TIP3P (Jorgensen et al. 1983) for Amber, Charmm, and OPLS; SPC (Berendsen et al. 1981) for Gromos; and SPC/E (Berendsen, Grigera, and Straatsma 1987) for the KBFF models. The MOH + HOH system was chosen as an example of a solution containing two polar molecules. Benzene is representative of aromatic amino acid sidechains, and mixtures of Ben + MOH display interesting features in the KBIs as a function of composition, namely aggregation of MOH at low x_{MOH}. Here, we also argue that FFs that reproduce the composition dependence of the KBIs for Ben + MOH mixtures are also more likely to correctly model the behavior of both solvent-exposed and buried phenylalanine. NMA was chosen because it is often used as a model for the peptide group. Low NMA compositions were studied because they are more representative of physiological protein backbone concentrations than high NMA compositions would appear. The current compositions may very roughly mimic peptide–peptide interactions in the presence of large amounts of water and may be very loosely analogous to the peptide interactions present in the denatured state. Finally, zwitterionic Gly (*aq*) was chosen because, despite being electrically neutral as a whole, Gly contains features similar to a mixture of salt-bridge forming ions. All of these choices were also driven by the availability of the required experimental data.

The analysis of the experimental data has been presented by us before (Weerasinghe and Smith 2005; Kang and Smith 2006; Gee 2010; Ploetz and Smith 2011a; Karunaweera et al. 2012). All simulations were performed using classical MD techniques and the GROMACS 4.5.3 simulation package (Hess et al. 2008). The systems involved cubic simulation cells of 10 nm in length simulated for 100 ns of production in the *NpT* ensemble. Further technical details can be found in our previous work (see Table 5.1) and will be supplemented in a future publication, which will additionally expand the set of thermodynamic and physical properties compared across the FFs. Note that perfect agreement with our previous results is not

TABLE 5.1
Currently Completed Kirkwood-Buff Derived Force Field (KBFF) Models

Solute	Solvent	Relevant Species	Reference
Acetone	Water	Cosolvent	(Weerasinghe and Smith 2003b)
Urea	Water	Cosolvent	(Weerasinghe and Smith 2003c)
NaCl	Water	Cosolvent	(Weerasinghe and Smith 2003d)
Guanidinium chloride	Water	Cosolvent, Arg	(Weerasinghe and Smith 2004)
Methanol	Water	Ser	(Weerasinghe and Smith 2005)
Amides	Water	Asn, Gln, peptide group, blocked termini	(Kang and Smith 2006)
Thiols, sulfides	Methanol	Met, Cys, disulfide	(Bentenitis, Cox, and Smith 2009)
Aromatics, aromatic alcohols	Methanol	Phe, Tyr	(Ploetz and Smith 2011a)
Heterocycles	Methanol or Water	Trp, His (charged and neutral)	(Ploetz and Smith 2011a)
1°, 2°, 3° alcohols	Water	Thr, cosolvent	(Jiao et al. 2012)
Amine salts, carboxylates	Water	Charged: Lys, Asp, Glu, termini	(Gee and Smith 2012)
Amines, carboxylic acids	Water	Cosolvent, Neutral: Lys, Asp, Glu, termini	(Dai, Weerasinghe, and Smith 2012)

achieved here due to much larger simulation boxes and much longer simulation times used in the current study. The simulated values of the KBIs taken for comparison with experiment were obtained by averaging the KBIs between 0.95 and 1.2 nm for MOH + HOH; 2.0 and 2.5 nm for Ben + MOH; and, 1.2 and 1.6 nm for NMA + HOH and Gly + HOH (Weerasinghe and Smith 2005; Kang and Smith 2006; Ploetz and Smith 2011a).

There are technical issues surrounding the determination of KBIs from simulations of closed systems. These are mentioned in Section 1.3.9 in Chapter 1 and described in more detail in Chapter 6 and to some extent in Chapter 7. We do not contend that the truncation (with averaging) type of approach described above is the most accurate. However, in the absence of more formal and accurate approaches, which were not available when we first investigated these integrals, we believe it has provided reasonable approximations to the true integrals. Certainly, it has allowed us to distinguish major flaws in many force fields, flaws that cannot simply be explained by inaccuracies in the integration procedure. Some examples of these flaws are described below.

Figure 5.1 displays the excess coordination numbers $N_{ij} = \rho_j\, G_{ij}$ for the four systems and the four FFs. All of the models performed reasonably well for MOH + HOH, but there was a wide variety of results for the other systems. N_{11} or N_{22} values of ~100 typically indicate a mixture that is approaching immiscibility (Matteoli and Lepori 1984). The GROMOS N_{11} values are, indeed, very large for the compositions studied. Furthermore, snapshots from the simulations gave a visual indication that the GROMOS NMA + HOH systems appeared to phase separate at both NMA compositions studied. Hence, the RDFs for the GROMOS NMA + HOH mixtures could not be considered as converged. Therefore, the values reported in Figure 5.1 are only shown for comparison purposes.

It is interesting to note that, while some FFs appeared to reproduce the N_{ij}s reasonably well for certain systems, none of the FFs were able to reproduce the N_{ij}s for all four systems. Additionally, in each aqueous solution, all deviations from experiment were in the direction of too much self-association between the solute molecules, that is, large positive N_{22} values. In many cases, the deviations from experiment were quite large. Clearly, in these cases the models are not capturing the correct balance of solute–solvent, solute–solute, and solvent–solvent interactions. It was exactly these types of results, albeit for different systems, that prompted us to investigate what the reasons were for the lack of consistency among the FFs, why the results were sometimes in such poor agreement with experiment, and if FFs could be systematically improved.

Due to the issues mentioned previously surrounding the unknown polarization of molecules in condensed phases, we believe condensed phase charge distributions cannot, in general, be determined from a simple scaling of the QM calculated electrostatic potential. In our opinion, optimization of the effective charge distribution should be performed when studying the properties of solution mixtures, not from quantum calculations for molecules in the gas phase. This is rarely attempted. Biomolecular FFs are usually designed to reproduce the properties of pure liquids and/or infinitely dilute solutes, with the assumption that mixtures of these molecules, in any ratio, will produce reasonable solution properties. For example, the most

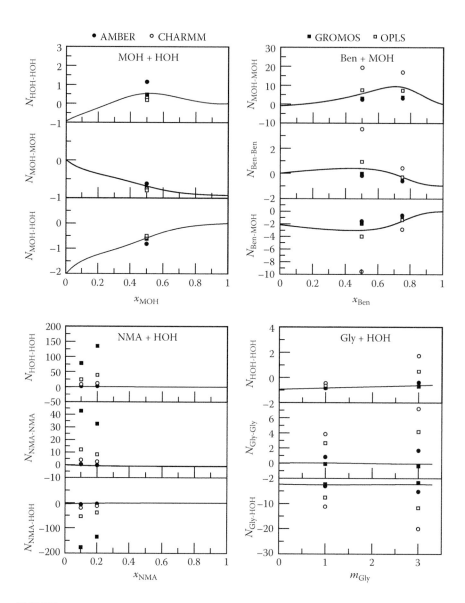

FIGURE 5.1 Comparisons of the excess coordination numbers for mixtures of methanol + water, benzene + methanol, N-methylacetamide + water, and zwitterionic glycine + water using four commonly used biomolecular force fields. Error bars are not shown for clarity (see Figure 5.2 for typical values). Please note the different y-axis scales. The GROMOS NMA + HOH points were calculated using unconverged RDFs.

recent GROMOS partial charges were parameterized to reproduce the hydration free enthalpies of amino acid analogs in water (Oostenbrink et al. 2004). However, it is clear that this FF does not produce the correct balance of solute–solute, solute–solvent, and solvent–solvent interactions for NMA + HOH at the compositions studied. We believe that it is more desirable to base the partial charge parameterization on properties of mixtures across a large range of compositions. This is not a particularly new concept, but finding data that is sensitive to changes in composition is just as important and has been difficult.

5.5 THE KIRKWOOD–BUFF DERIVED FORCE FIELD APPROACH

We have seen that current FFs do not necessarily reproduce the KBIs, and that generally the properties of solution mixtures are not taken into account during the parameterization procedure. Because different FFs produce such widely varying KBIs, and because the greatest variation in their parameters comes in the form of their charge distributions (Weerasinghe and Smith 2003a), it seems that the simulated KBIs are very sensitive to the choice of the charge distribution and that this distribution could be optimized to reproduce the KBIs. Armed with this philosophy, we set out to create a peptide/protein FF, known as the *Kirkwood-Buff derived force field* (KBFF), which attempts to reproduce the KBIs for small molecules, representative of the building blocks of proteins, and their mixtures with solvents such as water and methanol. The focus is centered on determining reasonable effective condensed phase charge distributions for polar solutes. More details of the philosophy can be found elsewhere (Ploetz, Bentenitis, and Smith 2010a; Weerasinghe et al. 2010). Table 5.1 shows the models we have parameterized to date.

Recently, approaches relying on KB theory have also been used by other groups for the parameterization or testing of several small molecule FFs (Lee and Van der Vegt 2005; Zhong and Patel 2010; Horinek and Netz 2011). Additionally, outside of the biological FF community, there has been significant work to develop FFs that reproduce solution activities, which is a key aim of the KBFF models. Notably the TraPPE models, developed by the Siepmann group, are specifically designed to model vapor–liquid coexistence curves (Maerzke et al. 2009). TraPPE has been developed primarily for thermodynamic and engineering purposes, and it cannot be directly applied to biomolecular systems in its present form.

Before indicating the type of results one can obtain using the KBFF philosophy, we briefly summarize the advantages and disadvantages of the approach.

Advantages of using KB theory in FF parameterization:

1. It is an exact solution theory, that is, it has no free parameters, requires no approximations, and makes no assumptions except that the system is in the thermodynamic limit.
2. Due to the complexities of the condensed phase, FF developers often must resort to using gas phase data to parameterize condensed phase FFs. KB theory allows condensed phase experimental data to be used.

3. The theory can be applied to solutions composed of any number of molecules of any size and complexity as long as the solution is stable with respect to composition.
4. It does not assume that the interactions are pairwise additive when expressing thermodynamic properties in terms of KBIs, but instead it takes into account the true many-body interactions in solutions.
5. One can decompose KBIs into distance-dependent effects, which is difficult to do experimentally.
6. KBIs are a sensitive probe of the molecular distributions observed for different solutions and can be used to distinguish between models with otherwise similar characteristics.

Disadvantages of using KB theory in the parameterization approach:

1. Relatively large simulation boxes (≥ 6 nm) may be necessary for systems with large molecules and/or in aggregating systems. The need for a large box may be anticipated if the experimental G_{ij}s show large deviations from their symmetric ideal values (Ploetz, Bentenitis, and Smith 2010b).
2. Relatively long simulation times (tens of nanoseconds) are needed in order for the RDFs and G_{ij}s to converge.
3. The search for a reliable charge distribution is more or less random. Our models have been developed based upon ~30 or fewer iterations (Bentenitis, Cox, and Smith 2009).
4. For some mixtures, the experimental data needed to conduct the KB inversion procedure is not available or is often unreliable.

The KBFF simulations followed the same protocols as the simulations for the other four FFs. Figure 5.2 displays the KBFF results compared to the results for the FF that came closest to reproducing the experimental N_{ij}s for a given system. To a certain extent the superior ability of the KBFF models to reproduce the KBIs is not surprising, as the parameters of the other FFs have not been optimized to reproduce this data, while the KBFF models were specifically parameterized to do so. But, we use it here as an indication of how this type of data can be improved with a reasonable degree of effort. We note that for MOH + HOH, GROMOS appears to have a slightly better agreement than KBFF. However, this may not be true across the full composition range. For the current comparison, we have picked the composition ($x_{MOH} = 0.5$) in which the KBFF model shows the largest deviation from experiment (Weerasinghe and Smith 2005).

The results indicate that several of the FFs are competitive with the KBFF models for one of the systems, but none exhibited the overall balance between solute–solute, solute–solvent, and solvent–solvent distributions for all the solutes as provided by the KBFF models. The NMA and water results are particularly concerning due to the central role of peptide–peptide interactions in determining the balance between folded and unfolded protein conformations. The observation that all the FFs display a significantly more positive solute–solute excess coordination number compared to

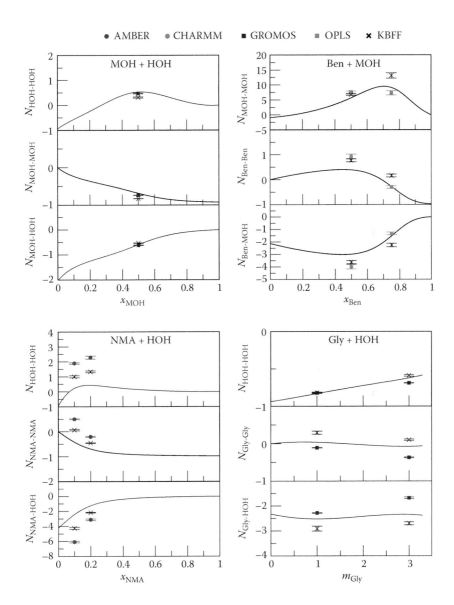

FIGURE 5.2 Comparisons of the excess coordination numbers for mixtures of methanol + water, benzene + methanol, N-methylacetamide + water, and zwitterionic glycine + water using the Kirkwood-Buff force field and the biomolecular force field that best reproduced the excess coordination numbers for each system. Please note the different y-axis scales. Error bars show the estimated standard deviation obtained from five 20-nanosecond subaverages. **(See color insert.)**

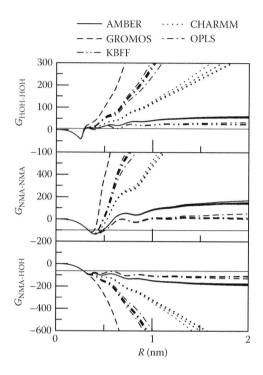

FIGURE 5.3 The Kirkwood-Buff integrals (G_{ij}, units of cm³/mol) as a function of integration distance for the *N*-methylacetamide (NMA) + water system at $x_{NMA} = 0.1$ for the five force fields. The thin horizontal line is the experimental value. The multiple lines correspond to five 20-nanosecond subaverages for each force field.

experiment also suggests a possible explanation for the imperfect thermodynamic behavior mentioned in Section 5.2.

Figure 5.3 displays the G_{ij}s as a function of distance for the NMA + HOH $x_{NMA} = 0.1$ system. While our choice of where to average the G_{ij}s is somewhat arbitrary, it is clear that the conclusions concerning the accuracy of the FFs are independent of the method used to obtain the exact values. For all FFs, the value of G_{11} and G_{22} are positive, indicating a degree of aggregation, however, the magnitude varies widely depending upon the FF. The values of the KBIs also tend to a constant limiting value for the systems in which the solute self-association was not too excessive. The absence of issues surrounding the integration of the RDFs is probably due to the large systems adopted here (involving between 10,000 and 33,000 molecules). The patterns displayed in Figure 5.3 for NMA and HOH was similar for all the systems studied here.

Figure 5.4 shows the RDFs for all five FFs at one composition for each mixture. Similar results were obtained for the other compositions. For MOH + HOH and Ben + MOH, there appears to be qualitative agreement among the RDFs for different FFs, despite their quantitative differences. This agreement is observed for both the short-ranged and the long-ranged structure. The NMA + HOH and Gly + HOH RDFs do not show qualitative agreement among the FFs. For NMA + HOH,

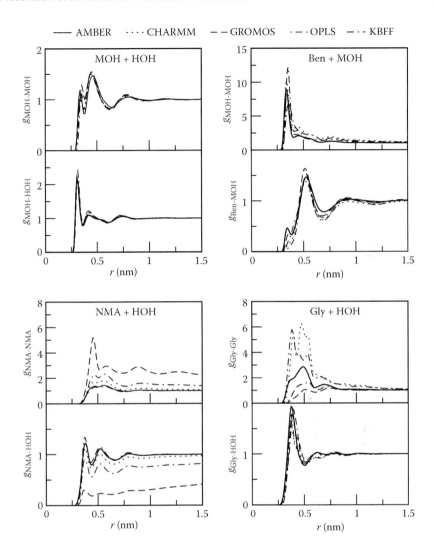

FIGURE 5.4 The center of mass based radial distribution functions (RDF) for five force fields for the mixtures methanol (MOH) + water (HOH) at $x_{MOH} = 0.5$, benzene (Ben) + methanol at $x_{Ben} = 0.75$, N-methylacetamide (NMA) + water at $x_{NMA} = 0.1$, and 1 m zwitterionic glycine (aq). Please note the different y-axis scales. The data represent the average RDF from 80 to 100 nanoseconds of production simulation after 10 nanoseconds of equilibration. The RDFs are converged with respect to time.

similarities are evident in the shape of the RDFs for all FFs excluding GROMOS. However, the height of the first maximum in the RDFs differs among all FFs. Note that the GROMOS NMA + HOH $g_{NMA-NMA}$ does not tend toward unity due to the excessive self-aggregation of NMA in this simulation. For the Gly + HOH system, the solute–solute RDF does not appear to display a similar shape among the five FFs. This indicates that the Gly-Gly pairing differs substantially between the FFs, although the FFs do provide the same excluded Gly-Gly distance. The long-range structural behavior is also different among the FFs for the Gly + HOH system, although not to the degree exhibited for NMA + HOH mixtures.

The simulated KBIs are dependent on the choice of water model used with each solute FF, but the effect is minimal compared to the differences observed for different solute FFs (Chitra and Smith 2001a; Weerasinghe and Smith 2003b, 2003c, 2003d, 2004, 2005; Kang and Smith 2006). Based upon our previous studies of the KBI dependence on the water model, we predict that the results would not be significantly different on switching to a different water model.

5.6   CONCLUSIONS

The key conclusions we have drawn from our present and past work include the following:

1. A KB analysis provides more data for testing and development of FFs. The KBIs are very sensitive to the charge distributions. The major advantage of the KBFF approach is that by developing an FF that reproduces the KBIs, it is assured that the corresponding solution thermodynamics (including the activity of the solute) will be well described. This then further ensures the correct balance between solvation and self-association of solutes.

2. The KBFF approach is designed to specifically reproduce the experimental KBIs for small solutes in solution mixtures. This is an important property to model correctly, but if it were only attainable by sacrificing agreement with experiment for other properties that are commonly reproducible, it would be of little use. Fortunately, this does not appear to be the case (Weerasinghe and Smith 2003b, 2003c, 2003d, 2004, 2005; Kang and Smith 2006; Bentenitis, Cox, and Smith 2009; Gee et al. 2011; Ploetz and Smith 2011a).

3. We consider the improvements made to be essential if one wants to quantitatively understand phenomena such as conformational or association equilibria in quantitative detail, despite the disadvantages of needing relatively long simulation times and relatively large system sizes during the parameterization procedure.

4. The actual charge distribution, not merely the solute polarity (dipole moment), appears to be what balances the solvation, thus it is not enough to simply scale up gas-phase charge distributions to mimic condensed-phase charge distributions (Weerasinghe and Smith 2003a).

5. In all aqueous solutions we have tested (excluding MOH + HOH, which was reasonable for all the FFs tested), whenever there has been a large deviation from the experimental KBIs, it has always been toward an overfavoring of the solute–solute interactions, and thus too much self-aggregation. For FFs, which overstabilize peptide–peptide group interactions, this likely leads to protein simulations that have low root mean square deviations (RMSDs) from experimentally determined structures, and thus may give a false appearance of being reliable. However, they would not be able to reproduce conformational equilibria accurately, resulting in a tendency to exaggerate the stability (melting temperature).

6. The KBIs are most sensitive to the accuracy of the activity data used during the analysis, and least sensitive to the accuracy of the compressibility data. Often it is even assumed that the compressibility of the solution is ideal, and this is a reasonable assumption in the absence of real compressibility data as is the use of isentropic compressibility data (Matteoli and Lepori 1984). Compressibility uncertainties as high as 50% have been shown to have insignificant effects on the values of the KBIs (Matteoli and Lepori 1984).

7. The KBIs should reach a plateau value at the distance that the RDFs tend toward unity. Although the procedure for choosing a distance over which to take the average of the KBIs is somewhat arbitrary, it is able to capture the significant differences in the degree of molecular association between different FFs.

8. Although in theory there should be multiple RDFs that, upon integration, could provide the same value of G_{ij}, and thus multiple solutions to the charge distribution, we have never found this to be a practical concern. In fact, it has been difficult to find one set of partial charges (or one set of RDFs), which reasonably reproduce the three KBIs.

In our opinion, the results presented here clearly show that conventional FFs can possess a significant imbalance of solute solvation and self-association in solutions. Fortunately, this problem can be overcome by adjusting the effective molecular charge distributions of the solutes during the parameterization procedure. In our opinion, the KBFF models represent an often dramatic improvement over the currently used protein FFs and should be seriously considered if one wishes to study conformational or association equilibria with reasonable accuracy.

ACKNOWLEDGMENTS

The project described was supported by grant R01GM079277 to Paul E. Smith from the National Institute of General Medical Sciences and grants DGE-0841414 and DGE-0750823 from the National Science Foundation to Elizabeth A. Ploetz. The content is solely the responsibility of the authors and does not necessarily represent the official views of the National Institute of General Medical Sciences, the National Institutes of Health, or the National Science Foundation.

6 Fluctuation Solution Theory Properties from Molecular Simulation

Jens Abildskov, Rasmus Wedberg, and John P. O'Connell

CONTENTS

Abstract: The thermodynamic properties obtained in the Fluctuation Solution Theory (FST) are based on spatial integrals of molecular total correlation functions (TCFs) between component pairs in the mixture. Molecular simulation, via either molecular dynamics (MD) or Monte Carlo (MC) calculations, can yield these correlation functions for model inter- and intramolecular potential functions. However, system-size limitations and statistical noise cause uncertainties in the functions at long range, and thus uncertainties or errors in the integrals. A number of methods such as truncation, distance shifting, long-range modeling, transforms, direct correlation function (DCF) matching, finite-size scaling, and adaptive resolution, have been explored to overcome these problems. This chapter reviews the issues and published work associated with using molecular simulation to obtain FST properties. The results suggest that molecular simulation should now be more fully utilized for obtaining quantitative FST thermodynamic properties of solutions.

6.1   INTRODUCTION

Although FST integrals may be obtained from simulations, most modeling has been done with empirical expressions for the integrals, which appear in the final exact expressions. We describe our work with these in Chapter 9 for both pure and mixed systems. However, since the radial distribution functions (RDFs) for the three molecular pairs of a binary mixture can be directly obtained from MD simulations, in principle, these may be integrated numerically to yield total correlation function integrals (TCFIs). With force fields capable of representing real behavior, real FST property variations can be computed. However, this task has proven to be more difficult than might have been expected. We are not aware of a fully reliable method for obtaining TCFIs from RDFs for polyatomic molecules. However, recent progress in calculating RDFs, as described in this chapter, should eventually lead to direct applications for real systems of interest, significantly expanding the knowledge and application of fluctuation solution theory. The focus here is on smaller molecules; simulations of proteins and larger substances have been described in Chapter 5.

6.2   BASICS

Unlike the atomic TCFs and pair RDFs introduced in Chapter 1, which are functions of only the spatial distance r between the centers of mass of the two molecules, molecular correlation functions depend on orientations, ω_1 and ω_2. The TCF can be resolved into isotropic and anisotropic parts (Gray and Gubbins 1984),

$$h_{ij}(r_{12},\omega_1,\omega_2) = h_{ij}(r_{12}) + h_{ij}^{(a)}(r_{12},\omega_1,\omega_2) \tag{6.1}$$

where $h_{ij}(r_{12}) = g_{ij}(r_{12}) - 1$ the isotropic part is obtained by averaging over angles,

$$h_{ij}(r_{12}) \equiv \left\langle h_{ij}(r_{12},\omega_1,\omega_2) \right\rangle_{\omega_1,\omega_2} \tag{6.2}$$

with

$$\langle \bullet \rangle_{\omega_1} \equiv \frac{1}{8\pi^2} \int d\omega_1 \equiv \frac{1}{8\pi^2} \int_0^{2\pi} d\phi \int_{-1}^1 d(\cos\theta_1) \int_0^{2\pi} d\chi_1 \qquad (6.3)$$

Thus, the anisotropic part is constructed to vanish upon averaging over orientations,

$$\left\langle h_{ij}^{(a)}(r_{12},\omega_1,\omega_2)\right\rangle_{\omega_1,\omega_2} = 0 \qquad (6.4)$$

For flexible molecules, the correlation functions are also functions of the molecular conformation. We consider small substances those for which conformation effects can be ignored. The molecular Ornstein–Zernicke (OZ) equation defines the molecular DCF, $c_{ij}(r_{12}, \omega_1, \omega_2)$ (Gray and Gubbins 1984),

$$h_{ij}(r_{12},\omega_1,\omega_2) = c_{ij}(r_{12},\omega_1,\omega_2) + \rho\sum_l x_l \int \left\langle h_{il}(r_{13},\omega_1,\omega_3)c_{lj}(r_{32},\omega_3,\omega_2)\right\rangle_{\omega_3} dr_3 \quad (6.5)$$

where ρ denotes the overall number density of the fluid and x_i is the number fraction of component i. In analogy with Equations 6.1 and 6.2, and Equation 6.4, the DCF can be written as a sum of isotropic and anisotropic parts,

$$c_{ij}(r_{12},\omega_1,\omega_2) = c_{ij}(r_{12}) + c_{ij}^{(a)}(r_{12},\omega_1,\omega_2) \qquad (6.6)$$

Substituting Equation 6.1 and Equation 6.6 into Equation 6.5 and angle averaging leads to

$$h_{ij}(r) = c_{ij}(r) + \rho\sum_l x_l \int h_{il}(r_{13})c_{lj}(r_{32})dr_3 +$$
$$r\sum_l x_l \int \left\langle \left\langle h_{il}^{(a)}(r_{13},\omega_1,\omega_3)\right\rangle_{\omega_1} \left\langle c_{lj}^{(a)}(r_{32},\omega_3,\omega_2)\right\rangle_{\omega_2}\right\rangle_{\omega_3} dr_3 \qquad (6.7)$$

Neglecting the last term in Equation 6.7 gives a simplified version of the OZ equation, in which the isotropic DCFs and TCFs are related without the anisotropic terms,

$$h_{ij}(r) = c_{ij}(r) + \rho\sum_l x_l \int h_{il}\left(\left|r-r'\right|\right)c_{lj}(r')dr' \qquad (6.8)$$

The various approaches to spatial integration of molecular simulation data described in Section 6.3 use this equation instead of the full OZ equation. While this may not be fully rigorous, it is supported by several different analyses. First, Equation 6.8 is exact in some integral equation theories of fluids with anisotropic interactions, such as the mean-spherical approximation and the generalized mean field theory (Gray and Gubbins 1984). Second, Wang et al. (1973) showed from MC simulations of Lennard–Jones (LJ) particles with significant dipole and quadrupole moments that anisotropic forces have limited effects on $h_{ij}(r)$. Also, Gubbins and O'Connell (1974) showed that, for dense fluids, compressibility data for water and argon could be scaled with only two parameters, meaning that the anisotropic effects were not apparent in the water data. In addition, several studies show successful corresponding-states scaling for the direct correlation function integrals (DCFIs) (Brelvi and O'Connell 1972, 1975a, 1975b; Campanella, Mathias, and O'Connell 1987; Huang and O'Connell 1987; Abildskov, Ellegaard, and O'Connell 2009, 2010a, 2010b), as described in detail by O'Connell (1994). Finally, the approximation of Equation 6.8 is the first term of the spherical harmonic expansions of the molecular correlation functions. Equation 6.8 can be systematically improved by considering the spherical harmonic expansions of the orientation-dependent TCFs and DCFs (Gray and Gubbins 1984). The significance of this improvement is currently unknown, however.

6.2.1　Equivalence of Ensembles

FST is based on the μVT ensemble, so the Kirkwood–Buff integrals (KBIs) are integrals over RDFs for an open system. However, simulations are most conveniently performed in the NpT, NVT, or NVE ensembles. MD simulations in the μVT ensemble are possible (Çagin and Pettitt 1991) but nontrivial due to the problems associated with inserting new particles (Beutler et al. 1994). For this and other reasons, simulations are normally done on closed systems, though rigorously, the corresponding KBIs are equal to 0 for unlike pairs and −1 for like pairs. The RDFs for μVT and NpT simulations differ by a term of the order of $1/N$, and the principle that $g_{ij}(r \rightarrow \infty) = 1$ is violated for closed systems (Ben-Naim 1990a). Fortunately, as illustrated in previous computational studies, RDFs in open and closed systems are extremely similar (Weerasinghe and Pettitt 1994). This means that while the original Kirkwood–Buff (KB) theory cannot be rigorously applied to a closed system, calculations converge to correct results with increasing N. Thus, it has become standard to use the *equivalence of ensembles* and to determine TCFIs using the MD simulations in the (NpT/NVT) ensembles rather than the μVT ensemble.

6.2.2　Integration

The usual approach to determine the RDF $g_{ab}(r)$ between the centers of mass of particle 1 of species a and particle 2 of species b separated by the distance r, is the accumulation of the number of particles b lying in the interval $[r, r + dr]$ from a given particle a, and for all available values of r within the central box. Numerical integration of molecular simulation RDFs is less straightforward. Theoretically, $h_{ij}(r)$ goes to zero when r goes to infinity. However, because the integral is evaluated numerically, convergence requires that $h_{ij}(r)$ goes faster to zero than r^2 goes to infinity. For practicality, since the upper limit of the integral is infinite, for a convergent integral,

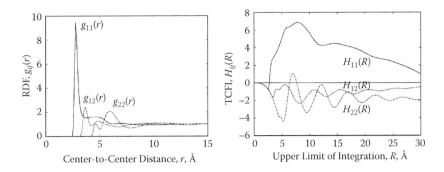

FIGURE 6.1   Left: RDFs from simulation of water (1) + t-butanol (2) at $x_1 = 0.65$, at 323 K and 1 atm. Right: Numerically evaluated (nonconverged) integrals $H_{ij}(R_{lim})$ from Equation 6.9.

there must be some upper separation distance, R_{lim}, beyond which the integrand no longer contributes significantly to the value of the integral. Thus, one defines

$$H_{ij}(R_{lim}) = \rho \int_0^{R_{lim}} r^2 h_{ij}(r)dr \tag{6.9}$$

with R_{lim} chosen sufficiently large that the integral is converged. This means that $H_{ij}(R_{lim})$ should be insensitive to larger values of R_{lim}. Experience shows, however, that $H_{ij}(R_{lim})$ frequently does not include all of the nonzero values of $r^2 h_{ij}(r)$ from an MD simulation (Salacuse, Denton, and Egelstaff 1996). An example of this is shown in Figure 6.1 (Wedberg 2011).

The lack of convergence is mainly due to the RDFs retaining subtle structure over relatively long distances that contribute to the integral. While using a larger system might minimize the effects of omitting such contributions, the result would be significantly increased computational time without necessarily ensuring accuracy or reliable extrapolation. There are techniques (Theodorou and Suter 1985; Nichols, Moore, and Wheeler 2009) to allow calculation of pair correlation functions for distances up to $\sqrt{3}/2$ times the box dimension. However, convergence may not be obtained, since the RDFs exhibiting substantial noise as the upper limit is approached (Salacuse, Denton, and Egelstaff 1996).

6.2.3   ASYMPTOTIC PROPERTIES OF RADIAL DISTRIBUTION FUNCTIONS AND POTENTIAL TRUNCATIONS

One clever approach to obtaining better convergence is to include asymptotic properties of the pair correlation functions (Lebowitz and Percus 1963). In particular, exact asymptotic expressions have been obtained by Attard and coworkers (Attard 1990; Attard et al. 1991), such as for dipolar fluids. Other work has extended simulation results for a system with a truncated potential to give those for the full potential (Lado 1964). The effects on pair distribution functions of potential truncations are important, for example, when the long-range tail of a potential must be truncated at some finite distance and the effect of the neglected part of the potential must be determined.

6.3   METHODS

The task of extending the pair distribution function based on theoretical consider-ations has been addressed many times (Verlet 1968; Galam and Hansen 1976; Jolly, Freasier, and Bearman 1976; Ceperley and Chester 1977; Dixon and Hutchinson 1977; Foiles, Ashcroft, and Reatto 1984). Often the goal has been to study the corre-lation functions themselves or to calculate structure factors, not to obtain properties. Here we will emphasize applications aimed toward representing thermodynamic properties of molecular fluids that do not have conformational variations. While many publications have been confined to atomic model fluids, such as LJ particles, we focus here on applications for real molecular systems and their mixtures.

6.3.1   DIRECT CORRELATION FUNCTION MATCHING

The method of Verlet (1968) is intended to correct correlation functions from simula-tion for the effects of finite-sized systems, such as those summarized by Salacuse, Denton, and Egelstaff (1996), and to extend computed correlation functions to long range. The method was originally used for a pure LJ fluid in a study of the DCF and the structure factor. Later it was extended to a LJ mixture (Jolly, Freasier, and Bearman 1976). With this method, pair TCFs are extended by forcing the corre-sponding DCF at large separations to be consistent with Equation 6.10,

$$\begin{pmatrix} h(r) \\ c(r) \end{pmatrix} = \begin{pmatrix} h_{MD}(r) & , & r \le R \\ a(u(r), h(r)), & r > R \end{pmatrix} \qquad (6.10)$$

Here $h_{MD}(r)$ is given by the simulation result and a is a closure relation giving the DCF in terms of the TCF, $h(r)$, and the potential, $u(r)$, evaluated at the same r. Verlet's method utilizes the simple structure and short range of c compared to h. The approach can use either the Mayer f-function or the Percus–Yevick relation to extend pair distribution functions obtained from simulations. For the pure LJ fluid, three different relations $a(u,h)$ have been commonly used: the Mayer function,

$$a(u(r), h(r)) = f(r) = \exp(-\beta u) - 1 \qquad (6.11)$$

the PY relation,

$$a(u(r), h(r)) = f(r)(1 + h)(1 - \exp(-\beta u)) \qquad (6.12)$$

and the first-order virial expansion of c (Wedberg et al. 2010),

$$a\big(u(r),h(r)\big) = f(r)\left[1 + \rho \int f\big(|r-r'|\big) f(r')dr'\right] \tag{6.13}$$

Note that these theories for c are all consistent with the asymptotic result $c(r) \rightarrow u(r)/kT$, when $r \rightarrow \infty$ (Lebowitz and Percus 1963). As shown in Wedberg et al. (2010), Equation 6.11 through Equation 6.13 usually yield similar results and there is no rigorous basis for selecting one over another. Calculations of isothermal compressibilities by integration of the extended pair correlation functions were reported only for three state conditions by Verlet (1968), and the statistical uncertainties were large. It is not clear whether these uncertainties were due to the quality of the simulations and their analysis, or to the assumptions made by the extension method. Furthermore, while the compressibilities were fairly reasonable for noble gases, agreement with experiment was not reported in quantitative terms, due to the focus of the paper being on other properties of the correlation functions.

For some time, the numerical solution of the liquid structure integral equations remained a challenge. Progress along these lines was made by the numerical implementation of Verlet's method, based on the Newton–Raphson method, known as the *Gillan scheme* for solving the OZ equation (Gillan 1979; Abernethy and Gillan 1980; Enciso 1985). Also, a factorization approach to the OZ equation was demonstrated by Jolly, Freasier, and Bearman (1976). Before describing applications of our variant of the Verlet method to systems resembling real molecular mixtures in Section 6.4, we review other approaches designed to improve convergence.

6.3.2 TRUNCATION

One of the simplest strategies is called *truncation* (Weerasinghe and Smith 2003b). With this method, the RDF from μVT simulations is approximated by the RDF from NpT simulations truncated at R_{lim}. This distance is chosen to be, "the range over which the intermolecular forces dominate the distribution of particles." Ideally, the truncated NpT RDF captures the major features of the μVT RDF, and its integral provides a good approximation to the desired KBIs. Truncated RDFs obtained from NpT simulations have been obtained for several different mixtures, and used to obtain properties, such as partial molar volumes (Lin and Wood 1996), and to express the KB equation in terms of local compositions (Mansoori and Ely 1985).

As described in Chapter 5, Smith and coworkers have employed the truncation method in order to develop accurate force fields for solutions, especially those with biochemicals and proteins. Over the past decade, a series of force-field development and validation studies have been published using KBIs (Chitra and Smith 2001a; Weerasinghe and Smith 2003d, 2004; Gee et al. 2011). The truncation approach can be successful if the TCFIs converge within the range of distance sampled by simulation. If not, the results will depend sensitively on the choice of R_{lim}. It is therefore common to average $H(R_{\text{lim}})$ with R_{lim} varying over a selected interval. There seem to be no general rules for selecting the interval, other than suggesting that it should cover one oscillation of the TCF.

6.3.3 DISTANCE SHIFTING

Distance shifting is another method employed with data on systems resembling real molecules. Both Perera and Sokolić (2004) and Hess and Van der Vegt (2009) attempt to correct the RDFs obtained from simulation by rescaling according to

$$g_{ij}(r) = \alpha_{ij}(r)g_{ij}^{(0)}(r) \tag{6.14}$$

Here α_{ij} is chosen in order to enforce that $g_{ij}(r)$ approaches unity at long distances. Perera and Sokolić (2004) presented NpT simulations of the water + acetone binary mixture. Although a correction of the order $1/N$ is required, the result $g_{ij}(r) \to 1$ (for $r \to \infty$) is often valid for simple fluids after a few molecular diameters. This can be realistic in simulation boxes with a few hundred particles. However, for systems characterized by microscopic aggregation, the RDFs decay in irregular fashion with the range of correlations in the RDF differing from that of pair interactions, even at conditions remote from a critical point. The apparent problem in evaluating these quantities is the upper bound of the integral relative to the range of the correlations described by the simulations. If the system is large enough, one may consider that the correct asymptotic behavior is attained at some cutoff R_{lim} smaller than the half-box length $L_{\text{box}}/2$. Accordingly, the KBI can be computed by replacing the infinite upper bound by R_{lim}, as in truncation methods. Though this may be satisfactory for simple fluids, it is probably incorrect for fluids with long-range correlations. If the upper bound is not large enough to capture the correct asymptotic behavior, it will lead to incorrect estimation of the KBIs. For water/solute systems, these correlations seem to extend for more than five to six water diameters, which is too large even for a system with $N = 1024$.

Since the $L_{\text{box}}/2$ values of all partial $g_{ij}(L_{\text{box}}/2)$ are always close to unity, Perera and Sokolić (2004) restored the correct asymptotic value at the natural half-box cutoff by shifting the value to unity. The expression used is

$$g_{ij}(r) = \alpha_{ij}(r)g_{ij}^{(0)}(r), \qquad \alpha_{ij}(r) = \frac{1}{1 + r/\left(L_{\text{box}}/2\right)\left[g_{ij}^{(0)}(L_{\text{box}}/2) - 1\right]} \tag{6.15}$$

where $g_{ij}^{(0)}(r)$ is the uncorrected RDF. This procedure leaves values of the RDF at contact nearly unchanged if $g_{ij}^{(0)}(L_{\text{box}}/2)$ is close to unity. Perera and Sokolić (2004) found that $N = 864$ is just enough to satisfy this condition, though $N = 2048$ is much better. The merit of this equation is the use of all $r < L_{\text{box}}/2$ values in the calculation of the KBI, to avoid artifacts in evaluations of the canonical ensemble KBIs.

Hess and Van der Vegt (2009) studied cation-binding affinity with carboxylate ions. They computed the excess coordination numbers, N_{ij}, defined in Section 1.1.5 in Chapter 1, for water (w) or cations (c) about cations,

$$N_{jc} = \rho_c 4\pi \int_0^\infty (g_{jc}(r) - 1)r^2 dr \tag{6.16}$$

where $j = w$, c and computed the chemical potential derivatives of Equation 1.49 from closed system NpT simulations. The finite-size correction is accounted for in a novel way. The idea is to consider a small part of a large system, such that this small part can be considered as open, and then evaluate the integral up to a finite distance where it has converged within the larger system. Since the RDFs can not converge exactly to unity, the Hess/Van der Vegt approach employs a scaling factor to account for the fact that the fluid composition far from a given molecule is different from the overall composition. The scaling factor is chosen, "such that the RDF becomes exactly 1." Assuming that $N_{jc}(R)$ is constant beyond a distance $R_{\lim}$, the RDF can be normalized to 1 by dividing it by the *observed* number of particles and multiplying by the *expected* number,

$$g_{jc}^* = \frac{N_j(1-V(R_{\lim})/V_{\text{box}})}{N_j(1-V(R_{\lim})/V_{\text{box}})-\Delta N_{jc}(R_{\lim})-\delta_{jc}} g_{jc} \qquad (6.17)$$

where N_j is the number of particles of species j, $V(R)$ is the volume of a sphere with radius R, and V_{box} is the volume of the simulation box, $\delta_{cc} = 1$ and $\delta_{wc} = 0$. For a system with 100 ion pairs, the scaling factor is around 1.005 (or one particle in 200). The correction for g_{wc} is two orders of magnitude smaller. Although the methods are straightforward to implement, they still require selecting an appropriate truncation distance, and there seems to be no systematic way to choose the value of $R_{\lim}$. This aspect can be a limitation, since in our experience the results are very sensitive to the selection of this value (Wedberg 2011).

6.3.4 LONG-RANGE MODELING

To avoid searching for an $R_{\lim}$ that gives the value of a converged KBI, an alternative is to find a model, and its parameters, for the long-range tail of the RDF to effectively extend the simulation results to infinite separation. The intention is to minimize sensitivity to the location where the simulation results are considered unreliable.

Matteoli and Mansoori (1995) gave a parametric expression for the RDFs of LJ fluids and their mixtures. That work arrived at a final form of the RDF based on the asymptotic conditions for zero density and infinite distance, as required by statistical thermodynamics, rather than rigorous geometrical and spatial considerations. Seven adjustable parameters were fitted to literature data on RDFs for each LJ fluid at different temperatures and densities. These were in turn expressed as functions of reduced temperature and density, so the complete parameterization used a total of 21 parameters. The capability of the expression to fit to RDFs of mixtures was checked against literature simulations of binary LJ mixtures with different diameters, molar fractions, and $\varepsilon_{AA}/\varepsilon_{BB}$ ratios. The agreement between calculated and simulation curves was satisfactory. The values of the reduced pressure and internal energy calculated by numerical integration of the completely parameterized equation compared reasonably well with literature MD simulations. This approach allows calculation by integration of related quantities such as compressibility, internal energy, pressure, and, using FST, the chemical potentials and partial molar volumes

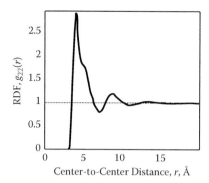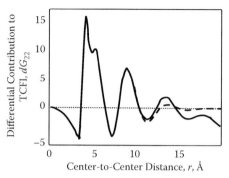

FIGURE 6.2 (a) Distribution function of benzene (1)/ethanol (2), $g_{22}(r)$, at $x_2 = 0.5$. (b) Differential contributions to the G_{22} integral, dG_{22}, at $x_2 = 0.5$. (From S. Christensen, G. H. Peters, F. Y. Hansen, J. P. O'Connell, and J. Abildskov, 2007b, Generation of Thermodynamic Data for Organic Liquid Mixtures from Molecular Simulations, *Molecular Simulation*, 33, 449. Reprinted with permission from Taylor & Francis.)

of the LJ mixture components for which RDF results are available. The works of Christensen et al. (2007a, 2007b, 2007c) treat molecular simulations of systems resembling real molecules.

As shown in Figure 6.2 (a) for the benzene(1)/ethanol(2) binary mixture, the solid line is $g_{22}(r)$, which deviates much less than 1% from unity near $r_{\max}$. However, divergence is clearly seen in Figure 6.2 (b), starting at $r \approx 15$ Å, where a local maximum in the integrand, dG_{22}, has a negative value. This is magnified in the TCFI because r^2 is a factor in the integrand, dG_{22}. To overcome this issue, a data reduction procedure based on tail modeling, in the sense of Matteoli and Mansoori (1995), was tried. Unfortunately, the assumptions underlying the Matteoli/Mansoori RDF are not valid in the direct correlation range (first peak) for the kinds of substances frequently encountered in chemical engineering applications, such as hydrogen bonding species. For example, the Matteoli correlation assumes that the variations of the RDF from unity decrease from peak to peak so a combination of the exponential and cosine functions can be used. However, this is not always the case with real molecules. For example, the ethanol–ethanol RDF at high concentrations of benzene shows an irregular multipeak behavior not seen in LJ mixtures. This reflects dilute solution association effects of the ethanol molecules. As a result, Christensen et al. abandoned the LJ expression and integrated the simulation results numerically for the range of direct interactions. Then, since the indirect part of the RDF does not change dramatically when different interaction potentials are used, especially when averaging over angles, the long-range shape of $g_{ij}(r)$ is simple and similar to that of hard spheres. Further, when a model expression is used, it is possible to analytically integrate the indirect g_{ij} out to $r_{\max}$, and ultimately to r_{nc}, beyond which there is no contribution to the TCFI. The model expression selected was

$$g_{ij}^{\text{indirect}}(r) = 1 + a \cdot \exp^{[-b(r-c)]} \sin(d(r-e)) \qquad (6.18)$$

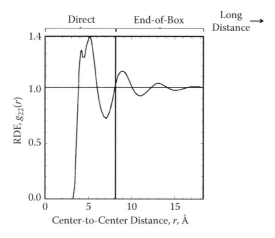

FIGURE 6.3   Spatial RDF–Blocks for integrating $h(r)$. "Direct" from simulation. "End-of-Box" is for matching simulation to fitted function such as Equation 6.18. "Long Distance" from fitted function.

The five parameters are determined by regression with the objective to reproduce $g_{ij}(r)$ from the third unity, r_{u3}, to r_{max}, with initial guesses found from

$$a_{init} = g_{ij}(r_{max,2}) - 1, \qquad b_{init} = \frac{-1}{r_{min,2} - r_{max,2}} \ln \frac{1 - g_{ij}(r_{min,2})}{g_{ij}(r_{max,2}) - 1}$$

$$c_{init} = r_{max,2}, \qquad d_{init} = \frac{\pi}{r_{u4} - r_{u3}}, \qquad e_{init} = r_{u3} \tag{6.19}$$

Here r_{u3} and r_{u4} are the radii of the third and fourth zeros of $h_{ij}(r)$, respectively, while $r_{max,2}$ is the radius of the second maximum of $g_{ij}(r)$, or the first maximum of the indirect interaction, located between the third and fourth zeros of $h_{ij}(r)$. The quantities $g_{ij}(r_{max,2})$ and $g_{ij}(r_{min,2})$ are the values of the second maximum and minimum, respectively. The dashed curve in Figure 6.2 is the result of such a regression. The contribution to H_{ij} from integration from r_{max} to r_{nc} is viewed as a long distance correction term.

Thus, the final form for TCFIs has three contributions as shown in Figure 6.3: (1) the direct interaction part of the RDF, H_{ij}^{direct}, integrated numerically; (2) the integral of a trial function from the end of the direct interaction to the maximum distance determined by the box size, H_{ij}^{box}; and (3) the long distance contribution, H_{ij}^{ld},

$$H_{ij} = \rho \underbrace{\int_0^{r_{u3}} r^2 (g_{ij}(r) - 1) dr}_{=H_{ij}^{direct}} + \rho \underbrace{\int_{r_{u3}}^{\infty} r^2 (g_{ij}(r) - 1) dr}_{=H_{ij}^{indirect} + H_{ij}^{ld}} = H_{ij}^{direct} + H_{ij}^{indirect} + H_{ij}^{ld} \tag{6.20}$$

The statistical uncertainty of the direct interaction part of the RDFs is normally negligible. Uncertainties are small for the first portion of the indirect part, though they could be significant at greater distances. Finally, the contribution of H_{ij}^{ld} to H_{ij} is the least, so its uncertainty can be ignored.

The 2006 International Fluid Properties Simulation Challenge (IFPSC) (http:// fluidproperties.org/) competitions were initiated in 2001 to stimulate and assess prediction methods for properties of industrially important fluids, by comparing methods, assessing the state of the art in simulation, and enhancing alignment of academic efforts with industrial needs. The 3rd IFPSC was held from March to September 2006. The focus of this contest was on the transferability of force fields and simulation methods for bubble pressures of mixtures of 1,1,1,2,3,3,3-heptafluoropropane (HFC-227ea refrigerant) and ethanol, based on limited data. In addition to their interesting pure component chemical properties, HFCs are often mixed with other fluids to be replacements for environmentally damaging chlorofluorocarbon refrigerants, and for cleaning solutions, fire retardants, and propellants. In the system of interest, hydrogen-bonding interactions could occur between the HFC-227ea proton and the ethanol hydroxyl, leading to attractive unlike interactions. It was expected that this system would not be well modeled by simple EOS mixing rules.

Entrants were provided with the experimental bubble points for 15 mixture compositions of 1,1,1,2,3,3,3-heptafluoropropane (HFC-227ea) and ethanol at 283.17 K, and properties of the pure materials at 343.13 K. The challenge was to compute bubble points for seven mixture compositions at 343.13 K, using any experimental data for the pure components but the only mixture points at 283.17 K. Entries using any theory/modeling/simulation method were accepted. Entries were judged based on the criterion,

$$SCORE = \frac{100}{7} \sum_{i=1}^{7} \left| \frac{P_{i,exp} - P_{i,calc}}{P_{i,calc}} \right| \qquad (6.21)$$

The experimental data for the mixtures at 343.13 K, measured at DuPont, were not released until all entries had been received. Four contestants used a range of different techniques including statistical-mechanical and molecular-simulation approaches that gave significantly more accurate predictions than from the semiempirical NRTL model (Prausnitz, Lichtenthaler, and Gomes de Azevedo 1999). The entry provided by Christensen et al. (2007a) was based on an NpT-MD simulation of the liquid phase at each mixture composition with FST linking the predicted microscopic structure (from the calculated RDF), to parameters that optimized a G^E-model for the liquid phase. It was assumed that the vapor phase was an ideal gas. For this contest, the Chemistry at HARvard Molecular Mechanics (CHARMM) force field was modified by revising the LJ parameters for the –CHF-part of the HFC-227ea molecule to fit experimental densities. The long-range modeling method of Christensen et al. (2007a) was sufficiently accurate to win (Case et al. 2007) in the State Conditions Transferability category. Their predictions of VLE behavior shown in Figure 6.4 gave a SCORE of 1.52. Activity coefficients found for this system are unusual in

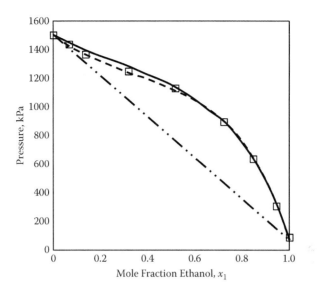

FIGURE 6.4 Ethanol/HFC-227ea pressure-composition diagram at 343 K. --□-- Experiment. (Data from F. H. Case, J. Brennan, A. Chaka, K. D. Dobbs, D. G. Friend, D. Frurip, P. A. Gordon, J. Moore, R. D. Mountain, J. Olson, R. B. Ross, M. Schiller, and V. K. Shen, 2007, The Third Industrial Fluid Properties Simulation Challenge, *Fluid Phase Equilibria*, 260, 153. With permission.) — From simulation. —••— Ideal solution. (Data from S. Christensen, G. H. Peters, F. Y. Hansen, J. P. O'Connell, and J. Abildskov, 2007b, Generation of Thermodynamic Data for Organic Liquid Mixtures from Molecular Simulations, *Molecular Simulation*, 33, 449; Modified from S. Christensen, G. H. Peters, F. Y. Hansen, J. P. O'Connell, and J. Abildskov, 2007c, State Conditions Transferability of Vapor-Liquid Equilibria via Fluctuation Solution Theory with Correlation Function Integrals from Molecular Dynamics Simulation, *Fluid Phase Equilibria*, 260, 169.)

behavior, as shown in Figure 6.5. The total pressures show positive deviations from Raoult's law, but the activity coefficients have a maximum for HFC-227ae and a minimum for ethanol near $x_1 = 0.45$.

The approach of Christensen was later modified by Wedberg, Peters, and Abildskov (2008). First, the 5-parameter form of Equation 6.18 was changed to a 4-parameter form,

$$g_{ij}^{indirect}(r) = 1 + a \cdot \exp^{[-b(r-c)]} \sin(d(r-c))   \qquad (6.22)$$

Next, a tail model corresponding to the antiderivative of the trial expression for $g(r)$ was fitted to the truncated numerical integral of $g(r)$ as a function of the upper integration limit. This tail model was then used to extrapolate $H(R_{lim})$ to $R_{lim} = \infty$, which yielded the value of the TCFI. Defining the running integrals of the RDF, $G(r)$,

$$G(r) = 4\pi \int_0^r r'^2 [g(r') - 1] dr', \quad H = \lim_{r \to \infty} \rho G   \qquad (6.23)$$

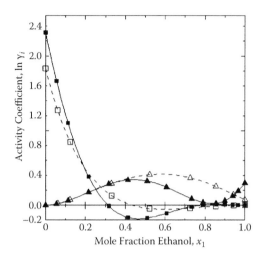

FIGURE 6.5   Activity coefficients (based on simulations) of (1) ethanol (□,■)/(2) HFC-227ea (Δ,▲) at 283.17K (—) and 343.13 K (- - -). (Modified from S. Christensen, G. H. Peters, F. Y. Hansen, J. P. O'Connell, and J. Abildskov, 2007c, State Conditions Transferability of Vapor-Liquid Equilibria via Fluctuation Solution Theory with Correlation Function Integrals from Molecular Dynamics Simulation, *Fluid Phase Equilibria*, 260, 169. Reprinted with permission from Elsevier.)

the antiderivative of Equation 6.23 used is

$$G_{smooth}(r) = -4\pi a \left[p(r)\cos(d(r-c)) + q(r)\sin(d(r-c)) \right] e^{-b(r-c)} + f$$

$$p(r) = \frac{d}{b^2+d^2} r^2 + \frac{4bd}{(b^2+d^2)^2} r - 2\frac{d^3-3b^2d}{(b^2+d^2)^3}$$

$$q(r) = \frac{b}{b^2+d^2} r^2 + \frac{2(b^2-d^2)}{(b^2+d^2)^2} r - 2\frac{3d^2b-b^3}{(b^2+d^2)^3}$$

(6.24)

The parameters for $G_{smooth}(r)$ were fitted to the sampled $G(r)$ for r ranging from r_{u3} to r_{max}, as defined above. The method was used to obtain isothermal compressibilities of five pure alkanes at three different state points and these were compared to the values derived from simulated overall density fluctuations. Results showed that the two approaches were fully consistent in values and uncertainties. Further, the computations converged in approximately the same simulation times. This suggests that computation of TCFIs is a route to isothermal compressibility, as accurate and fast as well-established benchmark techniques, with the advantage that it can be used in any ensemble (Puliti, Paolucci, and Sen 2011). Note that this approach has only been successfully tested on pure fluids. The main limitation of both methods (Christensen et al. 2007a, 2007b, 2007c; Wedberg, Peters, and Abildskov 2008) is that they apply only to systems where the TCF tails can be approximated by the model equations, which may not be true in general.

6.3.5 TRANSFORMS

Nichols, Moore, and Wheeler (2009) developed a method using finite Fourier-series expansions of molecular concentration fluctuations in order to reduce systematic errors from the simulation boundary conditions. The procedure was validated and compared to a truncation method for a nonideal binary liquid mixture of LJ particles tuned to imitate the system CF_4 and CH_4. A fluctuation expression is applied to a portion of the total volume within a closed simulation such as NVT. Rather than the sampling volume being spherical and centered on a single moving molecule, the sampling volume is a region with one or more rectangular slabs that is stationary with respect to the simulation cell. This leads to two alternative expressions,

$$S_{ij}(q) = x_i \delta_{ij} + x_i x_j \rho \int_0^\infty \frac{\sin(qr)}{qr} \big[g_{ij}(r) - 1 \big] 4\pi r^2 dr \tag{6.25}$$

and

$$S_{ij}(\mathbf{q}) = \frac{1}{N} \big\langle \delta N_i(\mathbf{q},t) \cdot \delta N_i(-\mathbf{q},t) \big\rangle = N \big\langle \psi_i(\mathbf{q},t) \cdot \psi_j(-\mathbf{q},t) \big\rangle \tag{6.26}$$

where $\psi_i(\mathbf{q},t)$ is the Fourier mass coefficient at the time t of the component i, and the wave vector $\mathbf{q}$ has components that are integer multiples of $2\pi/L_{box}$. The TCFIs are found from the structure factors via $\rho G_{ij} = S_{ij}(0) - 1$.

For an infinite system, the definitions of S_{ij} in Equations 6.25 and 6.26 are equivalent. However, for finite systems, truncation of the integrals can lead to errors in the integral of Equation 6.25. The advantage of Equation 6.26 relative to Equation 6.25 is that the structure factor can be based on a discrete Cartesian-based Fourier transform, rather than a continuous, spherically symmetric Fourier transform; Equation 6.26 is evaluated at various values of q, which are most accurate at larger values. The q-dependent S_{ij} are then extrapolated to $q = 0$ by fitting them to polynomials, the range of q and the polynomial order being selected empirically. Thus, it is not $g_{ij}(r)$ that is corrected; it is done via the structure factors related to the RDFs using the radial Fourier transform. The sampling volumes of the method do not truncate intermolecular correlations at a particular radial distance and no assumption is made that $g_{ij} \to 1$ at large intermolecular separations. In effect, this approach is consistent with periodic boundary conditions, but immune to long-range truncation effects. The inaccuracies in simulation RDFs at large r yield unreliable $S_{ij}(0)$ because of system size, but extrapolation to $q = 0$ provides the correction. Good results are obtained for LJ mixtures, but the method has apparently not yet been tested for molecular fluids.

6.3.6 FINITE-SIZE SCALING

The approach of Schnell et al. (2011) is to sample small nonperiodic systems in a (periodic) simulation box (reservoir) and then scale the results. The simulation system has sides of L_t in each dimension. Small systems are randomly selected subvolumes, denoted by L_{n-1}, L_n, and L_{n+1}, and can exchange energy and particles with the rest of the system. While the reservoir may not be a grand-canonical ensemble, the small systems will be when $L_n \ll L_t$. Finite-size scaling of KBIs determined from different-sized small systems is done by fitting and extrapolating them to the thermodynamic limit essentially using straight lines. The values compare well with those from integrating the RDF using a truncation method. In all cases, visual inspection is needed to identify where the subvolume results follow straight lines, but this is straightforward. While this approach has not been extensively tested, there is promise for such computations for practical applications with relative insensitivity to system size.

6.3.7 ADAPTIVE RESOLUTION

Quite recently, Mukherji et al. (2012) have proposed an adaptive resolution simulation method ('AdResS'). In a way resembling the distance-shifting method of Hess and Van der Vegt (2009), the 'AdResS' method divides the simulation domain into a small region of high-resolution (atomistic) molecules and a large region of (coarse-grained) lower resolution. Mukherji et al. analyze their method with a mixture resembling methanol and water by comparing the results with all-atom simulations and existing experimental results. Excellent agreement is found with the much larger and computationally more expensive all-atom simulations as well as with data. As with other methods that need to select an integration limit such that the integral converges to a plateau value or oscillates in a well-controlled way around a mean value, the 'AdResS' method also selects a high-resolution region width. The results indicate that a radius of 2 nm is adequate for the specific system of methanol and water, but a general approach to selecting the radius is uncertain.

6.4 DIRECT CORRELATION FUNCTION MATCHING— MIXTURES OF POLYATOMIC MOLECULES

Recently we (Wedberg et al. 2010, 2011a, 2011b; Wedberg 2011) have investigated more thoroughly the extension method of Verlet (1968) for mixtures with polyatomic molecules. This section gives a more complete description of our implementation, including some results not previously published. We have analyzed MD simulations of both pure (Wedberg et al. 2010) and mixed (Wedberg et al. 2011a, 2011b) LJ and Stockmayer fluids for wide ranges of state conditions and compared the results with the truncation method of Weerasinghe and Smith (2003b) and the distance-shifting method of Hess and Van der Vegt (2009).

6.4.1 METHOD

For potentials that decay faster than r^{-3}, the asymptotic behavior of the DCF is (Lebowitz and Percus 1963)

$$c(r) = -\beta u(r) + O\left\{ [\beta u(r)]^2 \right\}$$

(6.27)

Here, u is the pair intermolecular potential. In order to extend the TCF obtained from simulation, one chooses a value R within the range for which $h(r)$ is sampled, and determines h and c according to Equation 6.10. These requirements, together with the OZ equation, define a closed-form integral equation, which can be solved in order to obtain $h(r)$ for $r > R$. Simultaneously, $c(r)$ is obtained for all r. In our implementation, h and c are discretized as linear splines and both functions are assumed to be zero beyond a finite distance R_c. For example, in the cases of LJ and Stockmayer particles, $R_c = 15\sigma$, where σ is the LJ diameter. For *real* fluids, other values are used. Selecting this truncation radius is straightforward since the integrals generally converge, as can be checked by plotting the running integral versus r. If not converged, one repeats the calculation with a larger R_c. The Fourier-transformed OZ equation (Lebowitz and Percus 1963) is employed to express c explicitly in terms of h. This transforms the integral equation into a system of nonlinear equations for which a numerical solution is found using Newton's method (Wedberg et al. 2011a). The Jacobian is evaluated analytically and 5–15 iterations are normally required for convergence. Commonly, the Wiener–Hopf factorization technique is applied when the DCF is computed numerically from the TCF or vice versa (Jolly, Freasier, and Bearman 1976; Gray and Gubbins 1984; Press et al. 1992; Ramirez, Mareschal, and Borgis 2005). For the present application, the three-dimensional Fourier-transformed OZ equation can be employed. Applying the Fourier transform to Equation 6.6 gives a product,

$$\hat{h}_{ij}(k) = \hat{c}_{ij}(k) + \rho \sum_{l=1} \hat{h}_{il}(k) x_l \hat{c}_{lj}(k) \tag{6.28}$$

where $\hat{h}_{ij}(k)$ denotes the Fourier transformation of $h_{ij}(r)$. Due to radial symmetry, this is the zeroth-order Hankel transform, defined by

$$\hat{h}_{ij}(k) = 4\pi \int_0^\infty dr r^2 \frac{\sin(kr)}{kr} h_{ij}(r) \tag{6.29}$$

This same formulation holds for $\hat{c}_{ij}(k)$. The function $h_{ij}(r)$ is recovered from the inverse Hankel transform and is given by

$$h_{ij}(r) = \frac{4\pi}{(2\pi)^3} \int_0^\infty dk k^2 \frac{\sin(kr)}{kr} \hat{h}_{ij}(k) \tag{6.30}$$

Utilizing that $\hat{h}_{12}(k) = \hat{h}_{21}(k)$ and $\hat{c}_{12}(k) = \hat{c}_{21}(k)$, Equation 6.28 can be written as a linear system,

$$\tilde{h}(k) = \left[\mathbf{I} + \rho \mathbf{H}(k) \right] \tilde{c}(k) \tag{6.31}$$

with

$$\tilde{\mathbf{h}}(k) = \begin{bmatrix} \hat{h}_{11}(k) \\ \hat{h}_{12}(k) \\ \hat{h}_{22}(k) \end{bmatrix}, \quad \tilde{\mathbf{c}}(k) = \begin{bmatrix} \hat{c}_{11}(k) \\ \hat{c}_{12}(k) \\ \hat{c}_{22}(k) \end{bmatrix}, \quad \mathbf{H}(k) = \begin{bmatrix} x_1\hat{h}_{11}(k) & x_2\hat{h}_{12}(k) & 0 \\ 0 & x_1\hat{h}_{11}(k) & x_2\hat{h}_{12}(k) \\ 0 & x_1\hat{h}_{12}(k) & x_2\hat{h}_{22}(k) \end{bmatrix}$$

$$(6.32)$$

where $\mathbf{I}$ denotes the identity matrix. Equations 6.32 assume that the fluid mixture has at most two components. Equation 6.29 through Equation 6.31 provides a route for computing $c_{ij}(r)$ given $h_{ij}(r)$. The function $h_{ij}(r)$ is Hankel-transformed to yield $\hat{h}_{ij}(k)$. The linear system in Equation 6.31 is then solved for $\hat{c}_{ij}(k)$ at each k, followed by applying the inverse Hankel transform to obtain $c_{ij}(r)$. Solution of the problem of Equation 6.10 requires that the long-range part of $h_{ij}(r)$ be adjusted until the long-range part of $c_{ij}(r)$ matches a trial function $t_{ij}(r)$. This is accomplished by a Newton iteration scheme, for which grids in r and k space are introduced,

$$r_\alpha \equiv \alpha \cdot \Delta r, \quad \alpha = 0, \dots, N$$

$$k_\beta \equiv \beta \cdot \Delta k, \quad \beta = 0, \dots, N$$

$$(6.33)$$

The upper cutoffs are $R_c = N \cdot \Delta r$ and $K_c = N \cdot \Delta k$ for the integrals in Equations 6.29 and 6.30, respectively. Note that R_c is not the sampling limit set by the simulation box dimensions, but is typically much larger. The TCFs, DCFs, and their Hankel transforms at an iteration step t are represented by discrete vectors,

$$h_{ij,\alpha}^{(t)} \equiv h_{ij}^{(t)}(r_\alpha), \quad c_{ij,\alpha}^{(t)} \equiv c_{ij}^{(t)}(r_\alpha), \quad \alpha = 1, \dots, N$$

$$\tilde{h}_{ij,\beta}^{(t)} \equiv \hat{h}_{ij}^{(t)}(k_\beta), \quad \tilde{c}_{ij,\beta}^{(t)} \equiv \hat{c}_{ij}^{(t)}(k_\beta), \quad \beta = 1, \dots, N$$

$$(6.34)$$

Equation 6.29 for the TCF is approximated by truncating the integral at R_c and using the trapezoidal rule,

$$\tilde{\mathbf{h}}_{ij}^{(t)} = \mathbf{T} \cdot \mathbf{h}_{ij}^{(t)}$$

$$(6.35)$$

Elements of the matrix $\mathbf{T}$ are

$$T_{\beta\alpha} = 4\pi \cdot \Delta r \cdot r_\alpha^2 \cdot \frac{\sin k_\beta r_\alpha}{k_\beta r_\alpha}\left(1 - \frac{\delta_{0\alpha} + \delta_{N\alpha}}{2}\right)$$

$$(6.36)$$

with $\alpha = 1, \dots, N$, $\beta = 1, \dots, N$, with $\sin(kr)/(kr)$ being unity if either k or r is zero. Equation 6.30 for the DCF is approximated in a similar way by truncating the integral at K_c,

$$\tilde{\mathbf{c}}_{ij}^{(t)} = \mathbf{U} \cdot \mathbf{c}_{ij}^{(t)}$$

$$(6.37)$$

with

$$U_{\beta\alpha} = \frac{4\pi}{(2\pi)^3} \cdot \Delta k \cdot k_\alpha^2 \cdot \frac{\sin k_\beta r_\alpha}{k_\beta r_\alpha} \left(1 - \frac{\delta_{0\beta} + \delta_{N\beta}}{2} \right) \tag{6.38}$$

with α and β as in Equation 6.36. As stated above, the middle step of converting $\hat{h}_{ij}^{(t)}$ to $\hat{c}_{ij}^{(t)}$ is carried out by solving the linear system of Equation 6.31 for each value of β. If n_{ij} denotes indexing such that $r_{n_{ij}} \leq R_{ij} \leq r_{n_{ij}+1}$, and $h_{ij}^{(t)}$ and $c_{ij}^{(t)}$ denote vectors containing the elements of $h_{ij}^{(t)}$ and $c_{ij}^{(t)}$, respectively, with $n_{ij}+1 \leq \alpha \leq N$, $h_{ij}^{(t)}$ is updated at each iteration step according to

$$\mathbf{h}_{ij}^{(t+1)} = \mathbf{h}_{ij}^{(t)} + \Delta\mathbf{h}_{ij}^{(t)} \tag{6.39}$$

Here, $\Delta\mathbf{h}_{ij}^{(t)}$ in Newton's method is found by solution of the linear system,

$$\begin{bmatrix} \mathbf{J}_{11}^{11} & \mathbf{J}_{12}^{11} & \mathbf{J}_{22}^{11} \\ \mathbf{J}_{11}^{12} & \mathbf{J}_{12}^{12} & \mathbf{J}_{22}^{12} \\ \mathbf{J}_{11}^{22} & \mathbf{J}_{12}^{22} & \mathbf{J}_{22}^{22} \end{bmatrix} \begin{bmatrix} \Delta\mathbf{h}_{11}^{(t)} \\ \Delta\mathbf{h}_{12}^{(t)} \\ \Delta\mathbf{h}_{22}^{(t)} \end{bmatrix} = \begin{bmatrix} \Delta\mathbf{c}_{11}^{(t)} \\ \Delta\mathbf{c}_{12}^{(t)} \\ \Delta\mathbf{c}_{22}^{(t)} \end{bmatrix} \tag{6.40}$$

where the right-hand side represents the difference between the approximation of the long-range DCF to be enforced and the currently computed DCF,

$$\Delta\mathbf{c}_{ij,\alpha}^{(t)} = t_{ij}(r_\alpha) - \mathbf{c}_{ij,\alpha}^{(t)}, \quad \alpha = n_{ij}+1, \ldots, N \tag{6.41}$$

The Jacobian has the elements,

$$\mathbf{J}_{\alpha\beta}^{ij} \equiv \begin{bmatrix} \dfrac{\partial c_{ij,n_{ij}+1}^{(t)}}{\partial h_{\alpha\beta,n_{\alpha\beta}+1}^{(t)}} & \cdots & \dfrac{\partial c_{ij,n_{ij}+1}^{(t)}}{\partial h_{\alpha\beta,N}^{(t)}} \\ \cdots & \cdots & \cdots \\ \dfrac{\partial c_{ij,N}^{(t)}}{\partial h_{\alpha\beta,n_{\alpha\beta}+1}^{(t)}} & \cdots & \dfrac{\partial c_{ij,N}^{(t)}}{\partial h_{\alpha\beta,N}^{(t)}} \end{bmatrix} \tag{6.42}$$

These are partial derivatives that can be expanded by the chain rule to,

$$\frac{\partial c_{ij,\alpha}}{\partial h_{ab,\alpha'}} = \sum_{\beta=0,\beta'=0}^{N} \frac{\partial c_{ij,\alpha}}{\partial \tilde{c}_{ij,\beta}} \frac{\partial \tilde{c}_{ij,\beta}}{\partial \tilde{h}_{ab,\beta'}} \frac{\partial \tilde{h}_{ab,\beta'}}{\partial h_{ab,\alpha'}} = \sum_{\beta=0}^{N} U_{\alpha\beta} \frac{\partial \tilde{c}_{ij,\beta}}{\partial \tilde{h}_{ab,\beta}} T_{\beta\alpha'} \tag{6.43}$$

The last equality is due to the result,

$$\frac{\partial \tilde{c}_{ij,\beta}}{\partial \tilde{h}_{ab,\beta'}} = \delta_{\beta\beta'} \frac{\partial \tilde{c}_{ij,\beta}}{\partial \tilde{h}_{ab,\beta}} \tag{6.44}$$

a consequence of Equation 6.31, Equation 6.39, and Equation 6.41. The partial derivatives are obtained from three linear systems derived from Equation 6.31 by differentiation with respect to $\hat{h}_{11,\beta}$, $\hat{h}_{12,\beta}$, and $\hat{h}_{22,\beta}$. The results are

$$
\begin{bmatrix}
\dfrac{\partial \tilde{c}_{11,\beta}}{\partial \tilde{h}_{11,\beta}} & \dfrac{\partial \tilde{c}_{11,\beta}}{\partial \tilde{h}_{12,\beta}} & \dfrac{\partial \tilde{c}_{11,\beta}}{\partial \tilde{h}_{22,\beta}} \\[2ex]
\dfrac{\partial \tilde{c}_{12,\beta}}{\partial \tilde{h}_{11,\beta}} & \dfrac{\partial \tilde{c}_{12,\beta}}{\partial \tilde{h}_{12,\beta}} & \dfrac{\partial \tilde{c}_{12,\beta}}{\partial \tilde{h}_{22,\beta}} \\[2ex]
\dfrac{\partial \tilde{c}_{22,\beta}}{\partial \tilde{h}_{11,\beta}} & \dfrac{\partial \tilde{c}_{22,\beta}}{\partial \tilde{h}_{12,\beta}} & \dfrac{\partial \tilde{c}_{22,\beta}}{\partial \tilde{h}_{22,\beta}}
\end{bmatrix}
= \left[\mathbf{I} + \rho \mathbf{H}(\beta \cdot \Delta k) \right]^{-1}
\begin{bmatrix}
1 - x_1 \rho \tilde{c}_{11,\beta} & -x_2 \rho \tilde{c}_{12,\beta} & 0 \\
-x_1 \rho \tilde{c}_{12,\beta} & 1 - x_2 \rho \tilde{c}_{22,\beta} & 0 \\
0 & -x_1 \rho \tilde{c}_{12,\beta} & 1 - x_2 \rho \tilde{c}_{22,\beta}
\end{bmatrix}
\tag{6.45}
$$

At each iteration step, these systems are solved for the partial derivatives, which then are used to evaluate the Jacobians in Equation 6.44.

The short-range parts of the calculated DCFs are not used within the iteration scheme, though the short-range part of the DCF obtained from the final iteration is considered in selecting the parameters R_{ij}. Initially, the discretized TCFs are set to $h^{(0)}_{ij,\alpha} = h_{\mathrm{MD},ij}(r_\alpha)$ for all r_α within the sampling range for $h_{\mathrm{MD},ij}(r)$, and $h^{(0)}_{ij,\alpha} = 0$ for larger α. The iteration is carried out until

$$
\sum_{i,j} \sum_{\alpha = n_{ij}+1}^{N} \left[\Delta c^{(t)}_{ij,\alpha} r_\alpha^2 \right]^2 < \eta
\tag{6.46}
$$

with $\eta = 10^{-4}$ or less. Typically, this is achieved after 5 to 15 iterations. For some systems, in particular those at high density where the functions $h_{ij}(r)$ have significant structure beyond the sampling range, the tail model by Christensen et al. (2007a, 2007b, 2007c) was used to estimate the long-range behavior for the initial guess $h^{(0)}_{ij,\alpha}$. Using this approach, the Newton iterations have converged for all systems we have studied to date. Issues of how to select the matching distance and the angle averaging of potentials are discussed in detail by Wedberg (2011).

6.4.2 RESULTS

We now discuss results from MD simulations that test the capabilities of the method. The KBIs are primarily verified by comparing the derivative properties obtained from the integration procedure with the same properties obtained from alternative analyses, or from simulation results in the literature. For the simulations of water/organic solvent mixtures, the derivative properties obtained by integration are also compared against values derived from correlations of experimental data. In this last case, consistency depends not only on the accuracy of the integration procedure, but also on the accuracy of the force field.

6.4.3 MODEL FLUIDS

The methodology was first tested on pure and mixed LJ and Stockmayer fluids (Wedberg et al. 2010) for several reasons. First, these fluids are well defined, so simulation results over wide temperature and density ranges could be acquired with limited computational effort. Second, the thermodynamic derivative properties obtained from the extended pair-distribution function can be validated against data derived from correlations of previous simulations. Third, the simple form of the interatomic potentials allows a basic test of the assumption that the OZ equation can be resolved into isotropic and anisotropic parts. For the Stockmayer fluid, this form of the OZ relation is inexact, becoming less accurate as the (reduced squared) dipole moment, μ^{*2}, increases. The accuracy of the properties obtained for large μ^{*2} could indicate validity of the isotropic OZ equation. In what follows, the physical quantities are in dimensionless values, where the quantities marked with an asterisk (*) have been reduced with respect to the LJ parameters ε (energy), σ (length), and the atomic mass. We give here later developments (Wedberg 2011; Wedberg et al. 2011b) than described by Wedberg et al. (2010), which have less reliable approximations for the DCF tail.

6.4.4 LJ FLUIDS

Figure 6.6 compares the isothermal compressibilities obtained from the method with those obtained from the equation of state (EOS) of Mecke et al. (1996) for LJ fluids. At all four temperatures, our results are qualitatively consistent with the EOS, with differences in the range of 1 to 5%. The comparisons are not as good at the two lowest temperatures, where the differences are as high as 6%.

The greatest disagreement is seen when $T^* = 1.5$ and ρ^* is 0.3 or 0.4, which are the state points closest to the critical point of $\rho_c^* = 0.304$ and $T_c^* = 1.316$ (Smit 1992). At those conditions, the differences are 7% and 8%, respectively. This disagreement is not surprising considering that the reduced bulk modulus ($\rho k_B T \kappa_T = 1 - C$) can be very small in this region. We conclude that our method is best suited for systems at liquid density ($\rho > 2\rho_c$). For $\rho^* = 0.7$ and $\rho^* = 0.8$, the agreement with the EOS was better at higher temperature (1% at $T^* = 2.5$) than at lower temperature (6% at $T^* = 0.85$). It is possible that derivatives of the Mecke EOS are less accurate at lower temperatures since the EOS does not reproduce the simulation pressures very well under these conditions and at higher temperatures. Nevertheless, though low temperatures seem to offer more of a challenge, the results obtained under those conditions may still be considered satisfactory. The standard error in T was less than 0.5%, indicating that the calculations were well converged.

6.4.5 LJ/STOCKMAYER MIXTURES

LJ/Stockmayer mixtures include "Stockmayer" atoms (2) with finite dipole moments and "LJ" atoms (1) with zero dipole moment. LJ–LJ and LJ–Stockmayer interactions

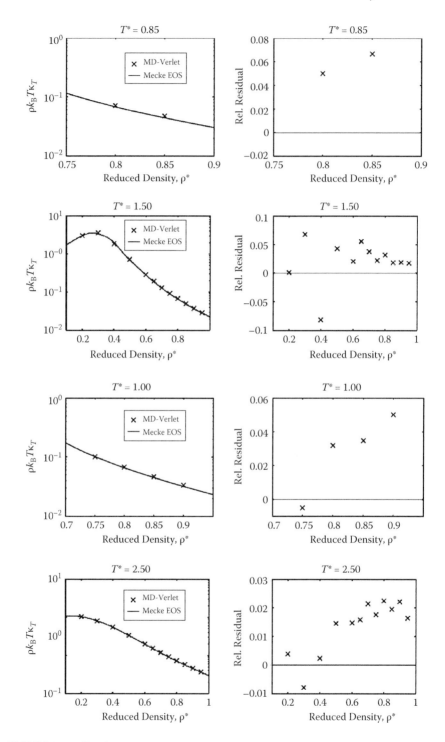

FIGURE 6.6   *(See facing page)*

thus follow the LJ potential, while Stockmayer–Stockmayer interactions include dipole–dipole interactions. The EOS of Gross and Vrabec (2006) describes mixtures of fluid particles with different dipole moments and thus can be used to obtain the isothermal compressibility. Values of the reduced bulk modulus obtained from our method are compared in Figure 6.7 with the Gross/Vrabec EOS. For $\mu^{*2} = 1$ (a), the Verlet values agree very well with the EOS; the differences are 1 to 1.5%. For the higher dipole moments, the agreement is still good when x_2 is small but deteriorates as x_2 increases (b) and (c). This becomes more pronounced for $\mu^{*2} = 3$ where the discrepancies are as large as 11% at Stockmayer-rich compositions. Since the Gross/ Vrabec EOS did not reproduce simulation pressures very well (Wedberg 2011) with errors increasing with larger μ^{*2} and x_2, the discrepancy may not be due only to the simulation results.

6.4.6 COMPARISON WITH EXISTING APPROACHES

The truncation (Weerasinghe and Smith 2003b) and distance-shifting (Hess and Van der Vegt 2009) methods were also employed to calculate $\rho k_B T \kappa_T$. Simple truncation requires averaging $H(R_{lim})$ over a specific interval, where $H(R_{lim})$ is the numerical TCFI as a function of the upper integration limit R_{lim}. It is not obvious how to choose these values. Here, the TCFIs have been averaged with R_{lim} in the interval $[2\sigma, 3\sigma]$, where σ roughly corresponds to the oscillation period of $h_{ij}(r)$. With the Hess method, the scaling factor α_{ij} was evaluated with $R = 4\sigma$. The RDFs were rescaled, but the integrals still did not converge within the sampled range. The integration of the rescaled RDFs was thus carried out as with the truncation method, but using larger truncation radii, based on the idea that corrected RDFs are more reliable at large separations. The TCFIs were averaged using truncation radii in the interval $[3.5\sigma, 4.5\sigma]$. The values obtained for the isothermal compressibility (Figure 6.7) demonstrate the limitations on simple truncation and distance-shifting methods. Truncation overestimates the compressibility by 10 to 15% while distance shifting underestimates it by 10 to 40%. These are sensitive to the choice of truncation radii, so it is possible that better results could have been obtained with other radii. The methods should also perform better with significantly larger simulation systems, but this was not tried here. The Verlet method yielded accurate results for LJ/Stockmayer mixtures as indicated by comparisons with benchmark values. While the method achieves better accuracy than simpler integration approaches, caution is advised regarding activity coefficient derivatives when a system is nearly ideal or when the mole fraction of a component is less than approximately 15%, as discussed by Wedberg et al. (2011a).

FIGURE 6.6   *(see facing page)* Values (x) for results (left) and relative residuals (right) from calculations of the reduced bulk modulus, $\rho k_B T \kappa_T$, for the pure LJ fluid at reduced temperatures $T^* = 0.85$, $T^* = 1.0$, $T^* = 1.5$, and $T^* = 2.5$. Lines derived from EOS of Mecke et al. are also shown. (From M. Mecke, A. Muller, J. Winkelmann, J. Vrabec, J. Fischer, R. Span, and W. Wagner, 1996, An Accurate Van der Waals-Type Equation of State for the Lennard-Jones Fluid, *International Journal of Thermophysics*, 17, 391, with permission from Springer.)

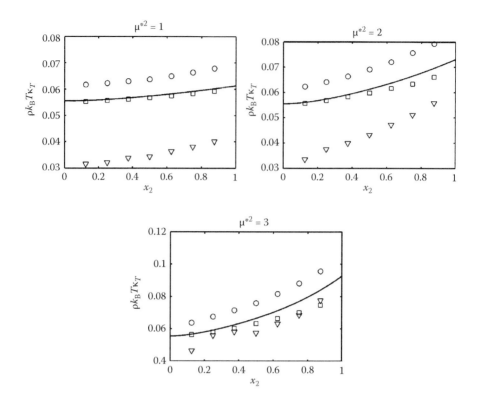

FIGURE 6.7 Values of $\rho k_B T \kappa_T$ for LJ/Stockmayer mixtures versus the mole fraction of Stockmayer particles, x_2, for dipole moments of (a) $\mu^{*2} = 1$, (b) $\mu^{*2} = 2$, and (c) $\mu^{*2} = 3$, derived from the Gross/Vrabec EOS (——) (From J. Gross and J. Vrabec, 2006, An Equation-of-State Contribution for Polar Components: Dipolar Molecules, *American Institute of Chemical Engineers Journal*, 52, 1194), compared with results from our MD-Verlet (□), truncation (o) (From S. Weerasinghe and P. E. Smith, 2003b, Kirkwood–Buff Derived Force Field for Mixtures of Acetone and Water, *Journal of Chemical Physics*, 118, 10663); and distance-shifting (∇) methods (From B. Hess and N. F. A. van der Vegt, 2009, Cation Specific Binding with Protein Surface Charges, *Proceedings of the National Academy of Sciences of the United States of America*, 106, 13296).

6.4.7 AQUEOUS ALCOHOL MIXTURES

The major goal is to establish an integration method that accurately predicts activity coefficient derivatives, partial molar volumes, and isothermal compressibilities from simulations of molecular mixtures with atom–atom interaction models. This section focuses on such applications with results compared to values derived from correlations of experimental data. It should be noted that the accuracy also depends on the validity of the molecular force fields and reliability of experimental data. As with the analysis of the LJ/Stockmayer mixtures, the simple truncation (Weerasinghe and Smith 2003b) and distance-shifting methods (Hess and Van der Vegt 2009) were employed to evaluate the same properties. Simple truncation averaging of the integral

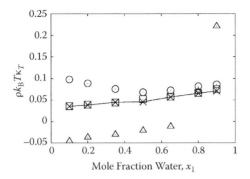

FIGURE 6.8 Isothermal compressibilities for the mixture water (1)/t-butanol (2). Results from our Verlet (□), truncation (○) (Results from S. Weerasinghe and P. E. Smith, 2003b, Kirkwood-Buff Derived Force Field for Mixtures of Acetone and Water, *Journal of Chemical Physics*, 118, 10663), and distance-shifting (Results from B. Hess and N. F. A. van der Vegt, 2009, Cation Specific Binding with Protein Surface Charges, *Proceedings of the National Academy of Sciences of the United States of America*, 106, 13296) (Δ) methods compared with values obtained from the fluctuation formula (—x—).

varied in the interval for R_{lim} from 1.0 to 1.5 nm. With the distance-shifting method, the scaling factors α_{ij} were evaluated from the calculated RDFs with the parameter $R = 2.0$ nm. Numerical integration of the rescaled RDFs did not converge within the sampling range, so the integrals of the rescaled TCFs were evaluated by truncation using intervals of 1.4–1.9 nm. The truncation radii employed for integration of the rescaled TCFs were larger than those used with the simple truncation approach since the rescaled TCFs probably were more accurate than the original TCFs for large r, as discussed in Section 6.3. For comparison, isothermal compressibilities were evaluated via the fluctuations of the simulation box volume. The results are shown in Figure 6.8. The Verlet method reproduced the fluctuation formula results to within 5%, while the simple truncation and distance-shifting methods were greatly in error. In fact, distance shifting yielded negative compressibilities. It is likely that reliable results by these methods require simulations of larger systems.

In order to validate the partial molar volumes obtained by the different integration methods, the excess molecular volume, V^{E}, was evaluated for the simulations at each composition according to

$$V_{\text{m}}^{\text{E}}(x_1) = V_{\text{m}}(x_1) - x_1 V_1^{\circ} - x_2 V_2^{\circ} \tag{6.47}$$

where $V_{\text{m}}(x_1)$ denotes the average molecular volume obtained at the composition x_1, the mole fraction of water. Also, V_1° and V_2° denote the average molar volumes of the corresponding pure components, obtained from separate simulations. The polynomial model of Handa and Benson (1979),

$$V_{\text{m}}^{\text{E}}(\text{model}) = x_1 x_2 \left[a_0 + a_1(x_2 - x_1) + a_2(x_2 - x_1)^2 \right] \tag{6.48}$$

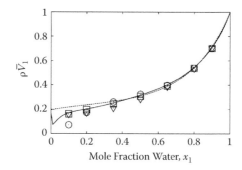

FIGURE 6.9 Relative partial molar volumes ($\rho\bar{V}_1$) for water (1)/t-butanol (2). The results from the Verlet (□), truncation (○) (Results from S. Weerasinghe and P. E. Smith, 2003b, Kirkwood-Buff Derived Force Field for Mixtures of Acetone and Water, *Journal of Chemical Physics*, 118, 10663), and distance-shifting (Results from B. Hess and N. F. A. van der Vegt, 2009, Cation Specific Binding with Protein Surface Charges, *Proceedings of the National Academy of Sciences of the United States of America*, 106, 13296) (Δ) methods for obtaining the TCFIs compared with results from full simulations smoothed with a quadratic polynomial (- - -) and with smoothed experimental data (———).

was fitted to the calculated values of the excess molar volume. Reduced partial molar volumes were evaluated by analytical differentiation of the model according to

$$\rho\bar{V}_1 = \rho V_m^E + \rho\left[\partial(NV_m^E)/\partial N_1\right]_{T,p,N_2} \tag{6.49}$$

In Figure 6.9, the results for $\rho\bar{V}_1$ are compared with TCFIs calculated by our Verlet, simple truncation, and distance-shifting methods. The partial molar volumes obtained from the correlations of simulation volumes are in very good agreement with those obtained from experimental correlations (Handa and Benson 1979). The results obtained via the three TCFI calculation methods agreed very well with both correlations. Furthermore, the three methods yielded similar results, though for dilute water systems, simple truncation underestimates the values relative to the other methods and experimental data.

Figure 6.10 shows activity coefficient derivatives over the whole composition range for experiment from three correlations and the Verlet method. A procedure for experimental data analysis was described by Wooley and O'Connell (1991), in which one extracts the isothermal compressibility, partial molar volumes, and activity coefficient derivatives from experimental data. The activity coefficient derivatives are obtained by fitting mixture vapor–liquid equilibrium data to obtain parameters for at least two different G^E models. Wooley and O'Connell employed the Wilson, non-random, two liquid (NRTL) and modified Margules (mM) models. Partial molar volumes are obtained from correlations of mixture densities (Handa and Benson 1979). Isothermal compressibilities are either taken from measurements or estimated with

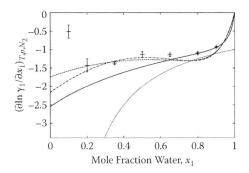

FIGURE 6.10   Composition derivative of the activity coefficient for water versus the water mole fraction x_1 of water (1)/t-butanol (2). The Verlet method (I) is compared with experimental data smoothed using the Wilson (———), NRTL (- - -), and mM (–•–) models. For phase stability, activity coefficient derivatives must lie everywhere above the curve $y = -1/x_1$ (•••).

the correlation of Huang and O'Connell (1987). Figure 6.10 also shows the relation $y = -1/x_1$; for complete miscibility, the activity coefficient derivative must always lie above this curve. The simulation value at $x_1 = 0.1$ may be unreliable due to diluteness. The results from the NRTL and mM correlations show immiscibility, which is not observed, while the simulations suggest complete miscibility. The Wilson correlation cannot give two liquid phases so it is more consistent with the simulations at dilute alcohol concentrations.

As described in Chapter 1, neither TCFIs nor DCFIs can be measured directly in experiments, though they can be derived from correlations of experimental data for other thermodynamic properties or integrals of X-ray or neutron scattering measurements. Figure 6.11 shows TCFIs from the correlated results of Figures 6.9 and 6.10 along with results from our Verlet method.

The TCFIs obtained by our Verlet method apparently converged at all compositions, indicating phase stability over the whole composition. They compare favorably with those from simple truncation (Weerasinghe and Smith 2003b), and the distance-shifting (Hess and Van der Vegt 2009) methods. When a simulated system is sufficiently large, the three methods can be expected to yield similar results, but the Verlet method is superior for smaller systems and when the RDFs have significant structure beyond the sampling limit. This is an important result, since the Verlet method might allow thermodynamic derivative properties to be accurately obtained from simulations of complex systems with relatively low computational effort.

6.5   FUTURE APPLICATIONS

Classical thermodynamics can only provide relations among properties; values must be found by experiment or computation. Given our current techniques, future applications of TCF integrations for properties may be achieved in the following areas.

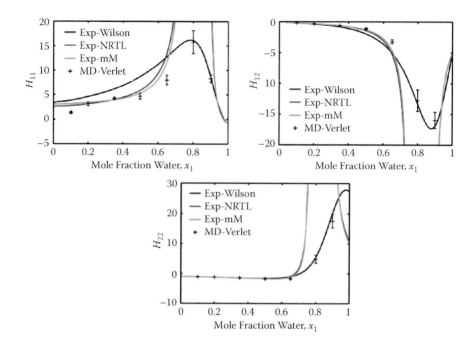

FIGURE 6.11 TCFIs for (a) water/water, (b) water/t-butanol, and (c) t-butanol/t-butanol obtained from simulation of water/t-butanol mixtures using the Verlet method (crosses) versus the water mole fraction x_1, compared with TCFIs obtained from experimental data using the Wooley/O'Connell procedure, where either the Wilson (black line), NRTL (red line), or mM (green line) models were employed for obtaining the activity coefficient derivatives. Note that the NRTL and mM model approaches infinity since they predict a phase split. (Calculated values from R. J. Wooley and J. P. O'Connell, 1991, A Database of Fluctuation Thermodynamic Properties and Molecular Correlation-Function Integrals for a Variety of Binary Liquids, *Fluid Phase Equilibria*, 66, 233.) **(See color insert.)**

6.5.1 Enzyme Solutions

Water activity is an important element of nonaqueous biocatalytic systems. Recently, we have explored different approaches to this property via MD simulation (Wedberg, Abildskov, and Peters 2012). Two main strategies to study how protein properties depend on water activity are termed *real-time* control and *a posteriori* analysis. The former comprises simulations of the protein in a nonaqueous medium in which the number of water molecules is adjusted to maintain a desired water activity. In the latter strategy, conventional MD simulations are carried out, but the water activity is calculated through postanalysis of the simulations. The study of Branco et al. (2009) is apparently the only work that explicitly considers water activity as a variable. However, their medium was assumed to be an ideal mixture. The greater challenge of nonideal media, such as aqueous organic solutions, has been addressed (Wedberg, Abildskov, and Peters 2012). Much more work needs to be done before establishing a standard method.

6.5.2 Diffusion and Reaction

The current works on properties have involved only equilibrium properties. Computed RDFs for homogeneous, but nonequilibrium, states could lead to local chemical potential gradients for diffusional driving forces and chemical reaction driving forces. Such computations would be unique and powerful for both thermodynamic and transport phenomena.

In particular, the Stefan–Maxwell constitutive equation for multicomponent diffusion in nonideal solutions (Curtiss and Bird 1999; Wheeler and Newman 2004a, 2004b) has driving forces derived from chemical potential gradients of all but one component. The nonideality adjustment has been obtained for mutual diffusion in binary mixtures using FST (Jolly and Bearman 1980; Schoen and Hoheisel 1984; Chitra and Smith 2001c). Simulations based on the methods described here could describe higher multicomponent systems, which are of significant interest.

6.6 CONCLUSIONS

The successes described in this chapter for both model and real mixtures indicate that molecular simulation methods for FST should now be ready for greater implementation and extension. Investigations to refine the various methods to compute KBIs are still ongoing. At this point, our extended Verlet method appears to be the most general and reliable approach to obtain thermodynamic properties, especially for dense systems. Its advantages of minimal computational effort and limited need for case-by-case judgment in analysis indicate its efficiency and robustness.

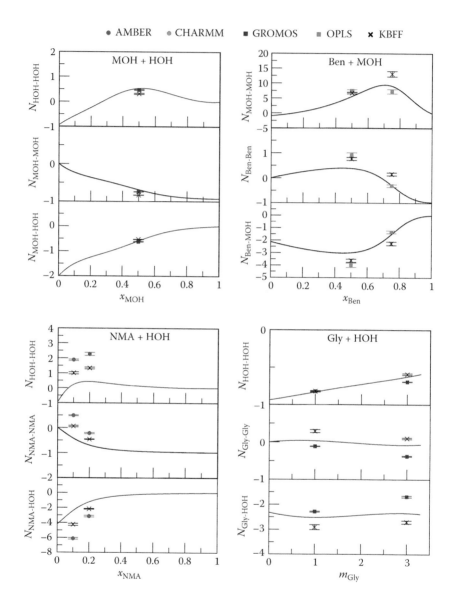

FIGURE 5.2 Comparisons of the excess coordination numbers for mixtures of methanol + water, benzene + methanol, N-methylacetamide + water, and zwitterionic glycine + water using the Kirkwood-Buff force field and the biomolecular force field that best reproduced the excess coordination numbers for each system. Please note the different y-axis scales. Error bars show the estimated standard deviation obtained from five 20-nanosecond subaverages.

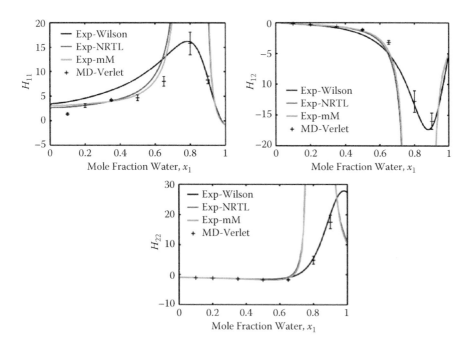

FIGURE 6.11 TCFIs for (a) water/water, (b) water/t-butanol, and (c) t-butanol/t-butanol obtained from simulation of water/t-butanol mixtures using the Verlet method (crosses) versus the water mole fraction x_1, compared with TCFIs obtained from experimental data using the Wooley/O'Connell procedure, where either the Wilson (black line), NRTL (red line), or mM (green line) models were employed for obtaining the activity coefficient derivatives. Note that the NRTL and mM model approaches infinity since they predict a phase split. (Calculated values from R. J. Wooley and J. P. O'Connell, 1991, A Database of Fluctuation Thermodynamic Properties and Molecular Correlation-Function Integrals for a Variety of Binary Liquids, *Fluid Phase Equilibria*, 66, 233.)

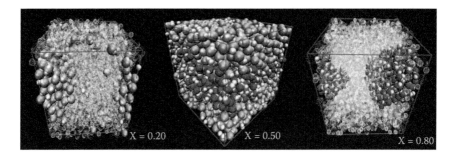

FIGURE 7.7 Snapshots of the λ = 1/3 mixture model which phase separates. The SPC/E water molecules are shown as red spheres (oxygen site) with two white spheres (hydrogen sites), while weak water is shown with the oxygen as a cyan sphere. Here, x is the mole fraction of weak water. For $x = 0.2$ and $x = 0.8$, the majority species is shown as ghost particles.

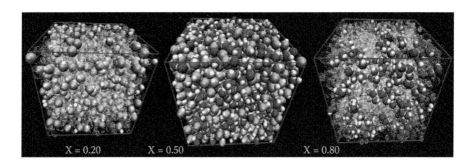

FIGURE 7.8 Snapshots of the λ = 4/5 mixture model which is a model for hydrophilic aqueous mixture. The representation conventions are as in Figure 7.7. Here, x is the mole fraction of weak water. For $x = 0.2$ and $x = 0.8$, the majority species is shown as ghost particles.

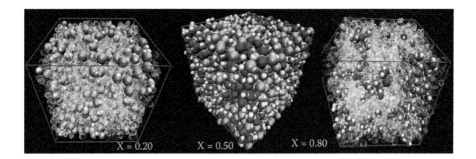

FIGURE 7.9 Snapshots of the λ = 2/3 mixture model which is a model for hydrophobic aqueous mixture. The representation conventions are as in Figure 7.7. Here, x is the mole fraction of weak water. For $x = 0.2$ and $x = 0.8$, the majority species is shown as ghost particles.

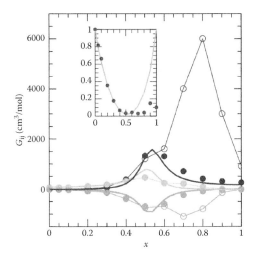

FIGURE 7.11 KBIs for the mixture model $\lambda = 2/3$ as a function of the weak-water mole fraction. The open dots are the uncorrected KBIs obtained by the RDFs and the filled dots are obtained through the TS approach described in the text. Blue is for G_{WW}, gold for G_{ww}, and green for G_{Ww} ("W" stands for SPC/E water and "w" for weak water). Dots are for data obtained from integration of the RDFs and the lines from the KB theory. The inset shows the concentration fluctuation function for a simple model (cyan) and correct version (red).

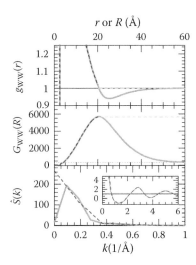

FIGURE 7.12 Illustration of the TS approach for molecular emulsions. Top panel shows a close-up of the water–water RDF at large distances, with red curve being the initial data from MD, gold curve is the abrupt continuation and green curve the TS continuation. Middle panel shows the corresponding RKBIs as a function of integration distance, and the lower panel the structure factor with a visible domain prepeak.

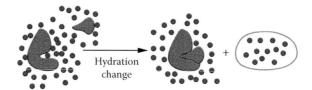

FIGURE 11.1 Changes of hydration accompany many biomolecular processes and reactions. The enumeration of the number of water molecules released (or adsorbed) for such reactions was a matter of debate.

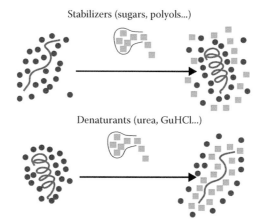

FIGURE 11.11 A schematic diagram summarizing the difference between protein stabilizers and denaturants in terms of their interactions with a protein.

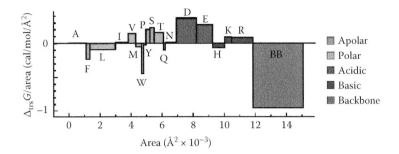

FIGURE 12.1 Free energy of transfer of the Nank4–7 denatured state plotted for each of the side chains and backbone contributions as a function of change in the accessible surface area from native (water) to denatured state (1 M urea).

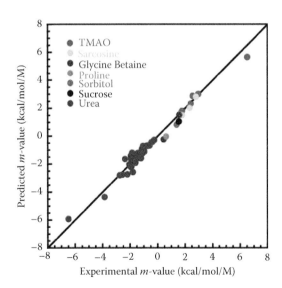

FIGURE 12.2 Predicted group transfer model *m*-values (related to the slope of the free energy change with respect to osmolyte concentration) versus that measured experimentally.

7 Concentration Fluctuations and Microheterogeneity in Aqueous Mixtures

New Developments in Analogy with Microemulsions

Aurélien Perera

CONTENTS

Abstract: In contrast with simple liquid mixtures, aqueous mixtures show enhanced concentration fluctuations that are experimentally detected through the Kirkwood–Buff integrals (KBIs). By revisiting the exact molecular Ornstein–Zernike (OZ) relation between correlation functions, we derive an expression for the asymptotical form of the correlation functions, which predicts the existence of specific domains, in addition to the usual Ornstein–Zernike correlation length associated with fluctuations. This allows one to reinterpret enhanced concentration fluctuations in terms of the existence of such specific domains, which we associate with the inherent microheterogeneity of these mixtures. In particular, the formalism allows us to predict the asymptotic decay of the correlation functions as incompletely obtained from finite size computer simulations, but in self-consistent agreement with the volumetric properties given by the same simulations. The analogy with microemulsions, as inspired from the theoretical developments, provides a unified way of describing fluctuations in mixtures in the context of condensed matter physics in general.

7.1   INTRODUCTION

Fluctuations in thermodynamics automatically imply the existence of an underlying structure that has created them. We know that such structure is comprised of molecules, and that their large number allows statistical studies, which, in turn, allow one to relate various statistical moments to macroscopic thermodynamic quantities. One of the purposes of the statistical theory of liquids (STL) is to provide such relations for liquids (Frisch and Lebowitz 1964; Gray and Gubbins 1984; Hansen and McDonald 2006). In such theories, many macroscopic quantities appear as limits at zero wave number of the Fourier transforms of statistical correlation functions. For example, the Kirkwood–Buff theory allows one to relate integrals of the pair density correlation functions to various thermo-physical properties such as the isothermal compressibility, the partial molar volumes, and the density derivatives of the chemical potentials (Kirkwood and Buff 1951). If one wants a connection between detailed correlations and integrated moments, one may ask about the nature of the wave-number dependence of these quantities. It turns out that the statistical theory of liquids allows an answer to such a question very precisely, which leads to new types of questions. The Ornstein–Zernike equation (Hansen and McDonald 2006), which is an exact equation of the STL, introduces the concept of correlation length ξ, which relates to the spatial extension of the density and/or concentration (the latter in the case of mixtures) fluctuations. This quantity cannot be accessed from pure thermodynamical quantities, and requires a statistical theory and the concept of wave-number-dependent quantities.

The same way fluctuations emerge from a background of constituent molecules, one may ask if a new type of object can also emerge from a fluctuating statistical collection of such molecules. The answer turns out to be yes, and the microheterogeneity of aqueous mixtures is precisely such a type of emergent new object (Perera et al. 2012). This emergence requires very specific interactions, such as, for example, the hydrogen bond interaction. In the absence of such interactions, the system is just subjected to density or concentration fluctuations. These fluctuations can only alter the system in one way: a phase transition. In the presence of specific interactions, however, some systems can undergo a new type of transition, sometimes called an *arrested phase transition*, which is *not* a true phase transition, whose exact statistical nature is still largely unknown, and remains an intense subject of research in domains related to glasses and gels (Foffi et al. 2005). The microheterogeneity of aqueous mixtures belongs to this latter category of statistical phenomena, and therefore differs from pure concentration fluctuations. This may be a puzzling concept at first since it is commonly admitted that microheterogeneity is just another manifestation of fluctuations (Franks and Ives 1966), since it appears to occur at zero-wave vector. It is by considering wave-number dependence that one can distinguish them properly. The fact that fluctuations and local emergent quasi-particles compete against one another is actually quite common in high-energy physics (Wen 2004). The particular state where both phenomena coexist is called the *Lifshitz state*, and is also found in microemulsions (Ciach and Gozdz 2001). Microemulsions are just a particular type of aqueous solutions (Davis 1993), where fluctuations and microheterogeneity scale up to micrometer sizes (Poland and Scheraga 1965), because of the larger size of the solute-surfactant molecules, which are usually large alcohol chains. Aqueous alcohol binary mixtures have attracted a great deal of attention because of their particular tendency to clustering and microsegregation (Dixit et al. 2002; Guo et al. 2003; Soper et al. 2006; Koga 2007).

It is the purpose of this chapter to sketch a statistical approach of wave-number-dependent fluctuations within the background of the STL, which goes beyond the correlation length concept, by incorporating in particular high order moments of the pair direct correlation function, at the level of description of molecular liquids. This approach permits one to unify the Teubner–Strey (TS) approach of microemulsions (Teubner and Strey 1987) to the usual Ornstein–Zernike approach of molecular emulsions in general.

7.2 STATISTICAL THEORY OF LIQUIDS: CONDITIONS FOR THE EXISTENCE OF MORPHOLOGY IN LIQUIDS

A liquid such as water is very different from a typical simple liquid such as liquid argon, for example. The hydrogen bonding ability of the water molecule allows it to build a network, which gives a very specific structure to liquid water that is missing in any simple liquid (Perera 2011). This hydrogen bonding energy is highly directional, and it leads to the exceptional properties of liquid water (Ball 2008). It turns out that some effective isotropic interactions can also mimic the specific structure of water, such as the various Jagla type interactions (Jagla 1999). All such isotropic effective interactions introduce a second length scale in the interaction to compensate for the spherical symmetry of this interaction (Perera 2009). This is a strong indication that there is a link between the symmetry breaking of the isotropic interaction and the appearance of morphology through additional length scales.

7.2.1 CORRELATIONS AND STABILITY

Let us sketch with a simple example what we mean by morphology of a liquid. Simple liquids and their mixture are essentially random disposition of the molecules; they have no morphology. On the other end, liquids that have undergone a full phase transition, such as liquid–vapor or liquid–liquid phase separation, have a clear interface, thus a simple morphology with one phase neatly separated from the other. Microsegregation is the example of liquid with nontrivial morphology. Microsegregated liquids look very much like simple liquids in a nonequilibrium situation stuck in the two-phase region. These are subject to nucleation processes that irreversibly drive small nuclei to full phases. This is the realm of fluctuations. Microsegregated liquids are also the siege of fluctuations, but it is the domains that fluctuate without driving a full phase separation. This situation makes it difficult to decide how to distinguish a phase-separating liquid from a microsegregated liquid. The STL provides a criterion for the mechanical stability of liquids that can be expressed in terms of the correlation functions.

7.2.1.1 Liquid State Theory for One Component Simple Liquid

For simplicity, we consider here a monoatomic liquid where two spherical particles interact through a distance-dependent interaction $u(r)$. The STL deals with the pair radial distribution function (RDF) $g(r)$ and the direct correlation function (DCF) $c(r)$. We will define these two functions in Section 7.3 in a more general context. For the simple liquid of interest, these two functions are related through two *exact* equations (Hansen and McDonald 2006), which are the OZ equation,

$$h(r) = c(r) + \rho \int d\vec{r}'h(|\vec{r} - \vec{r}|)c(r')$$

(7.1)

where $h(r) = g(r) - 1$, and the closure equation,

$$g(r) = \exp\left[-\beta u(r) + h(r) - c(r) + b(r)\right]$$

(7.2)

and the new function $b(r)$ is the so-called bridge function which contains contributions from higher rank correlations than pair correlations (Perera 2009). The OZ equation is more convenient to handle when it is Fourier transformed, with the definition

$$\hat{f}(\mathbf{k}) = \int d\mathbf{r}\exp(i\mathbf{k}\cdot\mathbf{r})f(\mathbf{r})$$

(7.3)

which transforms Equation 7.1 into,

$$\hat{S}(\mathbf{k}) = 1 + \rho\hat{h}(\mathbf{k}) = \frac{1}{1 - \rho\hat{c}(\mathbf{k})} \tag{7.4}$$

where $\hat{S}(\mathbf{k})$ is the structure factor. The isothermal compressibility κ_T, which is related to the fluctuation in the number of particles N in an open system, is also related to the denominator of Equation 7.4 through the relation (Hansen and McDonald 2006),

$$\frac{\langle N^2 \rangle - \langle N \rangle^2}{\langle N \rangle} = \kappa_T^* = \hat{S}(0) = \frac{1}{1 - \rho\hat{c}(0)} \tag{7.5}$$

where $\kappa_T^* = \rho k_B T \kappa_T$ is the reduced isothermal compressibility. Equation 7.5 is the typical form of the fluctuation–dissipation theorem, which relates fluctuations to the appropriate integral of correlation functions. This theorem is at the heart of statistical mechanics and provides the precious link between the macroscopic thermodynamics and the microscopic properties at small wavelengths. This is however not the end of the story, since the structure factor contains rich information as one approaches large wavelengths through zero wave-numbers k. The key relation to this is the exact result that states that the direct correlation function has the following asymptotic behavior,

$$\lim_{r \to \infty} c(r) = -\beta u(r) \tag{7.6}$$

Since all known interactions are analytic integratible functions, this means that, from Equation 7.3, the following Taylor expansion exists,

$$\tilde{c}(k) = c_0 + k^2 c_2 + k^4 c_4 + o\left(k^6\right) \tag{7.7}$$

where the moments c_n are defined as,

$$c_n = \frac{(-1)^{n/2}}{(n+1)!} \int d\mathbf{r}\, \mathbf{r}^n c(\mathbf{r}) \tag{7.8}$$

If we retain the expansion to second order in Equation 7.7, and examine the corresponding limit of the OZ Equation 7.4, we find,

$$\lim_{k \to \infty} \hat{S}(k) = \frac{A}{k^2 + \xi^{-2}} \tag{7.9}$$

where $A = -1/\rho c_2$ and $\xi = \sqrt{-\rho c_2 \kappa_T}$ is the correlation length. Equation 7.9 gives the following Yukawa form for the universal decay of the pair correlation function $h(r)$,

$$\lim h(r) = \frac{A}{r} \exp\left(\frac{-r}{\xi}\right) \tag{7.10}$$

The asymptotic relations (Equations 7.9 and 7.10) are exact for any one-component liquid with spherical interactions, except when the Taylor expansion (Equation 7.7) fails to hold, which is near a critical point (Fisher 1964). At the critical point, the DCF becomes a nonanalytic function, and the correlations develop a nonexponential algebraic decay in $1/r^{1+\eta}$ where $\eta \approx 0.041$ is a critical exponent (Fisher 1964). In all that follows, we will stay away from criticality, hence the OZ forms, Equations 7.9 and 7.10, are essentially exact. When the critical point is approached, the isothermal compressibility diverges, and the analysis above shows that the correlation length diverges as the square root of the compressibility. The divergence of ξ leads to a Coulomb decay of the pair correlation in Equation 7.10, but the correct exponent is slightly faster than pure Coulomb decay.

7.2.1.2   Liquid State Theory for Binary Mixtures of Simple Liquids

The formalism described above applies equally to any mixtures of simple liquids. The form of the decay is very similar to that obtained above, except that the prefactor A, and the correlation length ξ, are complicated expressions. The important point is the following: while the prefactor depends on the species pairs, the correlation length is the same for all species pairs. In other words, all pair correlation functions decay in an identical fashion. We will demonstrate this in a very general case in Section 7.3. Here, we examine the case of isotropic interactions and binary mixtures. If we label the two species as 1 and 2, we have 3 independent interactions: $u_{11}(r)$ and $u_{22}(r)$ between like species 1 and 2, respectively, and $u_{12}(r) = u_{21}(r)$ between unlike species. Similarly, there will be three RDFs: $g_{11}(r)$, $g_{22}(r)$, and $g_{12}(r)$, and three DCFs: $c_{11}(r)$, $c_{22}(r)$, and $c_{12}(r)$. All these functions are related to each other through the two same exact equations, the OZ equation written in the Fourier space as

$$\hat{h}_{ij}(k) = \hat{c}_{ij}(k) + \sum \rho_n \hat{h}_{in}(k) \hat{c}_{nj}(k) \tag{7.11}$$

and the closure equations,

$$g_{ij}(r) = \exp\left(-\beta u_{ij}(r) + h_{ij}(r) - c_{ij}(r) + b_{ij}(r)\right) \tag{7.12}$$

The OZ equation contains the partial number densities that are defined as $\rho_i = N_i/V$, where N_i is the number of molecules of species i within the volume V. One can also define the mole fraction of species i through $\rho_i = x_i\rho$. The OZ equation can be conveniently cast into the following matrix form,

$$\hat{S}\left(I - \hat{C}\right) = I \tag{7.13}$$

where $\mathbf{I}$ is the identity matrix, $\hat{\mathbf{S}}$ is the matrix of the structure factors with elements $\hat{S}_{ij}(k) = \delta_{ij} + \hat{H}_{ij}(k)$, and the matrices $\hat{\mathbf{H}}$ and $\hat{\mathbf{C}}$ are generically defined through $\hat{A}_{ij}(k) = \sqrt{\rho_i \rho_j}\,\hat{a}(k)$. Equation 7.13 looks just like Equation 7.4 and this can be used to get to the conclusions similar to those of the one component simple liquid in the previous paragraph.

First of all, Equation 7.6 holds for any interaction pairs. Therefore, one can expand the DCFs similarly to Equation 7.7,

$$\hat{c}_{ij}(k) = c_{0;ij} + k^2 c_{2;ij} + k^4 c_{4;ij} + o\left(k^6\right) \tag{7.14}$$

which can be written in matrix form as

$$\hat{C} = C_0 + k^2 C_2 + k^4 C_4 + o\left(k^6\right) \tag{7.15}$$

with the moment matrix defined through elements

$$C_{n;ij} = \sqrt{\rho_i \rho_j}\, c_{n;ij} = \sqrt{\rho_i \rho_j}\, \frac{(-1)^{n/2}}{(n+1)!} \int d\mathbf{r}\, r^n c_{ij}(\mathbf{r}) \tag{7.16}$$

which is analogous to Equation 7.8. When applied to a two-component system, one gets from Equation 7.13 the same form as Equation 7.8 for the small-k behavior of the structure factors,

$$\lim_{k \to 0} \hat{S}(k) = \frac{A}{k^2 + \xi^{-2}} \tag{7.17}$$

where

$$A_{ij} = \frac{\delta_{\underline{i}\underline{j}} - \hat{c}_{0;\underline{i}\underline{j}}}{-\gamma} \tag{7.18}$$

with the notation $\underline{i}$ for swapping species (i.e., $\underline{1} = 2$ and $\underline{2} = 1$),

$$\xi = \sqrt{\frac{-\gamma}{|\mathbf{I} - \mathbf{C}_0|}} \tag{7.19}$$

and

$$\gamma = (1 - C_{0;11}) C_{2;22} + (1 - C_{0;22}) C_{2;11} + C_{0;12} C_{2;12} \tag{7.20}$$

One sees that there is only one correlation length and this is equally true for any number of components. Following Equation 7.17, the decay of the pair correlations has the same Yukawa form as in Equation 7.10,

$$\lim h_{ij}(r) = \frac{A_{ij}}{r} \exp\left(\frac{-r}{\xi}\right) \tag{7.21}$$

The isothermal compressibility of a mixture is given by the expression

$$\kappa_T = \frac{\left|\hat{\mathbf{S}}(0)\right|}{\sum_{i,j}\sqrt{x_i x_j}\ \hat{\mathbf{S}}^{ij}} \tag{7.22}$$

an expression that trivially differs from that found in Hansen and McDonald because of the definition of the structure factor we have adopted above, a choice that keeps compatibility with the form of the OZ equation (Hansen and McDonald 2006). By using the OZ equation, we find an equivalent expression in terms of the DCFs,

$$\kappa_T^* = \frac{1}{1-\rho\sum x_i x_j \hat{c}_{ij}(0)} \tag{7.23}$$

which is identical to that reported in Hansen and McDonald (2006). In particular, we see that for a mixture, the isothermal compressibility is not related to the $\hat{\mathbf{c}}$ instability of the isotropic phase. The latter is governed by the determinant $\left|\mathbf{I}-\hat{\mathbf{C}}(0)\right|$ = $1/\left|\hat{\mathbf{S}}(0)\right|$, which is also related to the fluctuation of the number of particles through the relations:

$$\hat{S}_{ij}(0) = \frac{\langle N_i N_j \rangle - \langle N_i \rangle \langle N_j \rangle}{\sqrt{\langle N_i \rangle \langle N_j \rangle}} \tag{7.24}$$

We also see that the correlation length in Equation 7.19 diverges with the divergence of the correlations through the term $\left|\hat{\mathbf{S}}(0)\right|$. Another important link with thermodynamics is provided by the Kirkwood–Buff theory (Kirkwood and Buff 1951),

$$\rho\sqrt{x_i x_j}\left(\frac{\partial \beta\mu_i}{\partial \rho_j}\right)_{T,V,\{x\}'} = \delta_{ij} - \rho\sqrt{x_i x_j}\hat{c}_{ij}(0) \tag{7.25}$$

Equation 7.9 and Equation 7.19 reveal an important feature that is completely missed in any fluctuation theory, that the correlation length is related to the second moment of the DCF (Fisher 1964), while all fluctuation theories concern only zero-order moments. Coming back to the morphology of liquids, we see that the correlation length describes the amount of correlation inside disorder, but it is clearly insufficient to describe morphology. Let us see how we can bring up the concept of morphology through the study of simple liquid models.

7.2.2 Correlations and Fluctuations in a Simple Liquid with Repulsive Interactions

To make things simple, let us consider the following pair interaction between spherical particles of diameter σ, with $a > 0$,

$$u(r) = \begin{cases} +\infty & r < \sigma \\ a \exp(-\dfrac{r}{\lambda}) & r \geq \sigma \end{cases} \tag{7.26}$$

We can use the analytical expression for the DCF from the Percus–Yevick (PY) theory (Hansen and McDonald 2006) to compute the direct correlations inside the core region $r < \sigma$, and just use Equation 7.6 outside the core. This is in the spirit of the mean spherical approximation (MSA) (Hansen and McDonald 2006). This approximation reproduces a structure factor in relatively good agreement with that of the hard spheres (HS) as computed from the PY approximation, as can be seen in Figure 7.1, which itself is in good agreement with the computers for dense fluids. The resulting first three moments of the DCF can be computed following Equation 7.8. These moments are shown in Figure 7.2. The zero-order moment c_0 is always negative while the second moment c_2 is always positive when $a > 0$, so the discussion above on the possibility of an unstable fluid phase does not apply for this

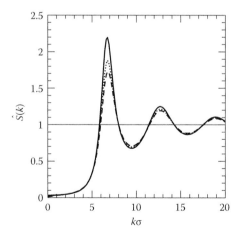

FIGURE 7.1   Structure factor for the hard sphere fluid versus wave vector times the diameter of the sphere σ. Full curve for PY approximation, dotted curve for the model used in the text with $a = 1$ and $\kappa = 1$, dashed curve for $a = 1$ and $\kappa = 2$.

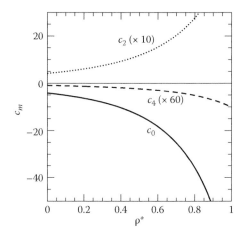

FIGURE 7.2 Moments of the direct correlation function for the HS fluid model with ($a = 1$, $\kappa = 1$) as a function of the reduced density defined as $\rho^* = (N/V)\sigma^3$, where N is the number of particles per volume V and σ is the diameter of the hard spherical particles. Full curve is for zero moment, dotted curve for second moment (up to a factor 10) and dashed curve for fourth moment (up to a factor 60).

type of system. Figure 7.2 confirms that hard-core models cannot predict diverging correlations. For repulsive interactions, the RDF is oscillatory at all densities: the correlations are dominated by packing effects. So such types of liquids have no morphology, so to speak, they are just random packing of spheres.

Mixture of hard and soft spheres can undergo a phase separation if the size ratio is large enough (Biben and Hansen 1991). However, the most recent results suggest that this phase separation occurs when the larger spheres crystallize (Coussaert and Baus 1997). Crystallization is an entirely different phenomenon and we are concerned only with liquids here. So we will not venture further into considering morphology of such repulsive interactions.

7.2.3 Correlations and Fluctuations in a Simple Liquid with Attractive Interactions

We consider now the attractive version of the above Yukawa interaction (Equation 7.26), with this time $a < 0$. Figure 7.3 shows the structure factors for different values of a, and one sees that density fluctuations can increase dramatically and eventually destabilize the phase toward a liquid–gas phase separation. Figure 7.4 shows the first three moments of the DCF as function of the density as well as the critical denominator $1 - \rho\hat{c}(0)$, which is seen to reach zero at the two spinodal points corresponding to the gas and liquid phases, respectively. It is also seen that the correlation length diverges at these points, indicating that these models are governed by density fluctuations. Similarly, one could devise interactions for two-component systems that would also undergo either a gas–liquid or a liquid–liquid phase separation, governed by the density and concentration fluctuations, respectively.

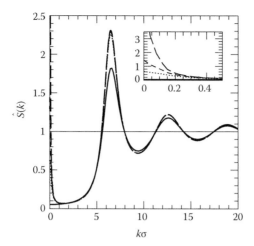

FIGURE 7.3   Structure factor for the model with attractive interactions. Full curve for ($a = 1$, $\kappa = 1$) as a reference, dotted curve for ($a = 1$, $\kappa = 1.5$), and dashed curve for ($a = 0.95$, $\kappa = 1.5$). The inset shows the small-k behavior.

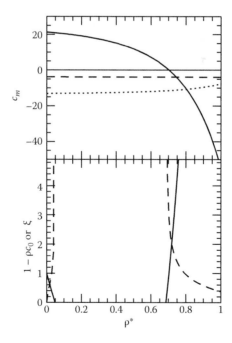

FIGURE 7.4   Upper panel, moments of the DCF for the attractive interaction model with parameters ($a = 1$, $\kappa = 1$), as a function of the density (full for $m = 0$, dotted for $m = 2$ and dashed for $m = 4$). Lower panel, behavior of the inverse compressibility (full) and correlation length (dotted) corresponding to the upper panel.

7.2.4 CORRELATIONS, FLUCTUATIONS, AND MORPHOLOGY IN A MODEL COMPLEX LIQUID WITH SPHERICAL INTERACTIONS

Let us now consider a model for which some microstructure can appear. It is clear from the previous examples that microstructure cannot appear from either a purely repulsive or a purely attractive model, when the length scale of the interaction is only that of the size of the particles. Such models are governed by packing effects and fluctuations. In order to make the model more complex, we need to mix both repulsion and attraction in order to introduce a new distance parameter within the range of the competition between attractive and repulsive interactions. The most realistic way to achieve this is to include an attractive interaction at short range and a repulsive attraction beyond. This way, we encourage particles to cluster, but not beyond a certain range. This type of model was in fact used to mimic the experimentally observed aggregation in globular protein solutions and colloid-polymer solutions (Stradner et al. 2004). It was observed that varying the salt concentration in the first solution and that of the polymers in the second, one could tune the ratio of attraction versus repulsion, hence producing aggregation of the globular protein or the colloids. The aggregation was monitored through the appearance of a prepeak in the neutron scattering experiments. This work triggered intense research to find an interaction mechanism that would produce such a prepeak. By modeling these two systems as an effective one-component system, with particles interacting through a short-range attraction and long-range repulsion, one can indeed reproduce the rich phase behavior of the experimental systems (Broccio et al. 2006; Archer and Wilding 2007).

The experimental systems are not a one-component system and contain, in particular, water, salt, and polymers. It is the effective interaction resulting from the contributions of the other components that provide the combination of short-range attraction and long-range repulsion. For now, we will not worry about how long-range repulsion can appear and we shall focus on the shape of the structure factor when particles aggregate into domains.

Let us consider a one-component system with hard spherical particles interacting through two additional Yukawa interactions,

$$u(r) = \begin{cases} +\infty & r < \sigma \\ \exp\left(-\dfrac{r}{\kappa_1}\right)\Big/ r - a\exp\left(-\dfrac{r}{\kappa_2}\right)\Big/ r & r \geq \sigma \end{cases} \qquad (7.27)$$

Such interactions are shown in Figure 7.5 in order to describe the purely attractive case, as well as the weak and strong repulsions at large distances. In order to understand how clustering can emerge from this type of interaction, we now treat the direct correlation function associated with this model in the mean spherical approximation fashion, which amounts to setting

$$c(r) = -\beta u(r) \qquad r \geq \sigma \qquad (7.28)$$

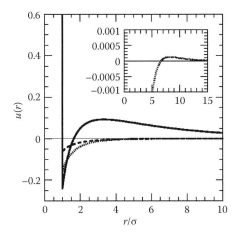

FIGURE 7.5   Yukawa interaction model. Full line for parameters ($a = 1.1$, $\kappa_1 = 2$, $\kappa_2 = 2$), dotted line for ($a = 1.3$, $\kappa_1 = 2$, $\kappa_2 = 1.85$), and dashed line for ($a = 1.85$, $\kappa_1 = 8$, $\kappa_2 = 2$).

and we will set the direct correlation inside the core to −1, which is the Mayer limit in the case of an ideal liquid (zero-density limit). A better MSA would be setting the core value to that of the Percus–Yevick approximation, just like we did in the previous paragraphs. The resulting direct correlation function in the Fourier space is

$$\hat{c}(k) = -\frac{4\pi}{k^3}\left[\sin(k) - k\cos(k)\right] - \hat{Y}_1(k) + a\hat{Y}_2(k) \tag{7.29}$$

with

$$\hat{Y}_i(k) = \frac{4\pi\kappa_i}{k\left(1 + k^2\kappa_i^2\right)}\left[\sin(k) + k\kappa_i\cos(k)\right]\exp\left(-\frac{1}{\kappa_i}\right) \tag{7.30}$$

The corresponding structure factors are plotted in Figure 7.6 for different values corresponding to three cases of small and tight clustering, large and soft clustering, and no clustering illustrated in Figure 7.5. In the latter case, it is seen that the structure factor exhibits high fluctuations as seen by the increase of $S(k)$ near $k = 0$, and due to the purely attractive global interaction. These are predomain forming fluctuations, a state that is called the *Lifshitz state* in several areas of physics ranging from microemulsions to high-energy physics (Ciach and Gozdz 2001). When a small repulsion is added, as seen in the inset, the small-k raise of the structure factor detaches into a prepeak, with still large fluctuations present in the system. This corresponds physically to large and soft clusters, where fluctuations rule the cluster formation. When the repulsion becomes predominant, the barrier between clusters increases and will stabilize them against fluctuations, making these latter very small

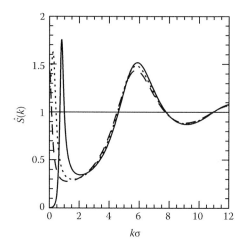

FIGURE 7.6   Structure factors for the two Yukawa interaction models of Figure 7.5.

as can be seen by the corresponding structure factor. In all these three cases, the main peak of the structure has remained unchanged and positioned at $k_m = 2\pi/\sigma$, corresponding to the size of the constituent particles. The prepeak witnesses the appearance of a new entity in the system, with size corresponding to $d = 2\pi/k_P$, where k_P is the wave vector associated with the prepeak. We claim that the microheterogeneity in aqueous mixtures is very similar to the physical phenomena we have just described, and corresponds to a situation close to the Lifshitz state or the large soft cluster case.

What is remarkable is that such a simple model is able to predict a prepeak in the structure factor in agreement with the physical requirement governed by the two parameters κ_i. The reason for the prepeak is, in fact, very simple. A single Yukawa potential provides a DCF that is essentially parabolic at small k, but two opposing Yukawa interactions can give a nonnegligible k^4 contribution because of the competition from the repulsive Yukawa. It is this contribution that gives a deviation from the standard OZ behavior we described in Equation 7.17 and favors the appearance of the higher-order k-vectors in the denominator. This k^4 term produces the prepeak and introduces an additional length scale, in addition to the OZ correlation length that we would have if we were restricted to the k^2 term only. This example shows how a simple one-component system can develop a non-OZ behavior. Hence, it indicates that fluctuations and domain size can appear for competing interaction.

In fact, the repulsive interaction is an artifact that we have introduced in order to mimic the screening effect of the salt or the polymer. In a full description, the k^4 term would appear naturally without the need to introduce a repulsive interaction. Is it possible to introduce such a term with only spherical interactions—through, for example, a binary mixture? It turns out that what is really required for morphology is an interaction with two distances. So, it is always possible to create such an interaction in model systems. The Jagla model for water-like systems is such a model (Jagla 1999). However, in real systems, the appearance of a second distance in the interaction seems to require a nonspherical symmetry breaking interaction. Water is a good example of such a nonspherical interaction that contains the potential for creating morphology. Aqueous mixtures are microheterogeneous because of water (Dixit et al. 2002; Allison et al. 2005). Similarly, polymers are highly nonspherical objects and their inclusion equally breaks the spherical symmetry of colloidal particles. The STL of nonspherical interactions is therefore required if one wants to further explore domain formation in realistic fluctuating systems.

7.3   STATISTICAL THEORY OF MOLECULAR LIQUIDS

Here, we formulate the molecular version of the STL for an arbitrary number of components, but the practical results will be restricted to binary mixtures.

7.3.1   A SURVEY OF THE FORMALISM

We consider a binary mixture made of two species of nonspherical molecules, and we will consider that each molecule is made of spherical sites, which is a convenient approximation for most realistic systems. Two such molecules will then interact through the sum of all their site–site interactions terms. We will consider the generic interaction where each site–site interaction is the sum of a Lennard–Jones (LJ) interaction and a Coulomb interaction if the sites bear partial charges localized at their centers. Each of the molecular site i_a belonging to species a is then characterized by its diameter σ_{i_a}, its LJ energy parameter ε_a, and its partial charge q_{i_a}. The total interaction between any two molecules, labeled 1 and 2, is then

$$u(1,2) = \sum u_{i_a j_b}\left(r_{i_a j_b}\right) = \sum 4\,\varepsilon_{i_a j_b}\left[\left(\frac{\sigma_{i_a j_b}}{r_{i_a j_b}}\right)^{12} - \left(\frac{\sigma_{i_a j_b}}{r_{i_a j_b}}\right)^{6}\right] - \frac{q_{i_a} q_{j_b}}{r_{i_a j_b}} \qquad (7.31)$$

where $r_{i_a j_b}$ is the distance between the two sites i and j on molecules of species a and b, respectively. The first term is the LJ term while the second is the Coulomb interaction term. The LJ energy and size parameters are usually defined empirically by the following rules, the arithmetic mean is called the *Lorentz rule* $\sigma_{i_a,j_b} = (\sigma_{i_a} + \sigma_{j_b})/2$ and the geometrical mean is called the *Berthelot rule* $\varepsilon_{i_a j_b} = \sqrt{\varepsilon_{i_a}\varepsilon_{j_b}}$. The interaction in Equation 7.31 can also be written in terms of the distance $\mathbf{r} = \mathbf{r}_2 - \mathbf{r}_1$ between the center of masses of any two molecules 1 and 2, as well as their respective orientations through the unit vectors, Ω_1 and Ω_2, each describing molecular orientations through the set of Euler angles $(\theta_i, \beta_i, \varphi_i)$ defined in the lab fixed frame (Gray and Gubbins 1984). Since any site i of a given molecule can be related to the center of mass at position $\mathbf{r}_i$ through the relation $\mathbf{r}_{i_a} = \mathbf{r}_i - \mathbf{I}_{i_a}(\Omega)$, where the last term is the position of the site in the reference frame attached to the molecule, one can indeed rewrite Equation 7.31 as

$$u(1,2) = u(\mathbf{r}, \Omega_1, \Omega_2) = u(r, \Omega_1, \Omega_2, \mathbf{r}/r) \qquad (7.32)$$

Both Equation 7.31 and Equation 7.32 contain the same information, but they lead to two different ways of handling the correlations. Indeed, one can define two types of pair correlation functions through the same statistical ensemble average: either the full angular distribution function,

$$g(1,2) = \left\langle \exp(-\beta u(1,2)) \right\rangle \tag{7.33}$$

or the site–site pair distribution functions,

$$g_{i_a j_b}(r) = \left\langle \exp\left(-\beta u_{i_a j_b}\left(r_{i_a j_b}\right)\right) \right\rangle \tag{7.34}$$

where the bracket indicates a canonical or grand canonical ensemble average. It is clear that the second expression contains less information than the previous one, because of the averaging of the partial interactions. The first formulation leads to a proper statistical mechanical formalism, with equivalent exact OZ and closure equations as in Equations 7.11 and 7.12, while the second formulation leads to approximate equivalent relations, in the context of the so-called Reference Interaction Site Models (RISM) (Gray and Gubbins 1984; Hansen and McDonald 2006). We will not venture into this second path here, since we rely on the form of the OZ equation in the molecular formulation, and that the equivalent site–site OZ (SSOZ) is plagued with various issues related to its inherent approximate nature (Kežić and Perera 2011). In particular, it is not possible to derive the correct critical behavior from a small-k expansion of the SSOZ. This is currently an open area of research as how to incorporate a proper diagrammatic formulation into the RISM-SSOZ formulation.

7.3.2 The Molecular OZ Approach to Microheterogeneity

7.3.2.1 The Molecular OZ Formalism

If we refer to the molecular formulation, then one can derive the molecular Ornstein–Zernike (MOZ) equation by following the standard STL formalism, which can be written in Fourier space as (Hansen and McDonald 2006)

$$\hat{h}_{ij}(1,2) = \hat{c}_{ij}(1,2) + \frac{1}{\omega} \sum_n \rho_n \int d\hat{\mathbf{h}}_{in}(1,3)\hat{\mathbf{c}}_{nj}(3,2) \tag{7.35}$$

with the same notation as in Equation 7.32 but with Fourier vector,

$$\hat{h}_{ij}(1,2) = \hat{h}_{ij}(\mathbf{k}, \mathbf{\Omega}_1, \mathbf{\Omega}_2) \tag{7.36}$$

and with $\omega = \int d\Omega$. Similarly, the exact molecular closure equation can be derived as

$$g_{ij}(1,2) = \exp\left[-\beta u(1,2) + h(1,2) - c(1,2) + b(1,2)\right] \tag{7.37}$$

The key point to handle these two equations was formulated by Blum and Torruela (1972) in a series of now landmark papers by using a rotational invariant expansion that would allow a separation between the radial and the angular dependence, the latter being handled through the Wigner elements $R_{\mu\nu}^{l}(\Omega)$, which are a generalization of the spherical harmonics to all three Euler angles (Gray and Gubbins 1984). This procedure creates an infinite series of "projections" $a_{ij:\mu\nu}^{mnl}(r)$ of the various correlation functions $a(1,2)$, both in r-space and Fourier space, for an expansion that is adapted to the rotationally invariant isotropic phase. The number integer of indexes $(m,n,l;\mu,\nu)$ depends on the symmetry of the phase. Without getting into such unnecessary complications, since we will be mainly considering macroscopic homogeneous and isotropic phase, the generic expansion reads:

$$a(1,2) = \sum a_{\mu\nu}^{mnl}(r)\varphi_{\mu\nu}^{mnl}(1,2) \tag{7.38}$$

where the "invariants" are defined as

$$\varphi_{\mu\nu}^{mnl}(1,2) = \varphi_{\mu\nu}^{mnl}(\Omega_1,\Omega_2,\mathbf{r}/r) = \sum_{\mu'\nu'\lambda'}\begin{pmatrix} m & n & l \\ \mu' & \nu' & \lambda' \end{pmatrix}R_{\mu\mu'}^{m}(\Omega_1)R_{\nu\nu'}^{n}(\Omega_2)R_{\lambda'0}^{l}(\mathbf{r}/r) \tag{7.39}$$

where the matrix element is a Racah 3-j symbol (Gray and Gubbins 1984), and $\underline{\lambda} = -\lambda$, a notation that we shall use throughout. The form of the invariant is entirely dictated by the symmetry of the phase. Note that the invariant does not carry the species pair indexing, since it depends only on the angles. This notation allows one to rewrite both exact Equation 7.35 and Equation 7.37 solely in terms of the projections $a_{ij:\mu\nu}^{mnl}(r)$, hence discarding all the angular dependence. The MOZ Equation 7.35 can be rewritten into the convenient matrix form, if one takes care of expressing Equation 7.35 in the intermolecular frame, where the vector $\hat{k}$ is aligned with the local z-axis. Since the mixture is isotropic and homogeneous, this operation does not alter the symmetry contained in the expansion of the type given in Equation 7.38, but it takes now a new form in the molecular frame (Perera and Patey 1988),

$$\hat{h}_{ij}(1,2) = \sum_{mn\chi\mu\nu} \tilde{h}_{ij:\mu\nu}^{mn\chi}(k)R_{\mu\chi}^{m}(\Omega_1)R_{\nu\underline{\chi}}^{n}(\Omega_2) \tag{7.40}$$

where the orientation dependence on $\mathbf{k}$ has been lifted. The new expansion coefficients $\hat{h}_{ij:\mu\nu}^{mn\chi}(k)$ are related to those in the lab fixed frame by a simple linear relation (Fries and Patey 1985),

$$\hat{h}_{ij:\mu\nu}^{mn\chi}(k) = \sum_{l}\begin{pmatrix} m & n & l \\ \chi & \underline{\chi} & 0 \end{pmatrix}\hat{h}_{ij:\mu\nu}^{mnl}(k) \tag{7.41}$$

In this new molecular new reference frame, the MOZ equation becomes a compact matrix equation (Kusalik and Patey 1988),

$$\hat{\mathbf{H}}_\chi = \hat{\mathbf{C}}_\chi \left[\mathbf{I} - (-1)^\chi \hat{\mathbf{C}}_c' \right]^{-1} \tag{7.42}$$

where each matrix is a $n_c \times n_c$ block matrix where n_c is the number of species in the mixture, and the new matrix $\hat{\mathbf{C}}_\chi'$ is related to the initial matrix $\hat{\mathbf{C}}_\chi$ by the following relation between their matrix elements,

$$\hat{C}_{ij;\mu\nu}^{t;mn\chi}(k) = (-1)^\mu \hat{C}_{ij;\mu\nu}^{mn\chi}(k) \tag{7.43}$$

For a two-component system, the typical 2×2 block matrix reads

$$\hat{\mathbf{H}}_\chi = \begin{pmatrix} \hat{\mathbf{H}}_{11} & \hat{\mathbf{H}}_{12} \\ \hat{\mathbf{H}}_{12} & \hat{\mathbf{H}}_{22} \end{pmatrix} \tag{7.44}$$

where each submatrix $\hat{\mathbf{H}}_{ij}$ is defined by elements $\sqrt{\rho_i\rho_j}\,\hat{h}_{ij;\mu\nu}^{mn\chi}(k)$. Equation 7.42 can be used to derive the small-k form of the MOZ equation. The relation equivalent to Equation 7.6 holds in the molecular case,

$$c_{ij}(1,2) \xrightarrow[r\to\infty]{} -\beta u(1,2) \tag{7.45}$$

which implies that the partial DCFs are Taylor expandable around $k = 0$, since the pair interactions are integrable at large separations. In order to manage the expansion, we need to take a look at the Fourier–Hankel transforms of the projections (Blum and Torruella 1972; Fries and Patey 1985). These are defined as

$$\hat{c}_{ij;\mu\nu}^{mnl}(k) = 4\pi i^l \int\limits_0^{+\infty} dr\ r^2 j_l(kr) c_{ij;\mu\nu}^{mnl}(r) \tag{7.46}$$

Since the spherical Bessel functions have the expansion (Abramowitz and Stegun 1970)

$$j_l(x) = \sum_{n=0} \frac{x^{l+2n}}{(2l+n+1)!!} \frac{(-1)^n}{2^{2n} n!} \tag{7.47}$$

We get the following small-k expansion for the DCFs,

$$\hat{c}_{ij;\mu\nu}^{mnl}(k) = \sum_{t=0} k^{l+2t} c_{ij;\mu\nu}^{mnl(t)} \tag{7.48}$$

with the coefficients defined as

$$c_{ij;\mu\nu}^{mnl(t)} = 4\pi i^l \frac{(-1)^t}{(2l+t+1)!!2^{2t}t!} \int_0^{+\infty} dr \, r^{l+2(1+t)} c_{ij;\mu\nu}^{mnl}(r) \tag{7.49}$$

Since, for symmetry reasons, l is always even, the expansion is even in powers of k (Blum and Torruella 1972). Therefore, we can expand the matrices $\hat{\mathbf{C}}_\chi$ and $\hat{\mathbf{C}}_\chi^t$ in even powers of k,

$$\hat{\mathbf{C}}_\chi = \hat{\mathbf{C}}_\chi^{(0)} + k^2 \hat{\mathbf{C}}_\chi^{(2)} + k^4 \hat{\mathbf{C}}_\chi^{(4)} + \mathbf{O}(k^6) \tag{7.50}$$

Formally, Equation 7.42 can be cast into the form,

$$\hat{\mathbf{H}}_\chi = \hat{\mathbf{C}}_\chi \hat{\mathbf{M}}_\chi^{-1} \tag{7.51}$$

with $\hat{\mathbf{M}}_\chi^{-1} = \mathbf{I} - (-1)^\chi \hat{\mathbf{C}}_\chi^t$, which has the formal solution,

$$\hat{\mathbf{H}}_\chi = \frac{1}{\left|\hat{\mathbf{M}}_\chi\right|} \hat{\mathbf{C}}_\chi \hat{\mathbf{M}}_\chi^{cof} \tag{7.52}$$

Since all the terms in the right-hand side have a Taylor expansion, we get the following small-k expansion of the MOZ equation,

$$\hat{\mathbf{H}}_\chi = \frac{1}{\left|\hat{\mathbf{M}}_\chi^{(0)}\right| + k^2 \Gamma_\chi^{(2)} + k^4 \Gamma_\chi^{(4)}} \hat{\mathbf{C}}_\chi^{(0)} \hat{\mathbf{M}}_\chi^{(0),cof} \tag{7.53}$$

where the coefficients in the denominator depend on the expansion structure of the matrix $\hat{\mathbf{M}}_\chi$. This is the working expression we shall use in what follows.

7.3.2.2 Fluctuation-Induced Phase Separation in Molecular Binary Mixtures

Equation 7.53 allows us to understand what happens if the term $\tilde{\Gamma}_\chi^{(4)}$ is negligible when compared with the two others. One gets a form comparable to Equation 7.9,

$$\hat{\mathbf{H}}_\chi = \frac{1}{\left|\hat{\mathbf{M}}_\chi^{(0)}\right| + k^2 \Gamma_\chi^{(2)}} \hat{\mathbf{C}}_\chi^{(0)} \hat{\mathbf{M}}_\chi^{(0),cof} \tag{7.54}$$

which gives the usual exponential decay to all the correlation functions as in Equation 7.10 and Equation 7.21 in terms of a *unique correlation length*, regardless of the number of components. Indeed, the denominator can be cast into the form,

$$\hat{\mathbf{H}}_\chi = \frac{1}{\Gamma_\chi^{(2)}} \frac{\hat{\mathbf{C}}_\chi^{(0)} \hat{\mathbf{M}}_\chi^{(0),cof}}{k^2 + \xi^{-2}} \tag{7.55}$$

with the correlation length defined as

$$\xi = \sqrt{\frac{\Gamma_\chi^{(2)}}{\left|\hat{\mathbf{M}}_\chi^{(0)}\right|}} \tag{7.56}$$

The correlation length will diverge when the instability of the homogeneous isotropic phase is reached, that is, when $|\hat{\mathbf{M}}_\chi^{(0)}| = 0$, and phase separation would occur. This is very similar to what we have studied in the case of simple fluids, and one sees that the underlying statistical mechanism is very similar from the OZ point of view. The specificity of the interactions would affect the state point when this phase separation is likely to occur, but the mechanism stays the same.

But the MOZ formalism allows the next term in the denominator to have a non-trivial form, precisely because of the multiple contributions from all the projections. It is precisely because of such contributions that aqueous mixtures, in particular, can exhibit domain segregation without phase separation. We examine now under which conditions this could occur.

7.3.2.3   Domain Formation and Fluctuations in Molecular Binary Mixtures: The Teubner–Strey Approach to Microemulsion

Equation 7.53 shows that all correlation functions decay in the same manner through a common denominator. This decay can be cast into the generic form directly from Equation 7.53 as

$$\lim_{k \to 0} \hat{h}_{ij:\mu\nu}^{mn\chi}(k) = \frac{a_{ij:\mu\nu}^{mn\chi}}{a_2 + c_1 k^2 + c_2 k^4} \tag{7.57}$$

By introducing the following variables, the correlation length ξ and the domain size d, in association with the coefficients (a_2, c_1, c_2) in the denominator above in the following manner,

$$a_2 = \left(\bar{d}^2 + \xi^2\right)^2$$

$$c_1 = 2\left(\bar{d}\xi\right)^2\left(\bar{d}^2 - \xi^2\right) \tag{7.58}$$

$$c_2 = \left(\bar{d}\xi\right)^4$$

with the definition $\bar{d} = d/(2\pi)$, one can show that the resulting form of Equation 7.57 has the following exact inverse Fourier transform,

$$\lim_{r \to +\infty} h_{ij:\mu\nu}^{mn\chi}(r) \propto a_{ij:\mu\nu}^{mn\chi}\frac{\exp(-r/\xi)}{r}\sin\left(r/\bar{d}\right) \tag{7.59}$$

In other words, all the correlations decay in the usual OZ fashion, but with modulation associated with the domain size. The domain size is a unique parameter, just like the correlation length. It is the same for all species, regardless of their respective concentrations in the mixture.

In the above manipulation of the MOZ form at small-k vectors, we have followed the approach used by Teubner and Strey for explaining the existence of prepeaks in microemulsions (Teubner and Strey 1987). In the initial approach of these authors, they started from considerations on the form of a Landau–Ginzburg free energy Hamiltonian, containing gradients of the local density, considered as the order parameter. This is a phenomenological approach, suitable when analyzing universal behavior (Chaikin and Lubensky 2000). The present derivation of the Teubner–Strey form of the structure factor in k-space, as well as the decay in r-space, is more microscopic in nature since it starts from the STL formulation of the correlation functions. The derivation of Teubner and Strey is specifically adapted to microemulsions, which are about three orders of magnitude larger in scale than simple aqueous mixtures. Indeed, the size of micelles or lamellas is on the micron scale (Tanford 1974) when the segregated domains in simple aqueous mixtures have sizes around a few nanometers. The fact that the STL is able to describe both types of phenomena in a unified manner is not a surprise, since both are aqueous mixtures, differing only by the number of components and the size of the solute molecules. However, the phenomenological Teubner–Strey formalism explains the existence of domain modulations in microemulsions, as witnessed by a prepeak of the water–water structure factor.

It is not *a priori* obvious that, despite mathematical similarities, the expressions in Equation 7.57 and Equation 7.59 might also describe domain formations and appearance of prepeaks in simple aqueous mixtures. Recent small angle scattering experiments in binary aqueous mixtures of monools, diols, and triols (D'Arrigo, Giordano, and Teixeira 2009), show that only the latter two show clear non-OZ type decays. But the problem is deeper than this simple remark. It is well known, particularly since the work of Matteoli and Lepori, that some aqueous mixtures exhibit large KBIs (Matteoli and Lepori 1984). If KBIs represent solely concentration fluctuations, then domain formation can be considered as a variant of a concentration fluctuation. Following this line of thinking, a micelle, or a lamella, would just be a manifestation of concentration fluctuations, which is not very reasonable to assume. The MOZ formalism shows the way out of this dilemma. It indicates that large peaks that appear in the structure factors are not necessarily due to increasing fluctuations, but to the appearance of domains. These domains can be considered as *intrinsic concentration fluctuations*, but they are truly new objects, just like micelles or lamellas. The appearance of such new objects is witnessed by a new peak in the structure factor, just like the main peaks associated with the presence of each constituent molecule. When the concentration or other parameters are varied, the increase of the peak at $k = 0$ ceases and a prepeak emerges instead. The crossover point is called the *Lifshitz point* in the microemulsion jargon (Ciach and Gozdz 2001), but this name is also encountered in highbrow physics, such as high-energy physics and string theories (Wen 2004). Indeed, in such theories, particles are seen to emerge from the quantum fluctuations of the vacuum (Wen 2004), just like specific domains emerge from the fluctuations in aqueous mixtures. The STL allows a classical analogue of these manifestations for simple and complex aqueous mixtures. Nonaqueous mixtures, such as acetone-methanol, for example (Perera et al. 2011), can also be the siege of such manifestations. The formalism developed in this paragraph shows the complexity of the interplay between fluctuations and domain formations. Needless to say, a complex form of the interactions is a prerequisite, and the hydrogen bonding interaction, being highly directional, favors the formation of domains that compete with concentration fluctuations to reveal a complex physics.

It would be highly desirable to derive the conditions for the appearance of a pre-peak or a Lifshitz state directly from STL methods, such as the integral equation techniques, for example. Unfortunately, these methods are still very much approximate and in particular cannot handle complex liquids such as water. Much progress in improving these methods is required before we can think of using Equation 7.53 in more details. In the absence of better theories, we can explore simple models that can show us how domains emerge from fluctuating mixtures.

7.3.2.4 Simple Model of Molecular Emulsions

By analogy with microemulsions, we define here simple aqueous mixtures to be molecular emulsions, in view of the insights from the previous section. We consider now a binary mixture of the SPC/E water model (Berendsen, Grigera, and Straatsma 1987) with various "weaker" forms of the same model, where we have scaled down all the partial charges by the same factor λ. We consider three typical cases, $\lambda = 1/3$, which is a typical strong hydrophobic solute such as benzene, for example, $\lambda = 2/3$, which represents a moderately hydrophobic solute, like an alcohol or acetone, for example, and finally $\lambda = 4/5$, which represents a typical hydrophilic solute such as dimethyl sulfoxide or formamide, for example. We have studied these mixtures through molecular dynamics simulation techniques, under ambient conditions of pressure and temperature. We have considered $N = 2048$ molecules, which is a relatively large number of molecules for such simple mixtures. Details are given in Perera, Mazighi, and Kežić (2012). One of the principal interests in studying such simplified models of solutes is to eliminate the solute size considerations. Arguments from the Scaled Particle Theory indicate that the small size of the water molecule might be responsible for most of the aqueous mixtures thermodynamical properties, such as the hydrophobic effect, for example (Pohorille and Pratt 1990). The present models eliminate such considerations, since all these solutes are similar to water, both in size and geometry of the interactions. Only the Lennard–Jones energy parameters in Equation 7.31 must be adjusted in order to obtain liquid state for the neat solute under ambient conditions.

The $\lambda = 1/3$ mixture model demixes for most concentrations, except under solute mole fraction $x = 0.1$. This can be clearly seen in the snapshots of Figure 7.7. This model is principally governed by concentration fluctuations that become critical and lead to full demixing. The $\lambda = 4/5$ mixture model is fully miscible at all concentrations, as can be seen from the snapshots of Figure 7.8. There is a visible microsegregation, and it is water that segregates from the solute since it forms more compact clusters. This can be noticed by comparing the snapshots of low water and low

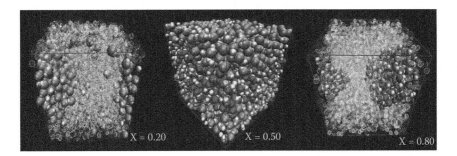

FIGURE 7.7 Snapshots of the $\lambda = 1/3$ mixture model which phase separates. The SPC/E water molecules are shown as red spheres (oxygen site) with two white spheres (hydrogen sites), while weak water is shown with the oxygen as a cyan sphere. Here, x is the mole fraction of weak water. For $x = 0.2$ and $x = 0.8$, the majority species is shown as ghost particles. **(See color insert.)**

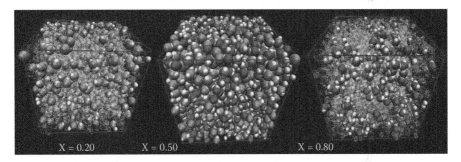

FIGURE 7.8 Snapshots of the $\lambda = 4/5$ mixture model which is a model for hydrophilic aqueous mixture. The representation conventions are as in Figure 7.7. Here, x is the mole fraction of weak water. For $x = 0.2$ and $x = 0.8$, the majority species is shown as ghost particles. **(See color insert.)**

solute contents systems: those with low water content show clear compact clusters, while those with low solute content show mostly dispersed solute. The KBIs for this mixture model can be computed directly by integrating the RDFs calculated from the simulations. It is important to underline that the RDFs from simulations do not approach unity at large separations, but to an asymptote that depends on both the system size and the concentration fluctuations of the species pair under consideration. This was demonstrated by Lebowitz and Percus for the case of the constant NVT canonical ensemble (Lebowitz and Percus 1961). In our constant NpT isobaric ensemble simulations, whenever the status of the asymptote was clear, we have noted that the RDFs also do not converge to unity. We have devised a simple mathematical operation that restores the correct unity asymptote without the need of evaluating explicitly the concentration fluctuations. Details are given in Kežić and Perera (2011) and Perera et al. (2011). A first approximation to the KBIs can also be determined through the KB formulas when the volumes of the mixture are known as a function of the solute mole fraction x, by assuming either an ideal form for the concentration fluctuation term $D(x) = 1$ (Equation 1.73), or a form that seems to suit most experimental data $D(x) = 1 - ax(1-x)$, where a is a constant parameter that can be adjusted to closely fit the available KBIs. For the present case, we used the KBIs determined by the simulations to determine the general shape of D. We found that all three KBIs provided very similar values of D. The results are shown in Figure 7.10. There is a reasonably good agreement between the direct and indirect evaluation of the KBIs. It is found that the KBIs are strongly nonideal, and look more like those for aqueous mixture of hydrophiles such as dimethyl sulfoxide or ethanolamine, for example. This confirms that the solute model is that of a hydrophilic molecule. In view of the small magnitude of the KBIs, one can say that such mixtures are mostly governed by concentration fluctuations and only to some extent by microheterogeneity.

The $\lambda = 2/3$ mixture model is a very good model of a strongly microheterogeneous mixture, as can be seen from the snapshots of Figure 7.9. It can be observed that water forms tighter clusters than the solute, and that the $x = 0.5$ mixture is clearly bicontinuous, with water and solute forming interlaced microaggregate domains, as observed in several realistic situations (Dixit et al. 2002; Dougan et al. 2004). Because of the rather large domains that are formed, the water–water RDFs $g_{WW}(r)$ and the solute–solute RDFs $g_{SS}(r)$ tend to be rather large at short range, indicating the strong correlation that build up between the like molecules that tend to microsegregate them. These large correlations tend to give very large running KBI (RKBI), a quantity that is defined as

$$G_{ij}(R) = 4\pi \int_0^R \left(g_{ij}(r) - 1 \right) r^2 dr \tag{7.60}$$

whose limit is precisely the KBI: $\lim_{R\to\infty} G_{ij}(R) = G_{ij}$. The RKBIs of the present model reach very large values, about several thousand after distances around 1 nm or so, precisely because of the large short-range correlations that build up at shorter ranges. As a result, the KBI evaluated with $N = 2048$ molecules are too high. Even though

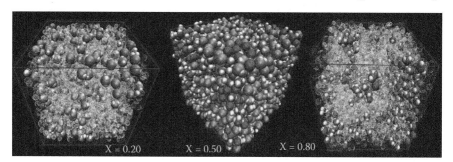

FIGURE 7.9 Snapshots of the $\lambda = 2/3$ mixture model which is a model for hydrophobic aqueous mixture. The representation conventions are as in Figure 7.7. Here, x is the mole fraction of weak water. For $x = 0.2$ and $x = 0.8$, the majority species is shown as ghost particles. **(See color insert.)**

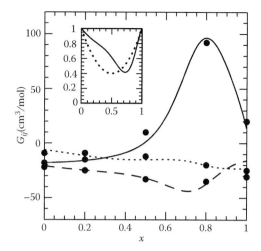

FIGURE 7.10   KBIs for the mixture model $\lambda = 4/5$ as a function of the weak-water mole fraction. Full line is for G_{WW}, dotted for G_{ww}, and dashed for G_{Ww} ("W" stands for SPC/E water and "w" for weak water). Dots are for data obtained from integrating of the RDFs and the lines from the KB theory. The inset shows the concentration fluctuation function $D(x)$ for a simple model (dotted) and correct version (full line).

these values are unphysically high, they can be considered as the final values for the G_{ij}. The principal reason for the ambiguity is the fact that the RDFs do not seem to reach a stable flat asymptote at the end of the simulation box. Clearly, larger system sizes seem required. The principal question is the nature of the relation between the large values obtained for the KBIs, as evaluated within the present system sizes, and the way these high RKBIs could settle down to smaller and more realistic values. This is where the idea of domain modulation seems the perfect solution to the problem. Indeed, if the RDFs are modulated by the domain size, they would oscillate at large distances, thus reducing the RKBI to smaller and more acceptable values. Figure 7.11 illustrates the problem of the KBIs in highly microheterogeneous systems. The dots are the KBIs evaluated by direct integration of the RDFs, by taking the last value of the RKBI even when the asymptotes are not stabilized. It is seen that for weak-water mole fraction above $x = 0.5$, the water–water and water–solute KBIs seem excessively large. If we use these KBIs to compute the $D(x)$ term, we find an unphysical behavior near $x = 1$, as illustrated in the inset of Figure 7.11. By enforcing these values to have the "proper" behavior illustrated by the continuous line, one can calculate the KBIs that would produce such correct behavior, which are found to be smaller. Examining the large distance behavior of the systems corresponding to $x > 0.5$, we find that the RDFs can be extrapolated by using a continuity solute that satisfies the TS requirements. This is illustrated in Figure 7.12 for $x = 0.8$. It is seen that the long-range extrapolation of the water–water RDF is very smooth and plausible, and leads to a smaller value of the KBI in excellent agreement with the expected value, as seen in Figure 7.11 where the new KBIs are shown. This calculation reveals

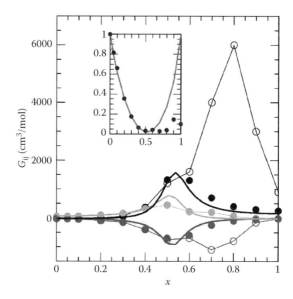

FIGURE 7.11   KBIs for the mixture model $\lambda = 2/3$ as a function of the weak-water mole fraction. The open dots are the uncorrected KBIs obtained by the RDFs and the filled dots are obtained through the TS approach described in the text. Blue is for G_{WW}, gold for G_{ww}, and green for G_{Ww} ("W" stands for SPC/E water and "w" for weak water). Dots are for data obtained from integration of the RDFs and the lines from the KB theory. The inset shows the concentration fluctuation function for a simple model (cyan) and correct version (red). **(See color insert.)**

that the domain modulation is certainly the solution of the dilemma of the high KBIs obtained in *small* size simulations (Chitra and Smith 2001a; Perera and Sokolic 2004; Lee and van der Vegt 2005; Zoranić et al. 2009; Perera et al. 2012).

This dilemma is posed in simulations of several realistic aqueous mixtures, and was often solved by readjusting the solute force field model in order to bring the RKBI to stable lower asymptotes within the considered simulation boxes (Weerasinghe and Smith 2003b; Lee and van der Vegt 2005; Ploetz, Bentenitis, and Smith 2010a). While there may be a vast number of force fields parameters that would allow reproducing a given panel of thermodynamic properties for any given system, it is not obvious that each such model would have the same morphology. In the case of simple aqueous mixtures, it all comes down to figuring out the role of fluctuations versus domain sizes, and this is certainly a clarified issue for more complex systems such as microemulsions. The idea that even simple aqueous systems can act as molecular emulsions is precisely to underline the role of domain formations instead of looking at them as fluctuations. The MOZ formalism shows that this can be achieved in principle provided we could have workable approximations of the direct correlation function for aqueous mixtures. The models discussed above are an indirect confirmation that the TS theory could be used to interpret the odd KBIs by rescaling the extent of the correlations with physical arguments based on domain formation just like in microemulsions.

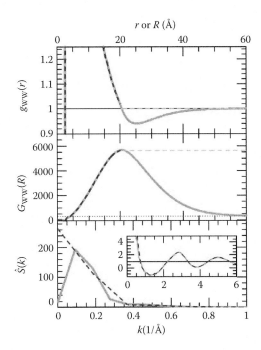

FIGURE 7.12 Illustration of the TS approach for molecular emulsions. Top panel shows a close-up of the water–water RDF at large distances, with red curve being the initial data from MD, gold curve is the abrupt continuation and green curve the TS continuation. Middle panel shows the corresponding RKBIs as a function of integration distance, and the lower panel the structure factor with a visible domain prepeak. **(See color insert.)**

7.4 CONCLUSIONS

In their introduction to nonionic aqueous mixtures, Rowlinson and Swinton state in their book on mixtures that these systems would require more insight than other types of mixtures (Rowlinson and Swinton 1982). In the early days of the developments of the theory of liquids, De Gennes reminded us that according to Landau, the great Soviet physicist, such a theory could not exist nor be useful (De Gennes 1977). Decades later, these two statements seem very much valid since the theory of aqueous mixtures is still full of mysteries and the corresponding theory of liquids is still in infancy. The microheterogeneity in aqueous mixtures seems a good place to understand liquids in a deeper way. Other challenges, such as understanding the role of water in biological systems, are appealing (Wiggins 2008), but we first need to understand aqueous mixtures.

The principal aim of this chapter was to show that complex liquid mixtures are characterized both by fluctuations and intrinsic microheterogeneity; the latter arises from the natural tendency of these systems to show microsegregated domains. These domains are not to be confused with large concentration fluctuations, such as those arising from phase-separating systems. The Kirkwood–Buff theory shows how the formalism of the statistical theory of liquids can be used to relate the integral of the correlation functions to concentration fluctuations and volumetric properties. But this theory does not give any clue concerning the correlations themselves. We show here that the STL allows us to retrieve important information about the correlations, such as the correlation length and the domain size, which gives us a more precise picture of the morphology of the mixture than one can deduce from KBIs alone. This insight appears to be important if one wants to interpret the vast amount of experimental data gathered on the KBIs of various mixtures (see Chapter 3).

In this chapter, we have shown how fluctuations and correlations are narrowly linked through the STL, first in the case of simple fluids and mixtures, then in complex fluids. We have illustrated how fluctuations are the sole feature of simple liquids and how they destabilize these systems toward a phase transition, the only morphology that is allowed in these systems. We have shown that morphology through clustering could be inserted into such simple models by including a supplementary length scale parameter through artificial long-range repulsive interactions. However, to implement such repulsion in realistic situations, one needs to incorporate angle-dependent interactions, and we have outlined the STL for such systems. Finally, we have illustrated the connection between microheterogeneity and domain modulation through an analogy with microemulsions with a simple aqueous mixture model. In order to go beyond this, one should be able to calculate from microscopic correlations how domain formation arises, by calculating both the correlation length and the domain sizes, as given in Equation 7.53, Equation 7.57, and Equation 7.58. This step would require that we have accurate correlation functions in the form of invariant projections. While such functions can be obtained through computer simulation, their true long-range behavior can only be obtained by theoretical methods, such as by solving MOZ with accurate closure equations. Preliminary work along such lines is in progress (Kežić and Perera 2011).

8 Solvation Phenomena in Dilute Solutions

Formal Results, Experimental Evidence, and Modeling Implications

Ariel A. Chialvo

CONTENTS

Abstract: We review the fundamentals underlying a general molecular-based formalism for the microscopic interpretation of the solvation phenomena involving sparingly soluble solutes in compressible media, an approach that hinges around the unambiguous splitting of the species correlation function integrals into short (finite)- and long (diverging)-ranged contributions at infinite dilution, where this condition is taken as the reference system for the derivation of composition expansions. Then, we invoke the formalism (a) to illustrate the well-behaved nature of the solvation contributions to the mechanical partial molecular properties of solutes at infinite dilution, (b) to guide the development of, and provide molecular-based support to, the macroscopic modeling of high-temperature dilute aqueous-electrolyte solutions, (c) to study solvation effects on the kinetic rate constants of reactions in near-critical solvents in an attempt to understand from a microscopic perspective the macroscopic evidence regarding the thermodynamic pressure effects, and (d) to interpret the microscopic mechanism behind synergistic solvation effects involving either cosolutes or cosolvents, and provide a molecular argument on the unsuitability of the van der Waals one-fluid (vdW-1f) mixing rules for the description of weakly attractive solutes in compressible solvents. Finally, we develop thermodynamically consistent perturbation expansions, around the infinite dilution reference, for the species residual properties in binary and ternary mixtures, and discuss the theoretical and modeling implications behind *ad hoc* first-order truncated expansions.

8.1 INTRODUCTION

The study of dilute fluid solutions has been essential to the foundation of solvation thermodynamics and the development of macroscopic modeling for the description of observed behavior and the correlation of experimental data. Typical studies of dilute solutions have dealt with the determination of limiting values for the solute activity coefficients and the corresponding slope of their composition dependence at infinite dilution (Jonah 1983, 1986), where these quantities played a crucial role in constraining the parameterization of excess Gibbs free-energy models (Van Ness and Abbott 1982; Lupis 1983; Wallas 1985; Chialvo 1990a).

While dilute fluid solutions are usually of theoretical and practical relevance regardless of the solvent conditions, the most appealing dilute systems are perhaps those comprising highly compressible environments, that is, at state conditions where small perturbations of either the system pressure or temperature can translate into significant changes in the system density, which in turn can substantially modify the species solvation behavior (Clifford 1999). Numerous technological advances in separation processes rely on our ability to tune the medium density within the highly compressible region to tailor the solvation behavior of species in solution for special applications including food processing (Brunner 2005), environmentally safe disposal of toxic wastes (Yesodharan 2002), chemical synthesis (Kruse and Dinjus 2007), as well as synthesis and processing of novel materials (Adschiri et al. 2011). Moreover, many geologic processes, including hydrothermal vents (Martin et al. 2008) and ore deposition (Richards 2011), take place in high-temperature aqueous media involving the simultaneous solvation of gases, nonpolar and ionic compounds whose proper macroscopic modeling requires a fundamental understanding of the associated microstructural changes of the fluid environments.

A common and frequently overlooked feature in the behavior of these dilute highly compressible environments is the coexistence of microscopic phenomena involving two rather different length scales (Chialvo and Debenedetti 1992; Chialvo and Cummings 1994). To understand their solvation behavior we must recognize the

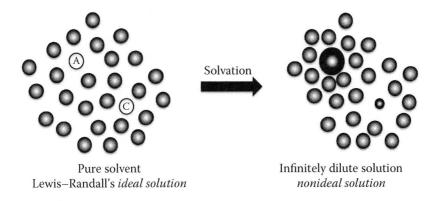

Pure solvent
Lewis–Randall's *ideal solution*

Infinitely dilute solution
nonideal solution

FIGURE 8.1 Schematic of the solvation process described in terms of the mutation of two solvent molecules labeled "A" and "C" into the anion and cation, where for simplicity we assume $v_+ = v_- = 1$.

mechanism underlying these phenomena; a short-ranged (SR) density perturbation induced by the insertion of a solute into the solvent medium, and the long-ranged (LR) propagation of this perturbation up to a distance dictated by the medium's correlation length (See Figure 8.1) (Chialvo and Debenedetti 1992; Debenedetti and Chialvo 1992).

The coexistence of these two length scales made difficult the early attempts to interpret the experimental evidence from, and the modeling of, systems characterized by highly compressible media, and gave rise to heated debates (Brennecke et al. 1990; Economou and Donohue 1990; McGuigan and Monson 1990). The source of controversy at the time was the failure to discriminate the solvation contributions from the overwhelming compressibility-driven contributions to the mechanical properties of the infinite dilute solutes. The debate was eased after the development of a rigorous solvation formalism for infinitely dilute solutions that identified and isolated the individual contributions from the two length scales (Chialvo and Cummings 1994; Chialvo, Kalyuzhnyi, and Cummings 1996; Chialvo et al. 1999, 2001). In fact, the coexistence of solvation and compressibility-driven phenomena, that is, what makes supercritical solutions challenging to model, becomes also the key to characterizing their thermodynamic properties in terms of the two distinctive length scales (Chialvo and Cummings 1994).

In Section 8.2 we discuss the main ideas behind the formalism and illustrate some of the features based on predictions from integral equation calculations involving simple binary mixtures modeled as Lennard–Jones systems (Section 8.2.1), to guide the development of, and provide molecular-based support to, the macroscopic modeling of high-temperature dilute aqueous-electrolyte solutions (Section 8.2.2), as well as to highlight the role played by the solvation effects on the pressure dependence of the kinetic rate constants of reactions in near-critical solvents (Section 8.2.3).

Nonvolatile species solubilities in highly compressible solvents might be significantly low, but usually much higher than the corresponding ideal gas solvent counterparts (typically referred to as the solubility enhancement); consequently, the actual systems are not infinitely dilute, that is, we must deal with finite compositions and their effect on the phase behavior. For that purpose, in Section 8.3 we introduce the main ideas behind the derivation of composition expansions for the thermodynamic properties of a dilute system whose infinite dilution reference systems are described by the solvation formalism of Section 8.2. Then we discuss the development of thermodynamically consistent perturbation expansions for the species residual properties in multicomponent dilute mixtures (Section 8.3.1), provide a microscopic interpretation of the expansion coefficients (Section 8.3.2), analyze special cases for which the thermodynamic consistency is well established (Section 8.3.3), discuss some modeling implications behind *ad hoc* first-order truncated expansions (Section 8.3.4), and interpret the microscopic mechanisms behind synergistic solvation effects involving cosolutes or cosolvents to provide a molecular argument on the unsuitability of the vdW-1f mixing rules for the description of weakly attractive solutes in compressible solvents (Section 8.3.5).

8.2 SOLVATION FORMALISM FOR INFINITELY DILUTE TERNARY SYSTEMS

The behavior of dilute solutes in compressible media has been traditionally associated with supercritical solubility enhancement and related phenomena, frequently analyzed in terms of the solute mechanical partial molecular properties at infinite dilution, that is, volume and enthalpy, as well as those of the pure solvent counterpart (Debenedetti and Kumar 1986, 1988). There are compelling reasons behind the use of these partial molecular properties at infinite dilution: (a) they are the pressure and temperature first derivatives of the corresponding chemical potentials, therefore, the choice of suitable approximations based on well-established limiting behavior for the pair correlation functions leads to the development of well-behaved and accurate correlations, and (b) they are rigorously related to volume integrals over the microstructure of the solvent around the infinite dilute solute according to the Kirkwood–Buff fluctuation formalism of mixtures (Kirkwood and Buff 1951). Consequently, the resulting macroscopic modeling will capture the relevant underlying physics (O'Connell, Sharygin, and Wood 1996). In fact, the solute partial molecular volume at infinite dilution has been often used to interpret the pressure dependence of the isothermal solubility of sparingly soluble species (Kumar and Johnston 1988), and not surprisingly, to make contact between the evolution of solubility (or the corresponding solute partial molecular fugacity coefficient) and the underlying changes in the solvent microstructure around the infinitely dilute solute (Debenedetti and Mohamed 1989) to facilitate the modeling or correlation of solubility data (Cochran, Lee, and Pfund 1990; O'Connell and Liu 1998).

Despite the referred advantages, the modeling of mechanical partial molecular properties of solutes at infinite dilution is intrinsically problematic, that is, they scale as the solvent isothermal compressibility (Levelt Sengers 1991), with finite prefactors associated with the magnitude of the molecular solute–solvent interaction asymmetry (Chialvo and Cummings 1995) as we will illustrate below. However, in an effort to facilitate the regression of experimental data, Wood and coworkers (Sedlbauer, Yezdimer, and Wood 1998) and O'Connell and coworkers (O'Connell,

Sharygin, and Wood 1996) manipulated specific combinations of partial molecular volumetric properties (i.e., compressibility-driven) that led naturally to well-behaved finite quantities associated with the solvation behavior of infinitely dilute solutes. These successful isothermal-density correlations suggested an underlying common mechanism that neutralized the original compressibility-driven contributions to the mechanical partial molecular properties of the solute at infinite dilution, and reinforced the need for its molecular-based interpretation (*vide infra* Section 8.2.2).

The main issue here is to achieve the unambiguous separation between solvation and compressibility-driven phenomena, based on the formal splitting of the total correlation functions into their corresponding direct and indirect contributions (Chialvo and Cummings 1994, 1995) according to the Ornstein–Zernike equation (Hansen and McDonald 1986), and then use the derived rigorous expressions as zeroth-order approximations, for example, reference systems, in the subsequent perturbation expansion of the composition-dependent thermodynamics properties of multicomponent dilute fluid mixtures (*vide infra* Section 8.3).

For that purpose, we can portray the solvation of a single ionic solute $C_{\nu+}^{z+} A_{\nu-}^{z-}$ in a pure solvent at fixed state conditions (constant T, and either constant p or ρ) as the formation of an infinitely dilute system in a *thought experiment* involving N solvent molecules in which a number ν of them ($\nu = \nu_+ = \nu_-$ where ν_+ and ν_- are the stoichiometric coefficients of the salt) can be distinguished by their *solute* labels. Thus, the initial conditions of the system represent an *ideal solution* in the sense of the Lewis–Randall rule (O'Connell and Haile 2005), that is, the residual properties of the $(N-\nu)$ solvent-labeled particles and those of the ν solute-labeled particles are identically the same (Chialvo 1990b; Chialvo 1993a). To form the desired infinitely dilute *nonideal solution*, the solvation process proceeds through the mutation (à la Kirkwood's coupling-parameter charging [Kirkwood 1936]) of the distinguishable ν solvent molecules into the final neutral ionic solute $C_{\nu+}^{z+} A_{\nu-}^{z-}$. The process in which the original ν *solute* species in the *ideal solution* (i.e., where all solute–solvent interactions are identical to the solute–solute and the solvent–solvent interactions) are converted into the neutral ionic species is driven by the free-energy difference, $\mu_2^{r,\infty}(T,p) - \nu\mu_1^{r,o}(T,p)$, where the superscript r denotes a residual quantity for a pure (o) or an infinitely dilute (∞) species at the specified state conditions, respectively. This driving force provides the sought link between the microstructural changes of the solvent around the mutating species and the macroscopic (thermodynamic) properties that best characterize the solvation process.

While this connection can be achieved in essentially four equivalent ways by interpreting the driving force of the solvation process (Chialvo et al. 2000b), from either a microscopic or a macroscopic standpoint, it is more instructive to invoke the one that explicitly illustrates the separation of length scales. To do so, we start from the exact thermodynamic expression (Modell and Reid 1983),

$$\mu_2^{r,\infty}(T,p) - \nu\mu_1^{r,o}(T,p) = \nu k_B T \ln\left(\overline{\phi}_2^{\infty} / \overline{\phi}_1^{o}\right)$$

$$= \int_0^{\rho(p)} \left(\overline{V}_2^{\infty} - \nu\overline{V}_1^{o}\right) \frac{d\rho}{\kappa_T \rho} \tag{8.1}$$

where $\bar{V}_1^o$ is the partial molecular volume of the pure solvent, $\bar{V}_1^o = \rho^{-1}$; κ_T is the corresponding solvent isothermal compressibility; and the partial molecular volume of the salt at infinite dilution $\bar{V}_2^\infty$ provides the link to the system structure through the total correlation function integrals (TCFI) as follows (Kusalik and Patey 1987; Chialvo et al. 1999),

$$\bar{V}_2^\infty = v\left(\bar{V}_1^o + G_{11}^o - G_{12}^\infty\right) \tag{8.2}$$

where G_{11}^o and G_{12}^∞ are the TCFI for the solvent–solvent and solute–solvent interactions, respectively.

Because $\bar{V}_2^\infty$ comprises a solvation contribution (associated with the solvent's local density perturbation) and a compressibility-driven contribution (associated with the propagation of the perturbation), it also offers a way to identify the short- and long-ranged contributions to this mechanical partial molecular property, and consequently, to discriminate the solvation process from the compressibility-driven phenomenon. In fact, after invoking the Ornstein–Zernike equation (Hansen and McDonald 1986), we can split the TCFIs into their direct and indirect contributions (Chialvo et al. 1999), that is,

$$\bar{V}_2^\infty = v\bar{V}_1^o\left(1 + C_{11}^o - C_{12}^\infty\right) + v\left(I_{11}^o - I_{12}^\infty\right) \tag{8.3}$$

where $I_{ij} = G_{ij} - \rho^{-1}C_{ij}$ represents the integral over the indirect pair correlation functions (ICFIs), for the ij-interactions (Chialvo and Cummings 1994). Then, it becomes clear that the first term in Equation 8.3 represents the solvation (finite) contribution to $\bar{V}_2^\infty$, $\bar{V}_2^\infty$(SR), that is,

$$\bar{V}_2^\infty(SR) = v\bar{V}_1^o\left(1 + C_{11}^o - C_{12}^\infty\right) \tag{8.4}$$

while the second term accounts for the compressibility-driven contribution, $\bar{V}_2^\infty$(LR), that is, (Chialvo et al. 1999)

$$\bar{V}_2^\infty(LR) = vk_BT\kappa_T C_{11}^o\left(C_{11}^o - C_{12}^\infty\right) \tag{8.5}$$

where C_{11}^o and C_{12}^∞ are the direct correlation function integrals (DCFIs) for the solvent–solvent and solute–solvent interactions, respectively (Kusalik and Patey 1987; O'Connell 1990).

The concept of short- and long-ranged contributions highlights the fact that the invoked correlation function integrals are either finite (DCFI) or diverging (ICFI) quantities, the result of volume integrals over either short- (direct) or long-ranged (indirect) correlation functions. Note also that both quantities involve the same finite factor $C_{11}^o - C_{12}^\infty$ linked to the solute–solvent molecular asymmetry, whose actual meaning can be revealed in a few insightful alternative representations of the structural perturbation of the solvent caused by the mutation of a solvent into the solute during the *thought experiment*. The first representation is an isothermal–isochoric rate of change of pressure caused by the structural perturbation of the solvent around the solute, $(\partial p/\partial x_2)_{T,\rho}^{\infty}$, that is, (Chialvo et al. 1999)

$$\left(C_{11}^{0} - C_{12}^{\infty}\right) = v^{-1}\kappa_{T}^{IG}\left(\partial p/\partial x_2\right)_{T,\rho}^{\infty} \qquad (8.6)$$

where $\kappa_{T}^{IG} = 1/\rho k_{B}T$ is the ideal gas isothermal compressibility and

$$\left(\partial p/\partial x_2\right)_{T,\rho}^{\infty} = \left(\bar{V}_2^{\infty} - v\bar{V}_1^{0}\right)/\left(\kappa_T \bar{V}_1^{0}\right) \qquad (8.7)$$

The second alternative representation is the finite volumetric response, or *solute-induced effect* as we called it earlier (Chialvo and Cummings 1994), that is,

$$\left(C_{11}^{0} - C_{12}^{\infty}\right) = \left(\bar{V}_2^{\infty}(\text{SR}) - v\bar{V}_1^{0}\right)/v\bar{V}_1^{0} \qquad (8.8)$$

where the numerator of the right-hand side of Equation 8.8, the volumetric *solute-induced effect*, can be explicitly written in terms of experimentally available volumetric properties as follows,

$$\left(\bar{V}_2^{\infty}(\text{SR}) - v\bar{V}_1^{0}\right) = \left(\kappa_T^{IG}/\kappa_T\right)\left(\bar{V}_2^{\infty} - v\bar{V}_1^{0}\right) \qquad (8.9)$$

Finally, a third alternative representation is based on the excess particle number $N_{2,\text{ex}}^{\infty} = \rho(G_{12}^{\infty} - G_{11}^{0})$ (Chialvo and Cummings 1995; Chialvo et al. 2000a), that is, the number of solvent molecules around the solute in excess of that around any solvent molecule (the Lewis–Randall ideal solution). Following the split given by Equation 8.3 through Equation 8.5, the solvation contribution to $N_{2,\text{ex}}^{\infty}$ becomes equal to $N_{2,\text{ex}}^{\infty}(\text{SR}) = -(C_{11}^{0} - C_{12}^{\infty})$ because $N_{2,\text{ex}}^{\infty} = -v^{-1}\kappa_T(\partial p/\partial x_2)_{T,\rho}^{\infty}$, consequently,

$$N_{2,\text{ex}}^{\infty}(\text{SR}) = \left(\kappa_T^{IG}/\kappa_T\right)N_{2,\text{ex}}^{\infty} \qquad (8.10)$$

and after recalling Equations 8.8 and 8.9, we can express Equation 8.10 in terms of experimentally available properties as follows,

$$N_{2,\text{ex}}^{\infty}(\text{SR}) = -\left(V_2^{\infty} - vV_1^{0}\right)/\left(vk_{B}T\kappa_T\right) \qquad (8.11)$$

Equation 8.7 through Equation 8.11 allow us to express the free-energy change in the *thought experiment*, Equation 8.1, in a few equivalent forms, depending on how we describe the structural perturbation of the solvent environment around the infinitely dilute solute, Equation 8.6, including: (a) in terms of the isothermal–isochoric rate of change of pressure, that is,

$$\mu_2^{r,\infty}(T,p) - v\mu_1^{r,0}(T,p) = \int_0^{\rho(p)}\left(\partial p/\partial x_2\right)_{T,\rho}^{\infty}\frac{d\rho}{\rho^2} \qquad (8.12)$$

an expression that highlights the finiteness of its integrand at any state condition, including the solvent's critical point where $(\partial p/\partial x_2)^\infty_{T_c,\rho_c}$ becomes known as *Krichevskii's parameter* (Levelt Sengers 1991); (b) through the solvation contribution to the excess particle number $N^\infty_{2,ex}(SR)$ (Chialvo et al. 2000b),

$$\mu_2^{r,\infty}(T,p) - \nu\mu_1^{r,o}(T,p) = -k_B T \int_0^{\rho(p)} N^\infty_{2,ex}(SR)\frac{d\rho}{\rho} \qquad (8.13)$$

to highlight the connection between the isothermal–isochoric pressure change associated with the perturbation of the solvent structure around the solute (Equation 8.6) and the corresponding effective change in the number of solvent molecules (Equation 8.10). Note that, since this quantity does not involve explicitly any size for the solvation shell, $N^\infty_{2,ex}(SR)$ can be considered as an *effective* solvation number (Chialvo et al. 1999), though it should not be confused with the conventional definition of solvation numbers based on the structural information of the first coordination shell; (c) in terms of the finite density perturbation of the solvent around the solute, that is,

$$\mu_2^{r,\infty}(T,p) - \nu\mu_1^{r,o}(T,p) = k_B T \int_0^{\rho(p)} \left(\bar{V}_2^\infty(SR) - \nu\bar{V}_1^o\right)d\rho \qquad (8.14)$$

The enthalpic and entropic solvation- and compressibility-driven counterparts to $\bar{V}_2^\infty(SR)$ and $\bar{V}_2^\infty(LR)$ can be derived straightforwardly by invoking the identity $\bar{H}_2^{r,\infty} = [\partial(\beta\mu_2^{r,\infty})/\partial\beta]_p$ (for details see Appendix C of Chialvo et al. (1999), that is,

$$\bar{H}_2^{r,\infty}(SR) = -\nu k_B T - \int_0^{\rho(p)} \left(\partial C_{12}^\infty/\partial\beta\right)_\rho\frac{d\rho}{\rho} - \beta\left(\partial p/\partial\beta\right)_\rho \bar{V}_2^\infty(SR) \qquad (8.15)$$

and

$$\bar{H}_2^{r,\infty}(LR) = -\beta\left(\partial p/\partial\beta\right)_\rho \bar{V}_2^\infty(LR) \qquad (8.16)$$

so that

$$\left[\bar{H}_2^{r,\infty}(SR) - \nu\bar{H}_1^{r,o}\right]_{T,p} = \int_0^{\rho(p)} \left[\partial\left(\bar{V}_2^\infty(SR) - \nu\bar{V}_1^o\right)/\partial\beta\right]_\rho d\rho$$
$$\qquad\qquad - \beta\left(\partial p/\partial\beta\right)_\rho\left(\bar{V}_2^\infty(SR) - \nu\bar{V}_1^o\right) \qquad (8.17)$$

Immediately, Equation 8.1 and Equation 8.17 lead to the corresponding entropic counterparts, that is,

$$\left[\bar{S}_2^{r,\infty}(\mathrm{SR}) - v\bar{S}_1^{r,o} \right]_{T,p} = -k_B \int_0^{\rho(p)} \left(\bar{V}_2^{\infty}(\mathrm{SR}) - v\bar{V}_1^{o} \right)_{T,p} d\rho$$

$$+ T^{-1} \int_0^{\rho(p)} \left[\partial \left(\bar{V}_2^{\infty}(\mathrm{SR}) - v\bar{V}_1^{o} \right) / \partial \beta \right]_{\rho} d\rho \qquad (8.18)$$

$$- \beta T^{-1} \left(\partial p / \partial \beta \right)_{\rho} \left[\bar{V}_2^{\infty}(\mathrm{SR}) - v\bar{V}_1^{o} \right]_{T,p}$$

with

$$S_2^{r,\infty}(\mathrm{LR}) = -\beta T^{-1} \left(\partial p / \partial \beta \right)_{\rho} V_2^{\infty}(\mathrm{LR}) \qquad (8.19)$$

Therefore, as for the case of the volumetric *solute-induced effect*, Equation 8.9, the other mechanical *solute-induced effects* can be written solely in terms of experimental data as follows,

$$\left(\bar{H}_2^{r,\infty}(\mathrm{SR}) - v\bar{H}_1^{r,o} \right)_{T,p} = \left(\bar{H}_2^{r,\infty} - v\bar{H}_1^{r,o} \right)_{T,p}$$

$$+ \beta \left(\partial p / \partial \beta \right)_{\rho} \left(\bar{V}_2^{\infty} - v\bar{V}_1^{o} \right) \left[1 - \left(\kappa_T^{\mathrm{IG}} / \kappa_T \right) \right] \qquad (8.20)$$

and

$$\left(\bar{S}_2^{r,\infty}(\mathrm{SR}) - v\bar{S}_1^{r,o} \right)_{T,p} = \left(\bar{S}_2^{r,\infty} - v\bar{S}_1^{r,o} \right)_{T,p}$$

$$+ \beta T^{-1} \left(\partial p / \partial \beta \right)_{\rho} \left(\bar{V}_2^{\infty} - v\bar{V}_1^{o} \right) \left[1 - \left(\kappa_T^{\mathrm{IG}} / \kappa_T \right) \right] \qquad (8.21)$$

where we have invoked the equivalent expression for $\bar{V}_2^{\infty}(\mathrm{LR})$ (see Equation 8.2 through Equation 8.5), that is,

$$\bar{V}_2^{\infty}(\mathrm{LR}) = \left(\bar{V}_2^{\infty} - v\bar{V}_1^{o} \right) \left[1 - \left(\kappa_T^{\mathrm{IG}} / \kappa_T \right) \right] \qquad (8.22)$$

Equations 8.17 and 8.18 suggest that the solvation process may be entirely characterized by the finite perturbation of the solvent structure due to the presence of the solute (also known as *solute-induced effects* [Chialvo and Cummings 1994]), that is,

$$\left(\mu_2^{r,\infty}(T,p) - v\mu_1^{r,o}(T,p) \right) = \left(\bar{H}_2^{r,\infty}(\mathrm{SR}) - v\bar{H}_1^{r,o} \right)_{T,p} - T \left(\bar{S}_2^{r,\infty}(\mathrm{SR}) - v\bar{S}_1^{r,o} \right)_{T,p} \qquad (8.23)$$

Moreover, according to Equation 8.16, Equation 8.19, and Equation 8.22, the long-range contributions to the enthalpic and entropic partial molecular properties of the infinitely dilute solute become

$$\bar{H}_2^{r,\infty}(\text{LR}) = -\beta\left(\partial p/\partial\beta\right)_\rho\left(\bar{V}_2^\infty - v\bar{V}_1^\circ\right)\left[1 - \left(\kappa_T^{IG}/\kappa_T\right)\right] \tag{8.24}$$

and

$$\bar{S}_2^{r,\infty}(\text{LR}) = -\beta T^{-1}\left(\partial p/\partial\beta\right)_\rho\left(\bar{V}_2^\infty - v\bar{V}_1^\circ\right)\left[1 - \left(\kappa_T^{IG}/\kappa_T\right)\right] \tag{8.25}$$

Thus, we can clearly see that the cancellation of these diverging quantities in Equation 8.23 results from the identity $\bar{S}_2^{r,\infty}(\text{LR}) = T^{-1}\bar{H}_2^{r,\infty}(\text{LR})$.

8.2.1 SOLUTE-INDUCED EFFECTS IN SIMPLE MODEL SYSTEMS

Here we illustrate the solvation formalism by integral equation calculations for binary mixtures described by the Lennard–Jones model (see Tables 8.1 and 8.2), and based on the Percus–Yevick approximation for the solution of the Ornstein–Zernike equations (Hansen and McDonald 1986) according to the approach proposed by McGuigan and Monson (McGuigan and Monson 1990). We focus on the *solute-induced effects* on the microstructure and the thermodynamic properties of infinitely dilute solutions of pyrene in carbon dioxide and Ne in Xe along the

TABLE 8.1
Lennard–Jones Parameters for the Model System Pyrene (2)/CO_2 (1)

ij-Interactions	$\sigma_{ij}(\text{Å})$	$\varepsilon_{ij}/k_B(\text{K})$
11	3.794	225.3
12	5.467	386.4
22	7.140	662.8

TABLE 8.2
Lennard–Jones Parameters for the Model System Ne (2)/Xe (1)

ij-Interactions	$\sigma_{ij}(\text{Å})$	$\varepsilon_{ij}/k_B(\text{K})$
11	4.047	231.0
12	3.433	87.0
22	2.820	32.8

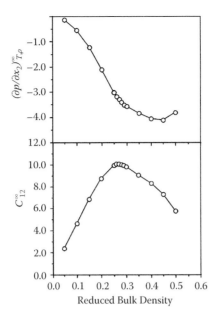

FIGURE 8.2   Solvent density dependence of $(\partial p/\partial x_2)^{\infty}_{T,\rho}$ and C^{∞}_{12} for the attractive Lennard–Jones pyrene (2)/CO$_2$ (1) system along the near-critical reduced isotherm $T_r = 1.02$. All quantities are made dimensionless using the solvent's Lennard–Jones parameters from Table 8.1.

near-critical reduced isotherm $T_r = T/T_c = 1.02$, within the solvent density range of $0.05 \leq \rho\sigma^3_{11} \leq 0.5$. These solutes behave as attractive and repulsive species, respectively, according to the classification of Debenedetti and Mohamed (1989), that is, $C^{\infty}_{12} > 1$ and $C^{\infty}_{12} < 1$, whose macroscopic manifestations correspond to those of nonvolatile and volatile species, that is, $(\partial p/\partial x_2)^{\infty}_{T,\rho} < 0$ and $(\partial p/\partial x_2)^{\infty}_{T,\rho} > 0$, respectively (Levelt Sengers 1991).

Figure 8.2 confirms the attractive nature of the pyrene-CO2 interactions, that is, $(\partial p/\partial x_{Py})^{\infty}_{T,\rho} < 0$ and $C^{\infty}_{PyCO_2} > 1$, conditions clearly reflected in the significantly large negative values of the corresponding solute's mechanical partial molecular quantities around the solvent's critical density (Figure 8.3). In Figure 8.4, we display the density dependence for the short-ranged contribution to the solute's mechanical partial molecular quantities $\bar{V}^{\infty}_{Py}(SR)$ and $\bar{H}^{r,\infty}_{Py}(SR)$, respectively, where we should highlight the significant differences in the size of these quantities in comparison with those in Figure 8.3. Note that, while $\bar{V}^{\infty}_{Py}(SR)$ displays a minimum on its density dependence at $\rho\sigma^3_{11} < \rho_c\sigma^3_{11} = 0.28$, the location of this minimum does not coincide with that of the $\bar{V}^{\infty}_{Py}$. Moreover, $\bar{H}^{r,\infty}_{Py}(SR)$ and the solute-induced effects, that is, $(\bar{V}^{\infty}_{Py}(SR) - \bar{V}^{o}_{CO_2})$ and $(\bar{H}^{r,\infty}_{Py}(SR) - \bar{H}^{r,o}_{CO_2})$ (Figure 8.5), exhibit monotonic density dependences.

In Figure 8.6, we validate the repulsive nature of the infinite dilute Ne in near-critical Xe, that is, $(\partial p/\partial x_{Ne})^{\infty}_{T,\rho} > 0$, and $C^{\infty}_{NeXe} < 1$, as well as by the positive values of the corresponding mechanical partial molecular properties of Ne (Figure 8.7). Note that $(\partial p/\partial x_{Ne})^{\infty}_{T,\rho}$ and C^{∞}_{NeXe} for the repulsive mixture exhibit monotonous density dependences, in contrast to the corresponding behavior of the attractive mixtures (Figure 8.2). Similar behavior is observed in the density dependence

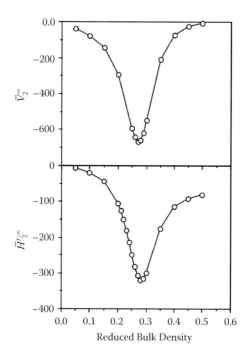

FIGURE 8.3 Solvent density dependence of $\bar{V}_2^{\infty}$ and $\bar{H}_2^{r,\infty}$ for the attractive Lennard–Jones pyrene (2)/CO_2 (1) system along the near-critical reduced isotherm $T_r = 1.02$. All quantities are made dimensionless using the solvent's Lennard–Jones parameters from Table 8.1.

for the short-ranged contribution to the solute's partial molecular volume $\bar{V}_{Ne}^{\infty}(SR)$ and enthalpy $\bar{H}_{Ne}^{r,\infty}(SR)$, Figures 8.7 and 8.8, as well as for the solute-induced local effects on the solvent's local density $(\bar{V}_{Ne}^{\infty}(SR) - \bar{V}_{Xe}^{0})$ and enthalpy $(H_{Ne}^{r,\infty}(SR) - \bar{H}_{Xe}^{r,0})$, Figure 8.9. The monotonicity of the density dependence of $(\partial p/\partial x_2)_{T,\rho}^{\infty}$ and C_{12}^{∞} should not be taken necessarily as a general rule for repulsive systems as we discussed elsewhere (Chialvo and Cummings 1994).

8.2.2 SOLUTE-INDUCED EFFECTS IN AQUEOUS ELECTROLYTE SYSTEMS AND THEIR MODELING

Considering that a primary goal of these molecular-based studies is to provide the fundamental understanding needed to develop successful engineering correlations, here we use the proposed solvation formalism as a versatile tool to interpret recent experimental data, and ultimately, to aid the choice of the best combination of properties for regression purposes. For example, Wood and collaborators (O'Connell, Sharygin, and Wood 1996; Gruszkiewicz and Wood 1997; Sedlbauer, Yezdimer, and Wood 1998) have found that the density dependence of the combined quantity $D_{21}^{\infty} \equiv (\kappa_T^{IG}/\kappa_T)\bar{V}_2^{\infty}$, for alkali halides in supercritical aqueous environments at supercritical water densities, exhibits an intriguingly weak temperature dependence in the temperature range $550 < T(K) < 725$. This behavior was obviously very appealing for

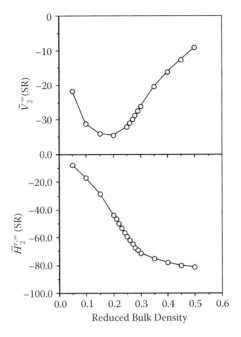

FIGURE 8.4 Solvent density dependence of $\bar{V}_2^{\infty}(SR)$ and $\bar{H}_2^{r,\infty}(SR)$ for the attractive Lennard–Jones pyrene (2)/CO_2 (1) system along the near-critical reduced isotherm $T_r = 1.02$. All quantities are made dimensionless using the solvent's Lennard–Jones parameters from Table 8.1.

regression purposes, for example, to estimate accurate values of $\bar{V}_2^{\infty}$ at other supercritical state conditions (Gruszkiewicz and Wood 1997) based only on the corresponding water properties, but it also suggested some relevant questions regarding the microscopic mechanism underlying such a behavior for $D_{21}^{\infty}(T,\rho)$.

To gain insights into the observed behavior of $D_{21}^{\infty}(T,\rho)$, we carried out integral equation (IE) calculations on model systems defined as charged hard-sphere ions immersed in an aqueous-like solvent, described as hard spheres with embedded point polarizabilities and permanent electrostatic multipole moments including quadrupole and octupoles, by solving the reference hypernetted-chain (RHNC) equations with solvent polarization effects treated at the self-consistent mean-field (SCMF) level (Chialvo and Cummings 1994, 1999; Chialvo et al. 2000c). In Figure 8.10, we display the IE-predicted solvent density dependence of $D_{21}^{\infty}(T,\rho)$ along three supercritical isotherms for infinitely dilute CsBr aqueous-like solutions (Chialvo et al. 2001), in comparison with the corresponding experimental data from Sedlbauer, Yezdimer, and Wood (1998) within the temperature and density ranges $604 < T(K) < 717$ and $0.26 < \rho$ (g/cm^3) < 0.60, respectively. The outstanding feature of the IE results is their lack of temperature dependence for supercritical densities, that is, $D_{21}^{\infty}(T,\rho) \cong D_{21}^{\infty}(\rho \geq \rho_c)$, a behavior that mirrors the experimental observations in other supercritical aqueous electrolyte and nonelectrolyte systems (O'Connell, Sharygin, and Wood 1996; Gruszkiewicz and Wood 1997; Sedlbauer, Yezdimer, and Wood 1998).

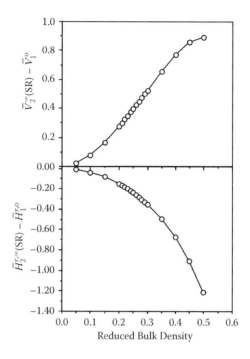

FIGURE 8.5   Solvent density dependence of $(\overline{V}_2^{\infty}(SR) - \overline{V}_1^{o})$ and $(\overline{H}_2^{r,\infty}(SR) - \overline{H}_1^{r,o})$ for the attractive Lennard–Jones pyrene (2)/CO_2 (1) system along the near-critical reduced isotherm $T_r = 1.02$. All quantities are made dimensionless using the solvent's Lennard–Jones parameters from Table 8.1.

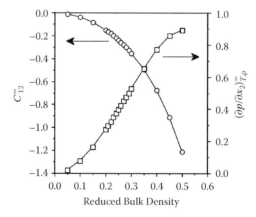

FIGURE 8.6   Solvent density dependence of $(\partial p/\partial x_2)_{T,\rho}^{\infty}$ and C_{12}^{∞} for the repulsive Lennard–Jones Ne (2)/Xe (1) system along the near-critical reduced isotherm $T_r = 1.02$. All quantities are made dimensionless using the solvent's Lennard–Jones parameters from Table 8.2.

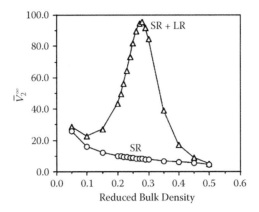

FIGURE 8.7 Solvent density dependence of $\bar{V}_2^{\infty}$ and $\bar{V}_2^{\infty}(SR)$ for the repulsive Lennard–Jones Ne(2)/Xe (1) system along the near-critical reduced isotherm $T_r = 1.02$. All quantities are made dimensionless using the solvent's Lennard–Jones parameters from Table 8.2.

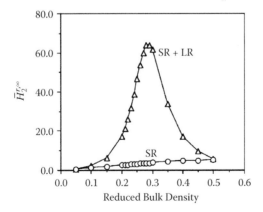

FIGURE 8.8 Solvent density dependence of $\bar{H}_2^{r,\infty}$ and $\bar{H}_2^{r,\infty}(SR)$ for the repulsive Lennard–Jones Ne (2)/Xe (1) system along the near-critical reduced isotherm $T_r = 1.02$. All quantities are made dimensionless using the solvent's Lennard–Jones parameters from Table 8.2.

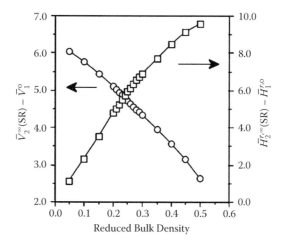

FIGURE 8.9 Solvent density dependence of $(\overline{V}_2^{\infty}(SR) - \overline{V}_1^{o})$ and $(\overline{H}_2^{r,\infty}(SR) - \overline{H}_1^{r,o})$ for the repulsive Lennard–Jones Ne (2)/Xe (1) system along the near-critical reduced isotherm $T_r = 1.02$. All quantities are made dimensionless using the solvent's Lennard–Jones parameters from Table 8.2.

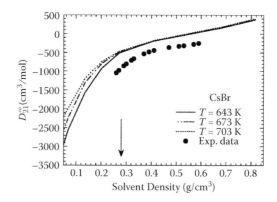

FIGURE 8.10 Behavior of $D_{21}^{\infty} = (\kappa_T^{IG}/\kappa_l)\overline{V}_2^{\infty}$ for an infinitely dilute CsBr aqueous solution as a function of the solvent density along three supercritical isotherms in comparison with experimental data. (Data from J. Sedlbauer, E. M. Yezdimer, and R. H. Wood, 1998, Partial Molar Volumes at Infinite Dilution in Aqueous Solutions of NaCl, LiCl, NaBr, and CsBr at Temperatures from 550 K to 725 K, *Journal of Chemical Thermodynamics*, 30, 3.) Arrow indicates the estimated critical density of the model solvent.

In order to interpret and offer an explanation for the weak temperature dependence exhibited by $D_{21}^{\infty}(T, \rho \geq \rho_c)$, we analyze its link with two closely related quantities, $N_{2,ex}^{\infty}(SR)$ and $\rho^{-1}(C_{12}^{\infty} - C_{11}^{o})$ (e.g., see Equation 8.6 and Equation 8.10), that also exhibit negligible temperature dependence for supercritical densities as illustrated in Figure 8.11. The link becomes clearer if we rewrite the quantity D_{21}^{∞}, after invoking Equation 8.4 through Equation 8.8, in the following four alternative forms,

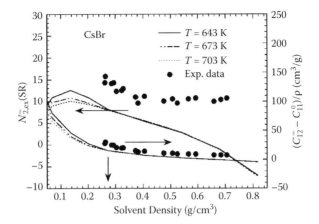

FIGURE 8.11 Behavior of $N_{2,ex}^\infty(SR)$ and $\rho^{-1}(C_{12}^\infty - C_{11}^0)$ for an infinitely dilute CsBr aqueous solution as a function of the solvent density along three supercritical isotherms in comparison with experimental data. (Data from J. Sedlbauer, E. M. Yezdimer, and R. H. Wood, 1998, Partial Molar Volumes at Infinite Dilution in Aqueous Solutions of NaCl, LiCl, NaBr, and CsBr at Temperatures from 550 K to 725 K, *Journal of Chemical Thermodynamics*, 30, 3.) Vertical arrow indicates the estimated critical density of the model solvent.

$$\begin{aligned} D_{21}^\infty &= D_{11}^0 + v\bar{V}_1^0\left(C_{11}^0 - C_{12}^\infty\right) \\ &= D_{11}^0 - \bar{V}_1^0 N_{2,ex}^\infty(SR) \\ &= D_{11}^0 + \kappa_T^{IG}\bar{V}_1^0\left(\partial p/\partial x_2\right)_{T,\rho}^\infty \\ &= D_{11}^0 + \bar{V}_2^\infty(SR) - v\bar{V}_1^0 \end{aligned} \tag{8.26}$$

with

$$D_{11}^0 = \bar{V}_1^0\left(\kappa_T^{IG}/\kappa_T\right) \tag{8.27}$$

Note that D_{11}^∞ can be thought of as the (Lewis–Randall) ideal solution counterpart of D_{21}^∞, that is, when all 22, and 12 interactions are identically the same as those of the pure solvent, 11. A closer look at Equations 8.26 and 8.27 in the context of the observed behavior for the density dependence of the solvation quantities $N_{2,ex}^\infty(SR)$ and $\rho^{-1}(C_{12}^\infty - C_{11}^0)$ suggests that a more appropriate regression quantity than $D_{21}^\infty(T,\rho)$ would be $\Delta D_{21}^\infty \equiv D_{21}^\infty - D_{11}^0$, since this quantity depends exclusively on the solute–solvent molecular asymmetries, that is,

$$\Delta D_{21}^\infty = v\bar{V}_1^0\left(C_{11}^0 - C_{12}^\infty\right) \tag{8.28}$$

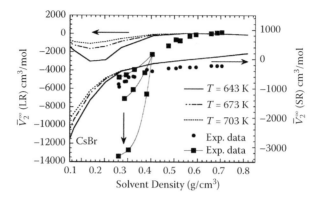

FIGURE 8.12 Behavior of $\bar{V}_2^\infty(SR)$ and $(\bar{V}_2^\infty(LR) \equiv \bar{V}_2^\infty - \bar{V}_2^\infty(SR))$ for an infinitely dilute CsBr aqueous solution as a function of the solvent density along three supercritical isotherms in comparison with experimental data. (Data from J. Sedlbauer, E. M. Yezdimer, and R. H. Wood, 1998, Partial Molar Volumes at Infinite Dilution in Aqueous Solutions of NaCl, LiCl, NaBr, and CsBr at Temperatures from 550 K to 725 K, *Journal of Chemical Thermodynamics*, 30, 3.) Vertical arrow indicates the estimated critical density of the model solvent. The three lines joining the experimental data correspond approximately to the isotherms of 669 K, 686 K, and 709 K from bottom to top, respectively.

In this sense, ΔD_{21}^∞ measures directly the relative structural changes undergone by the solvent around the solute, and can be linked to the alternative macroscopic quantities (also known as *solute-induced effects*) as follows,

$$\Delta D_{21}^\infty = -\bar{V}_1^\circ N_{2,ex}^\infty(SR)$$

$$= \kappa_T^{IG} \bar{V}_1^\circ \left(\partial p / \partial x_2\right)_{T,\rho}^\infty \qquad (8.29)$$

$$= \bar{V}_2^\infty(SR) - v\bar{V}_1^\circ$$

Finally, the rather strong temperature dependence of $D_{21}^\infty(T, \rho < \rho_c)$ is not an unexpected behavior if we consider the limiting behavior of $N_{2,ex}^\infty(SR)$ as the solvent density approaches zero, (see Section 8.3.3.2 for more details);

$$N_{2,ex}^\infty(SR) \approx 2v\rho^3(k_B T)^2 \left(B_{12}(T) - B_{11}(T)\right) + \dots \qquad (8.30)$$

an expression that clearly captures the strong temperature and density dependence predicted by the IE calculations of the model systems and observed experimentally, as illustrated in Figure 8.12 for high-temperature infinite dilute aqueous CsBr solutions.

8.2.3 Solvation Effects on the Kinetic Rate Constant

Chemical reactions in highly compressible (near-critical) media have been the focus of intense research for the last quarter of the century because of their applications in new "green" technologies such as waste treatment, material processing, biomass to fuel conversion, as well as electrochemical and enzymatic reactions (Williams and Clifford 2000; Brunner 2004; Hutchenson, Scurto, and Subramaniam 2009; Knez 2009). While the most compelling reason to carry out reactions in highly compressible media is the opportunity to tune the solvation power of the medium (i.e., species chemical potentials) and attenuate transport limitations by adjustment of the system pressure and/or temperature (Bruno and Ely 1991; Clifford 1999; Licence and Poliakoff 2008), there has been considerable controversy regarding the existence of *unusual* behavior (occasionally referred to as *anomalies*) in reaction kinetics at near and supercritical conditions (e.g., Brennecke 1993; Savage et al. 1995). In fact, it has been conjectured that the isothermal kinetic rate constants of some reactions carried out in near-critical solvent could not be ascribed solely to thermodynamic pressure effects, but rather to local composition enhancement and density augmentation effects around reactants (Brennecke 1993; Ellington, Park, and Brennecke 1994; Reaves and Roberts 1999), a claim centered around the lack of agreement between experimental measurements and predicted values from transition state theory (TST) using equation of state calculations.

Instead of arguing about the validity of the above conjectures, here we invoke the solvation formalism, Section 8.2, to rationalize some experimental findings and their interpretations by drawing explicit links between the (macroscopic) thermodynamic pressure effect on the kinetic rate constant and the (microscopic) species solvation behavior in a highly compressible medium. To that end, we study the solvent effect (or, more precisely, the solvation effect) on the kinetic rate constant within the framework of the TST (Hynes 1985; Steinfeld, Francisco, and Hase 1989), and its thermodynamic formulation that allows us to link it to changes of Gibbs free energy of activation.

Our analysis is based on the following simple reaction,

$$\nu_R R + \nu_S S \quad \underset{}{\overset{\text{solvent}}{\rightleftharpoons}} \quad \mathfrak{R}^{\ddagger} \quad \longrightarrow \quad \text{products} \tag{8.31}$$

where R is a solute reactant, S is a reactive cosolvent, $\mathfrak{R}^{\ddagger}$ is the activated complex in equilibrium with the reactants in solution, and ν_i are the corresponding stoichiometry coefficients. According to the TST, the reaction rate constant is given by (Hynes 1985)

$$k^{\text{TST}} = \frac{k_B T K_o}{\hbar} K_c^{\ddagger} \tag{8.32}$$

where $\hbar$ is the Planck constant/2π, K_o is a factor to provide the correct units for k^{TST}, and $K_c^{\ddagger}$ is the molar-based equilibrium constant for the activation process described by Equation 8.31. By expressing $K_c^{\ddagger}$ in terms of the species fugacities, the pressure effect on the kinetic rate constant becomes (for more details see Appendix A of Chialvo, Cummings, and Kalyuzhnyi [1998])

$$\left(\partial \ln k^{\text{TST}} / \partial p \right)_{T,x} = -\Delta \bar{V} \kappa_T + \left(\nu_R \bar{V}_R^{\infty} + \nu_S \bar{V}_S^{\infty} - \nu_{\mathfrak{R}} \bar{V}_{\mathfrak{R}}^{\infty} \right) / k_B T \tag{8.33}$$

where $\Delta \overline{V}^{\ddagger} = v_R \overline{V}_R^{\infty} + v_S \overline{V}_S^{\infty} - v_{\Re} \overline{V}_{\Re}^{\infty}$ is the actual activation volume, and $\Delta v = v_R + v_S - v_{\Re}$. While Equation 8.33 provides a macroscopic expression for the pressure derivative of the TST kinetic rate constant, it also allows us to introduce the solvation effects in a precisely microscopic way, by invoking the solvation formalism (Chialvo, Cummings, and Kalyuzhnyi 1998), that is,

$$\left(\partial \ln k^{\text{TST}}/\partial p\right)_{T,x} = \rho \kappa_T \left[v_R \left(\overline{V}_R^{\infty}(\text{SR}) - \overline{V}_R^{\infty(\text{IG})}(\text{SR})\right) + v_s \left(\overline{V}_S^{\infty}(\text{SR}) - \overline{V}_S^{\infty(\text{IG})}(\text{SR})\right) \right.$$
$$\left. - v_{\Re} \left(\overline{V}_{\Re}^{\infty}(\text{SR}) - \overline{V}_{\Re}^{\infty(\text{IG})}(\text{SR})\right) \right]$$

(8.34)

Thus, the pressure coefficient of the kinetic rate constant can be factorized into two terms, one involving the solvent's isothermal compressibility, and the other containing short-ranged solvation contributions to the species i in solutions relative to their ideal gas contributions, $\overline{V}_i^{\infty}(\text{SR}) - \overline{V}_i^{\infty(\text{IG})}$, where all these finite quantities can be explicitly written in terms of experimentally obtained volumetric properties (*vide supra* Section 8.2).

In passing, we should note that any "anomalous" behavior of this pressure effect in a highly compressible medium is simply the manifestation of the large solvent compressibility. In fact, we can remove the anomalous behavior by simply invoking the density rather than pressure effect, that is,

$$\left(\partial \ln k^{\text{TST}}/\partial \rho\right)_{T,x} = \left[v_R \left(\overline{V}_R^{\infty}(\text{SR}) - \overline{V}_R^{\infty(\text{IG})}(\text{SR})\right) + v_s \left(\overline{V}_S^{\infty}(\text{SR}) - \overline{V}_S^{\infty(\text{IG})}(\text{SR})\right) \right.$$
$$\left. - v_{\Re} \left(\overline{V}_{\Re}^{\infty}(\text{SR}) - \overline{V}_{\Re}^{\infty(\text{IG})}(\text{SR})\right) \right]$$

(8.35)

an expression that clearly highlights the effect of the differential solvation of the species on the kinetic rate constant. These two equations allow us to extract some valuable information regarding special scenarios that we can take as reference systems, including the case of ideal gas reactants, ideal gas solvent, and/or a combination of both; for example, if the reactants behave as ideal gas species then, from Equations 8.34 and 8.35 we have that $(\partial \ln k^{\text{TST}}/\partial p)_{T,x} = 0$ and $(\partial \ln k^{\text{TST}}/\partial p)_{T,x} = 0$.

Now, we interpret the effect of species-solvent molecular asymmetries on the pressure dependence of the kinetic rate constants for reacting systems studied by Roberts et al. (1995) according to the solvation formalism. The system under consideration consists of triplet benzophenone (^{3}BP) as an infinitely dilute reactant, O_2 as an infinitely dilute reactive cosolvent, and the infinitely dilute transition state (TS) species all immersed in near critical CO_2 solvent, where all species are described in terms of Lennard–Jones interactions (see Table 8.3) and unlike-pair interactions based on the Lorentz–Berthelot combining rules.

To highlight the central role played by the partial molecular volume of species 2 in solution $\overline{V}_i^{\infty}$, and especially the solvation counterpart $\overline{V}_i^{\infty}(\text{SR})$, we display in Figure 8.13 their density dependence for the three species at infinite dilution along the near-critical isotherm $T_r = 1.01$, where 100 volume units represent ~2.9 L·mol^{-1}, therefore the species partial molecular volumes at these near-critical conditions vary

TABLE 8.3
Lennard–Jones Potential Parameters for the Reacting Quaternary Mixture

ij-Interactions	$\sigma_{ij}(\text{Å})$	$\varepsilon_{ij}/k_B(\text{K})$
CO_2-CO_2	3.643	232.2
O_2-O_2	3.355	119.0
^3BP-^3BP	6.580	625.7
TS-TS[a]	7.291	681.7

[a] Parameters determined from the critical properties given in C. B., Roberts, J. W. Zhang, J. E. Chateauneuf, and J. F. Brennecke, 1995, Laser Flash-Photolysis and Integral-Equation Theory to Investigate Reactions of Dilute Solutes with Oxygen in Supercritical Fluids, *Journal of the American Chemical Society*, 117, 6553 and in C. R. Robinson, S. G. Sligar, M. L. Johnson, and G. K. Ackers, 1995, Hydrostatic and Osmotic Pressure as Tools to Study Macromolecular Recognition, *Energetics of Biological Macromolecules*, 259, 395.

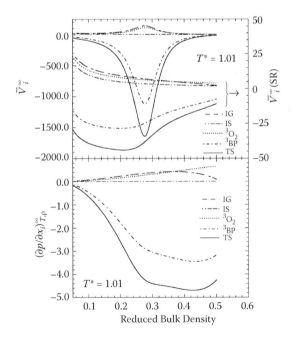

FIGURE 8.13 Solvent density dependence of the different species (i) partial molecular volume $\overline{V}_i^{\infty}$, its solvation counterparts $\overline{V}_i^{\infty}(\text{SR})$, and the corresponding $(\partial p/\partial x_i)_{T,\rho}^{\infty}$ for the reactive system $^3BP + {^3O_2} + CO_2 \rightleftharpoons TS$ along the near-critical reduced isotherm $T_r = 1.01$. All quantities are made dimensionless using the solvent's Lennard–Jones parameters from Table 8.3.

from ~4.3 L·mol⁻¹ (e.g., an IG solute) to ~47.9 L·mol⁻¹ (TS complex) while the solvation counterparts go from ~0.2 L·mol⁻¹ to ~1.1 L·mol⁻¹.

For illustration purposes, we have also included in these graphs the properties of two limiting cases, a *solute* in the ideal solution (IS), that is, the pure solvent, and the ideal gas solute (IG). These cases are very convenient references because they encompass two extreme molecular asymmetries, that is, nonzero intermolecular interactions with null solute–solvent asymmetry, and zero solute–solvent interactions, respectively. As the solute–solvent interactions become less repulsive, $(\partial p/\partial x_i^\infty)_{T,\rho,\{x\}'}^\infty > 0$ approaches its value for the ideal gas solute, $(\partial p/\partial x_i^\infty)_{T,\rho,\{x\}'}^\infty = \rho k_B T - \kappa_T^{-1}$, then reaches zero for the ideal solution, and finally turns negative as the solute–solvent molecular asymmetries become more significant.

The effect of the solute–solvent molecular asymmetries on the isothermal-pressure (or density) derivative of the kinetic rate constant might be highlighted by considering the sensitivity of the species solvation properties to small perturbations of the unlike-pair interactions, for example, deviations from some reference combining rules (Chialvo 1991; Vlcek, Chialvo, and Cole 2011). Since we have used the Lorentz–Berthelot (LB) combining rules throughout the reported IE calculations, here we illustrate how small deviations of the unlike-pair interactions from the corresponding LB rules can translate into significant changes in the magnitude and sign of either $(\partial \ln k^{TST}/\partial p)_{T,x}$ or $(\partial \ln k^{TST}/\partial \rho)_{T,x}$. In fact, as illustrated in Figure 8.14, a 4% decrease from the Berthelot rule, $\xi = (\varepsilon_{1i}/\varepsilon_{1i}^B) \approx 0.96$ and a 7% decrease from the Lorentz rule, $\eta = (\sigma_{1i}/\sigma_{1i}^L) \approx 0.93$ for the TS-CO_2 pair interactions is enough to compensate for the ³BP-CO_2 pair interaction asymmetry, and consequently, to cross the threshold $(\partial \ln k^{TST}/\partial p)_{T,x} \approx (\partial \ln k^{TST}/\partial \rho)_{T,x} \approx 0$ for the reaction ³BP + ³O₂ + CO_2 ⇔ TS along the near-critical isotherm $T_r = 1.01$.

Although most experimental data about supercritical solvation phenomena are reported using the system pressure as the dependent variable, obviously because this is the manipulated variable in the experiments, the natural dependent variable is actually the system density. In fact, the system density is typically estimated using the total pressure as input for an equation of state or standard thermodynamic tables, after assuming an effectively infinitely dilute system. While this assumption might be accurate enough for highly dilute solutions under normal state conditions, this approach might result in significant deviations from the actual solution densities in highly compressible systems. For the sake of argument, let us analyze the change of pressure upon the addition of small by finite amounts of solute species in an otherwise pure solvent,

$$p(T,x_2) = p(T,x_2 = 0) + \sum_i (\partial p/\partial x_i)_{T,\rho}^\infty x_i + \cdots \text{h.o.t} \qquad (8.36)$$

According to the reported IE calculations, $(\partial p/\partial x_i)_{T,\rho}^\infty$ is of the order of 1–3 kbar in the near critical solvent for the solutes under study, and considering the global mole fraction, $\Sigma_i x_i \equiv x_2 < 0.01$, we can estimate that $p(T,x_2 \neq 0) - p(T,x_2 = 0) \approx 30$ bar. Now, if we assumed that the system was at infinite dilution under the measured pressure $p(T,x_2 \neq 0)$, as usually done by the experimentalist (Kimura and Yoshimura

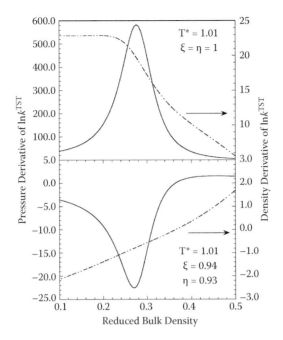

FIGURE 8.14 Comparison between the density dependence of $(\partial \ln k^{TST}/\partial p)_{T,x}$ and $(\partial \ln k^{TST}/\partial \rho)_{T,x}$ for the reactive system $^3BP + {}^3O_2 + CO_2 \rightleftharpoons TS$ along the near-critical reduced isotherm $T_r = 1.01$, when all unlike-pair interactions are described by the Lorentz–Berthelot (top) or perturbed Lorentz–Berthelot combining rules (bottom). All quantities are made dimensionless using the solvent's Lennard–Jones parameters from Table 8.3.

1992; Weinstein et al. 1996; Clifford et al. 1998; Reaves and Roberts 1999), then the estimated density for the pure solvent would be either smaller or larger than the actual pure solvent density $\rho = [\bar{V}_1^o]^{-1}_{p(T,x_2=0)}$. In fact, according to the isothermal density dependence of $(\partial p/\partial x_i)^{\infty}_{T,\rho}$, it becomes obvious that the magnitude of the density difference $\Delta \rho = [\bar{V}_1^o]^{-1}_{p(T,x_2 \neq 0)} - [\bar{V}_1^o]^{-1}_{p(T,x_2=0)}$ will be most significant within the same density region where $(\partial p/\partial x_i)^{\infty}_{T,\rho}$ exhibits the largest magnitude, typically but not necessarily, around the near-critical solvent density (e.g., Figure 8.13). Unfortunately, the consequences of these density differences have been usually, and mistakenly, taken as evidence of an anomalous solvation behavior ascribed to some unclear near-critical phenomena.

8.3 SOLVATION PHENOMENA IN MULTICOMPONENT DILUTE SOLUTIONS

One feature that makes dilute near-critical solutions technologically relevant and theoretically challenging is the condition of high dilution, albeit finite solute concentration. For modeling purposes, it is always advantageous to invoke a well-characterized model to play the role of the reference system, from which we can attempt to determine the properties of the real system by a perturbation expansion around the chosen reference. For the solvation behavior of sparingly soluble solutes, the obvious choice for that reference is the infinitely dilute solution at the same temperature and pressure. This is a natural choice in that the systems of interest are dilute solutions, whose properties can be obtained from suitable free-energy composition expansions, based on fugacity or activity coefficients (Van Ness and Abbott 1982; Prausnitz, Lichtenthaler, and Gomes de Azevedo 1986), around the infinite dilution counterpart, for which we have a full characterization based on the solvation formalism discussed in Section 8.2.

Truncated composition expansions have been frequently used in fluid phase equilibrium calculations, especially for binary mixtures (O'Connell and Haile 2005) for which compliance with the Gibbs–Duhem equation also implies exactness of the corresponding differentials. However, care must be exercised when dealing with multicomponent systems because they might become thermodynamically inconsistent whenever the truncated expressions do not fulfill the condition of exactness, also known as the *Maxwell relations* (O'Connell and Haile 2005). In other words, if the truncated expansions are not state functions, their predictions will depend on the path used for the integration of the differential expressions; consequently, the validity of any related theoretical developments and corresponding conclusions will become questionable (e.g., see Appendix F of Chialvo et al. [2008]). Moreover, the identification and calculation of the expansion coefficients, including their precise microscopic meanings in the context of the reference system, are as significant as the compliance of the thermodynamic consistency.

In what follows, we describe succinctly the development of isothermal–isobaric second-order truncated composition expansions of the relevant partial molecular properties of species in ternary systems, provide a molecular-based interpretation of the expansion coefficients, derive some special case systems, show how the expansions reduce to well-known results for the zero-density limit, and discuss some modeling implications.

8.3.1 SECOND-ORDER TRUNCATED EXPANSIONS FOR TERNARY DILUTE SOLUTIONS

Our goal here is to develop a self-consistent second-order truncated expansion of the residual chemical potential of species in ternary systems involving a compressible solvent and two solutes at high (but finite) dilution. For the sake of simplicity, we consider here a ternary mixture at constant pressure and temperature that comprises N_1 solvent molecules, N_2 solute molecules of species 2, and N_3 cosolute molecules of species 3, respectively, for which the corresponding residual chemical potentials, that is, partial molecular fugacity coefficients (O'Connell and Haile 2005) at finite composition are expanded around the infinite dilution condition for the two solutes as follows,

$$\ln \overline{\phi}_1(x_2, x_3) = \ln \overline{\phi}_1^0 + a_1 x_2 + b_1 x_3 + c_1 x_2^2 + d_1 x_3^2 + e_1 x_2 x_3$$

$$\ln \overline{\phi}_2(x_2, x_3) = \ln \overline{\phi}_2^\infty + a_2 x_2 + b_2 x_3 + c_2 x_2^2 + d_2 x_3^2 + e_2 x_2 x_3 \qquad (8.37)$$

$$\ln \overline{\phi}_3(x_2, x_3) = \ln \overline{\phi}_3^\infty + a_3 x_2 + b_3 x_3 + c_3 x_2^2 + d_3 x_3^2 + e_3 x_2 x_3$$

where we have chosen x_2 and x_3 as the independent variables, $\mu_i^{r,\otimes}(x_2, x_3)_{p,T} = k_B T \ln \overline{\phi}_i^{\otimes}(x_2, x_3)$ is the isothermal–isobaric residual chemical potential of species i, and $\otimes$ denotes either the pure component, or infinite dilution condition. Obviously, the 15 expansion coefficients are not all independent, since they are connected by two thermodynamic constraints, that is, they must satisfy simultaneously the Gibbs–Duhem (GD) equation and the exactness (Maxwell relations, MR) of the mixture's residual Gibbs free-energy function $G^r(p, T, N_1, N_2, N_3)$ (O'Connell and Haile 2005).

By following the procedure described in detail elsewhere (Chialvo et al. 2008), we arrive at the desired expressions,

$$\ln \bar{\phi}_1(x_2, x_3) = \ln \bar{\phi}_1^\circ + 0.5 k_{22} x_2^2 + 0.5 k_{33} x_3^2 + k_{23} x_2 x_3$$

$$\ln \bar{\phi}_2(x_2, x_3) = \ln \bar{\phi}_2^\infty - k_{22} x_2 - k_{23} x_3 + 0.5 k_{22} x_2^2 + 0.5 k_{33} x_3^2 + k_{23} x_2 x_3 \quad (8.38)$$

$$\ln \bar{\phi}_3(x_2, x_3) = \ln \bar{\phi}_3^\infty - k_{23} x_2 - k_{33} x_3 + 0.5 k_{22} x_2^2 + 0.5 k_{33} x_3^2 + k_{23} x_2 x_3$$

where the coefficients $k_{ij}(p,T)$ can be interpreted in terms of the infinite dilution limiting slopes, that is,

$$k_{ij}(p,T) \equiv -\left(\partial \ln \bar{\phi}_i / \partial x_j\right)^\infty_{p,T,x_k} \quad (8.39)$$

whose microscopic interpretation will be given below in Section 8.3.2. A close inspection of Equation 8.38 allows us to recast the solute properties in a more compact form, that is,

$$\ln \bar{\phi}_2(x_2, x_3) = \ln \bar{\phi}_2^\infty - k_{22} x_2 - k_{23} x_3 + \ln\left[\bar{\phi}_1(x_2, x_3) / \bar{\phi}_1^\circ\right]$$
$$\quad (8.40)$$
$$\ln \bar{\phi}_3(x_2, x_3) = \ln \bar{\phi}_3^\infty - k_{23} x_2 - k_{33} x_3 + \ln\left[\bar{\phi}_1(x_2, x_3) / \bar{\phi}_1^\circ\right]$$

where we have separated the linear from the quadratic composition dependence, a feature that will become handy in our discussion (Section 8.3.4) of the validity of current first-order truncated expansions. According to Equation 8.40, the composition dependence of the residual Gibbs free energy of a dilute ternary system becomes

$$g^r(p, T, x_2, x_3) = G^r(p, T, N_1, N_2, N_3) / (N_1 + N_2 + N_3)$$

$$= k_B T \left[\ln \bar{\phi}_1^\circ + x_2 \ln\left(\bar{\phi}_2^\infty / \bar{\phi}_1^\circ\right) + x_3 \left(\bar{\phi}_3^\infty / \bar{\phi}_1^\circ\right) - \right. \quad (8.41)$$

$$\left. 0.5 k_{22} x_2^2 - 0.5 k_{33} x_3^2 - k_{23} x_2 x_3 \right]$$

Because Equation 8.38 through Equation 8.40 are thermodynamically consistent, the corresponding pressure and temperature derivatives, that is, the partial molecular volumetric, enthalpic, and entropic counterparts will automatically satisfy the thermodynamic constraints discussed above. For example, by recalling that

$(\partial \ln \bar{\phi}_i / \partial p)_{T,x_j} = \beta(\bar{V}_i - \bar{V}_i^{IG}) = \beta \bar{V}_i^r$ (O'Connell and Haile 2005) and after invoking Equation 8.38, it follows that,

$$\bar{V}_1^r(x_2, x_3) = \bar{V}_1^{r,o} + k_B T \left[0.5\left(\partial k_{22}/\partial p\right)_T x_2^2 + 0.5\left(\partial k_{33}/\partial p\right)_T x_3^2 + \left(\partial k_{23}/\partial p\right)_T x_2 x_3 \right]$$

$$\bar{V}_2^r(x_2, x_3) = \bar{V}_2^{r,\infty} + k_B T \left[0.5\left(\partial k_{22}/\partial p\right)_T x_2^2 + 0.5\left(\partial k_{33}/\partial p\right)_T x_3^2 + \left(\partial k_{23}/\partial p\right)_T x_2 x_3 - \right.$$
$$\left. \left(\partial k_{22}/\partial p\right)_T x_2 - \left(\partial k_{23}/\partial p\right)_T x_3 \right] \qquad (8.42)$$

$$\bar{V}_3^r(x_2, x_3) = \bar{V}_3^{r,\infty} + k_B T \left[0.5\left(\partial k_{22}/\partial p\right)_T x_2^2 + 0.5\left(\partial k_{33}/\partial p\right)_T x_3^2 + \left(\partial k_{23}/\partial p\right)_T x_2 x_3 - \right.$$
$$\left. \left(\partial k_{23}/\partial p\right)_T x_2 - \left(\partial k_{33}/\partial p\right)_T x_3 \right]$$

Alternatively, the solute partial molecular volumes can be written in the same form as Equation 8.40 for the corresponding partial molecular fugacity coefficients, that is,

$$\bar{V}_2^r(x_2, x_3) = \bar{V}_2^{r,\infty} - k_B T \left[\left(\partial k_{22}/\partial p\right)_T x_2 + \left(\partial k_{23}/\partial p\right)_T x_3 \right] + \left[\bar{V}_1^r(x_2, x_3) - \bar{V}_1^{r,o} \right]_{T,p}$$
$$\bar{V}_3^r(x_2, x_3) = \bar{V}_3^{r,\infty} - k_B T \left[\left(\partial k_{23}/\partial p\right)_T x_2 + \left(\partial k_{33}/\partial p\right)_T x_3 \right] + \left[\bar{V}_1^r(x_2, x_3) - \bar{V}_1^{r,o} \right]_{T,p} \qquad (8.43)$$

where $[\bar{V}_i^r(x_2, x_3) - \bar{V}_1^{r,o}]_{T,p}$, involving the quadratic composition dependence, and $(\partial k_{23}/\partial p)_T$ are the quantities that couple the two residual partial molecular volumes in Equation 8.43. Consequently, the system's residual molar volume becomes

$$\upsilon^r(T, p, x_2, x_3) = V^r(p, T, N_1, N_2, N_3)/(N_1 + N_2 + N_3)$$
$$= \bar{V}_1^{r,o} + x_2\left(\bar{V}_2^{r,\infty} - \bar{V}_1^{r,o}\right)_{T,p} + x_3\left(\bar{V}_3^{r,\infty} - \bar{V}_1^{r,o}\right)_{T,p} - \qquad (8.44)$$
$$k_B T \left[0.5\left(\partial k_{22}/\partial p\right)_T x_2^2 + 0.5\left(\partial k_{33}/\partial p\right)_T x_3^2 + \left(\partial k_{23}/\partial p\right)_T x_2 x_3 \right]$$

Likewise, by recalling that $(\partial \ln \bar{\phi}_i / \partial T)_{p,x_j} = -(\beta/T)(\bar{H}_i - \bar{H}_i^{IG}) = -\bar{H}_i^r/k_B T^2$ (O'Connell and Haile 2005) and after invoking Equation 8.38, it follows,

$$\bar{H}_1^r(x_2, x_3) = \bar{H}_1^{r,o} - k_B T^2 \left[0.5\left(\partial k_{22}/\partial T\right)_p x_2^2 + 0.5\left(\partial k_{33}/\partial T\right)_p x_3^2 + \left(\partial k_{23}/\partial T\right)_p x_2 x_3 \right]$$

$$\bar{H}_2^r(x_2, x_3) = \bar{H}_2^{r,\infty} - k_B T^2 \left[0.5\left(\partial k_{22}/\partial T\right)_p x_2^2 + 0.5\left(\partial k_{33}/\partial T\right)_p x_3^2 + \left(\partial k_{23}/\partial T\right)_p x_2 x_3 - \right.$$
$$\left. \left(\partial k_{22}/\partial T\right)_p x_2 - \left(\partial k_{23}/\partial T\right)_p x_3 \right] \qquad (8.45)$$

$$\bar{H}_3^r(x_2,x_3) = \bar{H}_3^{r,\infty} - k_B T^2 \left[0.5\left(\partial k_{22}/\partial T\right)_p x_2^2 + 0.5\left(\partial k_{33}/\partial T\right)_p x_3^2 + \left(\partial k_{23}/\partial T\right)_p x_2 x_3 - \right.$$

$$\left. \left(\partial k_{23}/\partial T\right)_p x_2 - \left(\partial k_{33}/\partial T\right)_p x_3 \right]$$

Alternatively, we can express the solute partial molecular enthalpy in the same form as Equation 8.40 and Equation 8.43, that is,

$$\bar{H}_2^r(x_2,x_3) = \bar{H}_2^{r,\infty} + k_B T^2 \left[\left(\partial k_{22}/\partial T\right)_p x_2 + \left(\partial k_{23}/\partial T\right)_p x_3 \right] + \left[\bar{H}_1^r(x_2,x_3) - \bar{H}_1^{r,o} \right]$$

$$\bar{H}_3^r(x_2,x_3) = \bar{H}_3^{r,\infty} + k_B T^2 \left[\left(\partial k_{23}/\partial T\right)_p x_2 + \left(\partial k_{33}/\partial T\right)_p x_3 \right] + \left[\bar{H}_1^r(x_2,x_3) - \bar{H}_1^{r,o} \right]$$

(8.46)

Similar to the case of the other residual properties, the quadratic-dependent term $[\bar{H}_1^r(x_2,x_3) - \bar{H}_1^{r,o}]_{T,p}$ and $(\partial k_{23}/\partial T)_p$ link the two expressions for the residual partial molecular properties in Equation 8.46, so that the system's residual molar enthalpy becomes,

$$h^r(T,p,x_2,x_3) = H^r(p,T,N_1,N_2,N_3)/(N_1 + N_2 + N_3)$$

$$= \bar{H}_1^{r,o} + x_2\left(\bar{H}_2^{r,\infty} - \bar{H}_1^{r,o}\right)_{T,p} + x_3\left(\bar{H}_3^{r,\infty} - \bar{H}_1^{r,o}\right)_{T,p} +$$

(8.47)

$$k_B T^2 \left[0.5\left(\partial k_{22}/\partial T\right)_p x_2^2 + 0.5\left(\partial k_{33}/\partial T\right)_p x_3^2 + \left(\partial k_{23}/\partial T\right)_p x_2 x_3 \right]$$

Obviously, the entropic contributions are simply obtained by substituting Equation 8.38 and Equation 8.45 into the definition of the species residual partial molecular entropy as illustrated in Appendix D of Chialvo et al. (2008).

8.3.2 MICROSCOPIC INTERPRETATION OF THE EXPANSION COEFFICIENTS

Now we make contact between the macroscopic coefficients $k_{ij}(T,p)$, defined by Equation 8.39, and the microstructural details of the reference system, that is, the solvation behavior of the species in the infinitely dilute system, by invoking the solvation formalism discussed in Section 8.2 (and references therein) from which we have that

$$k_{ij}(T,p) = \left(G_{11}^0 + G_{ij}^\infty - G_{1i}^\infty - G_{1j}^\infty\right)/\bar{V}_1^0 \quad i,j = 2,3$$

(8.48)

Moreover, as discussed extensively elsewhere (Chialvo and Cummings 1994, 1999; Chialvo et al. 2000c), and according to the argument developed previously in Section 8.2, we can split the TCFI of an infinitely dilute solution into direct (DCFI) and indirect (ICFI) contributions as follows (Chialvo 1993b),

$$G_{ij}^{\infty} = \bar{V}_1^{\circ} C_{ij}^{\infty} + \kappa_T k_B T C_{1i}^{\infty} C_{1j}^{\infty} \quad i,j = 2,3$$

$$G_{1j}^{\otimes} = \kappa_T k_B T C_{1j}^{\otimes} \quad\quad\quad\quad j = 1,3$$

(8.49)

where species 1 always denotes the solvent, and the superscript $\otimes$ indicates either a pure, $j = 1$, or an infinite dilution, $j \neq 1$, condition. Therefore, by invoking Equation 8.49, we can also split k_{ij}, Equation 8.48, into a solvation $k_{ij}^{\text{solvation}}$ and a compressibility-driven $k_{ij}^{\kappa_T}$ contribution, that is,

$$k_{ij}(T,p) = \underbrace{C_{ij}^{\infty}}_{\text{solvation}} + \underbrace{\left(\kappa_T/\kappa_T^{\text{IG}}\right)\left(C_{11}^0 + C_{1i}^{\infty} C_{1j}^{\infty} - C_{1i}^{\infty} - C_{1j}^{\infty}\right)}_{\text{compressibility-driven}} \quad i,j = 2,3 \quad (8.50)$$

$$= k_{ij}^{\text{solvation}} + k_{ij}^{\kappa_T}$$

Obviously, $k_{ij}^{\kappa_T}$ becomes the dominant contribution in the compressible region of the solvent, where it scales as κ_T, with a sign-defining prefactor that depends on the solvation behavior of the solute species. In fact, as previously shown elsewhere (Chialvo 1993b), the behavior of $k_{ij}(T,p)$ in a highly compressible medium depends on whether $i = j$ or $i \neq j$ because at those conditions we expect that

$$k_{ii}(T,p) \rightarrow \kappa_T \left(1 - C_{1i}^{\infty}\right)^2$$

$$\sim \kappa_T \left[\left(\partial p/\partial x_i\right)_{T,\rho}^{\infty}\right]^2 \sim \kappa_T \left[N_{i,\text{ex}}^{\infty}(\text{SR})\right]^2 \quad i = 2,3 \quad (8.51)$$

and

$$k_{23}(T,p) \rightarrow \kappa_T \left(1 - C_{12}^{\infty}\right)\left(1 - C_{13}^{\infty}\right)$$

$$\sim \kappa_T \left(\partial p/\partial x_2\right)_{T,\rho}^{\infty} \left(\partial p/\partial x_3\right)_{T,\rho}^{\infty} \sim \kappa_T N_{2,\text{ex}}^{\infty}(\text{SR}) N_{3,\text{ex}}^{\infty}(\text{SR}) \quad (8.52)$$

In other words, Equations 8.51 and 8.52 indicate that from a microscopic view-point, $k_{ii}(T,p) > 0$ regardless of the sign of C_{1i}^{∞} ($i = 2,3$), while the sign of $k_{23}(T,p)$ depends on whether both C_{1i}^{∞} ($i = 2,3$) are simultaneously larger or smaller than unity. In particular, according to the Debenedetti–Mohamed (DM) classification (Debenedetti and Mohamed 1989), we expect that

$$k_{23}(p,T) > 0 \begin{cases} C_{12}^{\infty} > 1 \quad \text{and} \quad C_{13}^{\infty} > 1 & \rightarrow \quad \text{attractive} \\ 0 \leq C_{12}^{\infty} \leq 1 \quad \text{and} \quad 0 \leq C_{13}^{\infty} \leq 1 & \rightarrow \quad \text{weakly attractive} \quad (8.53) \\ C_{12}^{\infty} < 0 \quad \text{and} \quad C_{13}^{\infty} < 0 & \rightarrow \quad \text{repulsive} \end{cases}$$

whose macroscopic counterpart can be written in terms of the corresponding $(\partial p/\partial x_i)_{T,\rho}^{\infty}$, $N_{i,\text{ex}}^{\infty}(\text{SR})$, or $(\bar{V}_i^{\infty}(\text{SR}) - \bar{V}_1^{\circ})$ quantities, for example,

$$k_{23}(p,T) > 0 \rightarrow \tag{8.54}$$

$$
\begin{cases}
\left(\partial p / \partial x_2\right)_{T,\rho}^{\infty} < 0 \quad \text{and} \quad \left(\partial p / \partial x_3\right)_{T,\rho}^{\infty} < 0 & \rightarrow \quad \text{nonvolatile} \\[2ex]
0 \le \left(\partial p / \partial x_2\right)_{T,\rho}^{\infty} \le \left(\kappa_T^{IG}\right)^{-1} \quad \text{and} \quad 0 \le \left(\partial p / \partial x_3\right)_{T,\rho}^{\infty} \le \left(\kappa_T^{IG}\right)^{-1} & \rightarrow \quad \text{volatile} \\[2ex]
\left(\partial p / \partial x_2\right)_{T,\rho}^{\infty} > \left(\kappa_T^{IG}\right)^{-1} \quad \text{and} \quad \left(\partial p / \partial x_3\right)_{T,\rho}^{\infty} > \left(\kappa_T^{IG}\right)^{-1} & \rightarrow \quad \text{volatile}
\end{cases}
$$

where we have invoked the classification of volatile and nonvolatile solutes proposed by Levelt Sengers based on the size of the corresponding Krichevskii parameter (Levelt Sengers 1991). Note that in Equation 8.54, we are mapping the three types from the DM classification in terms of the corresponding $(\partial p / \partial x_i)_{T,\rho}^{\infty}$; consequently, we split the condition $(\partial p / \partial x_i)_{T,\rho}^{\infty} > 0$ into two subtle but significant subcases, based on the relative magnitude of $(\partial p / \partial x_i)_{T,\rho}^{\infty}$ in comparison with $(\kappa_T^{IG})^{-1}$.

8.3.3 SPECIAL CASE SYSTEMS FOR THE SECOND-ORDER TRUNCATED EXPANSION

At this point, we would like to illustrate how the expressions derived in Section 8.3.1 reduce to analogous forms for special system cases for which the thermodynamic consistency is already well established.

8.3.3.1 Dilute Binary Systems

Starting from the expressions of the ternary mixture, we can derive the corresponding binary system counterparts involving a solvent (species 1) and a solute (species 2). This is achieved by assuming that the solute (species 2) and the cosolute (species 3) in the original ternary system in Section 8.3.1 behave identically, that is, the two species are indistinguishable from an intermolecular interaction viewpoint. Consequently, by invoking Equation 8.48 it becomes obvious that this *pseudo*ternary system would be characterized by $k_{22} = k_{33} = k_{23}$, so that from Equation 8.38, the partial molecular fugacity coefficients for the binary system become equal to

$$\ln \overline{\phi}_1(z) = \ln \overline{\phi}_1^{\circ} + 0.5 k_{22} z^2 \tag{8.55}$$

$$
\begin{aligned}
\ln \overline{\phi}_2(z) &= \ln \overline{\phi}_2^{\infty} - k_{22}\left(z - 0.5 z^2\right) \\
&= \ln \overline{\phi}_2^{\infty} - k_{22} z + \ln\left[\overline{\phi}_1(z) / \overline{\phi}_1^{\circ}\right]
\end{aligned}
\tag{8.56}
$$

where the dummy mole fraction $z = x_2 + x_3$ denotes the mole fraction of the solute, that is, $z = 1 - x_1$. Thus, from Equation 8.41, and Equations 8.55 and 8.56, the residual Gibbs free energy of the binary system becomes

$$g^r(p,T,z) = k_B T \left[\ln \overline{\phi}_1^{\circ} + z \ln\left(\overline{\phi}_2^{\infty} / \overline{\phi}_1^{\circ}\right) - 0.5 k_{22} z^2 \right] \tag{8.57}$$

whose corresponding volumetric, enthalpic, and entropic expressions are given in full details elsewhere (Chialvo et al. 2008).

8.3.3.2 Mixtures of Imperfect Gases

Low-density gaseous systems are frequently described by virial-based equations of state, such as the case of the mixtures of imperfect gases whose behavior can be described accurately at the level of their second virial coefficients (Van Ness and Abbott 1982). These systems become an instructive example to illustrate that the consistency of the derived expressions in Section 8.3.1 is independent of the type of intermolecular forces involved. In fact, within the density range where $Z = 1 + Bp/k_BT$ is an adequate representation of the system behavior, with $B(T) = \Sigma_{i,j} x_i x_j B_{ij}(T)$ the total correlation function integrals $G_{ij}(T,p,x_k)$ become proportional to the second virial coefficients $B_{ij}(T)$ (Hansen and McDonald 1986; Ben-Naim 2006), that is,

$$\lim_{\rho \to 0} G_{ij}(p,T,x_k) = 4\pi \int_0^\infty \left[\exp(-\beta u_{ij}(r)) - 1 \right] r^2 dr$$

$$= -2B_{ij}(T)$$

(8.58)

where we have invoked the definition $g_{ij}(r;p,T,x_k) = \exp(-\beta \omega_{ij}(r;p,T,x_k))$, and the fact that the zero-density limit of the potential of mean force, $\lim_{\rho \to 0} \omega_{ij}(r;p,T,x_k) = u_{ij}(r)$ is the corresponding intermolecular pair potential $u_{ij}(r)$. Under these conditions, the molecular expression of the expansion coefficient $k_{ij}(T,p)$, Equation 8.48, reduces to the following simple combination of second virial coefficients, that is,

$$k_{ij}(T,p) = -2p\left(B_{11} + B_{ij} - B_{1i} - B_{1j}\right)/k_BT \qquad i,j = 2,3$$

(8.59)

Moreover, by invoking the statistical mechanic expression for the second virial coefficient of the mixture written in Van Ness' compact form (Van Ness and Abbott 1982), $B(T) = \Sigma_i x_i B_{ii}(T) + \Sigma_{i>j} x_i x_j \delta_{ij}(T)$ with $\delta_{ij}(T) = 2B_{ij} - B_{ii} - B_{jj}$, the expansion coefficients in Equation 8.38 for ternary mixtures of imperfect gases reduce to

$$k_{ij}(T,p) = p\left(\delta_{1i} + \delta_{1j} - \delta_{ij}\right)/k_BT \qquad i,j = 2,3$$

(8.60)

Consequently, by substitution of Equation 8.60 into Equation 8.38, we obtain the desired expressions for the mixture of imperfect gases, that is,

$$\ln \bar{\phi}_1(x_2,x_3) = \ln \bar{\phi}_1^\circ + (p/k_BT)\left[0.5\delta_{12}x_2^2 + 0.5\delta_{13}x_3^2 + (\delta_{12} + \delta_{13} - \delta_{23})x_2x_3\right]$$

$$\ln \bar{\phi}_2(x_2,x_3) = \ln \bar{\phi}_2^\infty - (p/k_BT)\left[\delta_{12}x_2 + (\delta_{12} + \delta_{13} - \delta_{23})x_3 - \right.$$

$$\left. 0.5\delta_{12}x_2^2 - 0.5\delta_{13}x_3^2 - (\delta_{12} + \delta_{13} - \delta_{23})x_2x_3\right]$$

(8.61)

$$\ln \bar{\phi}_3(x_2, x_3) = \ln \bar{\phi}_3^\infty - (p/k_B T)\big[(\delta_{12} + \delta_{13} - \delta_{23})x_2 + \delta_{13}x_3 -$$
$$0.5\delta_{12}x_2^2 - 0.5\delta_{13}x_3^2 - (\delta_{12} + \delta_{13} - \delta_{23})x_2 x_3\big]$$

Note that, after some straightforward manipulation, for example, Appendix E of Chialvo et al. (2008), these equations are precisely the expressions (4-121a–c) derived by Van Ness and Abbott (1982) for ternary mixtures obeying the equation of state $Z = 1 + Bp/k_B T$. Likewise, we can derive the corresponding volumetric, enthalpic, and entropic expressions for the mixture of imperfect gases after considering that (Chialvo et al. 2008)

$$(\partial k_{ij}/\partial p)_T = (\delta_{1i} + \delta_{1j} - \delta_{ij})/k_B T$$
$$(\partial k_{ij}/\partial T)_p = p\big[d(\delta_{1i} + \delta_{1j} - \delta_{ij})/dT\big]/k_B T - p(\delta_{1i} + \delta_{1j} - \delta_{ij})/k_B T^2$$
(8.62)

8.3.4 Highlights and Discussion on the Composition Expansions

The proposed second-order composition expansion highlights two features; namely, (a) that the expressions for the solvent's partial molecular properties do not contain first-order terms (e.g., Equation 8.38, Equation 8.42, and Equation 8.45), and consequently, (b) that the quadratic composition dependence of the corresponding solute and cosolute properties actually comes from that of the solvent, that is, the first expression in each of Equation 8.38, Equation 8.42, and Equation 8.45. These features are transferred to the special system cases, such as binary systems and the corresponding mixtures of imperfect gases analyzed in Section 8.3.3.1 to Section 8.3.3.2, and point to often-overlooked inconsistencies in first-order composition expansions.

Numerous publications have dealt with solvation phenomena involving dilute solutes in compressible media (Chialvo 1993b; Munoz and Chimowitz 1993; Li, Chimowitz, and Munoz 1995; Munoz, Li, and Chimowitz 1995; Ruckenstein and Shulgin 2001b, 2002a; Shulgin and Ruckenstein 2002a), based on the first-order truncated composition expansion for the solute partial molecular fugacity coefficients developed by Jonah and Cochran (1994) as a multicomponent generalization of Debenedetti and Kumar analysis for binary systems (Debenedetti and Kumar 1986). The common denominator in all these studies is the following first-order approximation,

$$\ln \bar{\phi}_2(x_2, x_3) \simeq \ln \bar{\phi}_2^\infty - k_{22}x_2 - k_{23}x_3$$
$$\ln \bar{\phi}_3(x_2, x_3) \simeq \ln \bar{\phi}_3^\infty - k_{23}x_2 - k_{33}x_3$$
(8.63)

which, according to Equation 8.40, would indicate that the validity of Equation 8.63 is tied to the condition $\bar{\phi}_1(x_2, x_3) = \bar{\phi}_1^\circ$. The immediate consequences of this condition are: (a) the Maxwell relations are only satisfied when $k_{22} = k_{33} = k_{23}$, that is, when the solute–solute, solute–cosolute, and cosolute–cosolute interactions are identically the same; consequently, when the system behaves effectively as a binary (e.g., Section 8.3.3.1); and (b) under the condition $k_{22} = k_{33} = k_{23}$, Equation 8.55 indicates that the only solution of $\ln[\bar{\phi}_1(x_2,x_3)/\bar{\phi}_1^\circ] = 0.5k_{22}(1-x_1)^2 = 0$ is given by $x_1 = 1.0$ since $k_{22} \neq 0$. In other words, the first-order approximation, Equation 8.63, is thermodynamically consistent only at infinite dilution, $x_2 = x_3 = 0$; consequently, the corresponding temperature and pressure derivatives will suffer from the same inconsistency (Chialvo et al. 2008).

8.3.5 SYNERGISTIC SOLVATION EFFECTS IN HIGHLY COMPRESSIBLE MEDIA AND THE VDW-1F MIXING RULES

The solubility enhancement of nonvolatile solutes in compressible solvents and the potential effect on it by other dilute species in solution have been the focus of much interest in the separation sciences (Bruno and Ely 1991; Kiran and Levelt Sengers 1994; Kiran, Debenedetti, and Peters 2000; Williams and Clifford 2000). These species are usually cosolutes, cosolvents, or a combination of both, leading to potential synergistic effects on their solubility, phenomena known as *cosolute* or *mixed-solute* (Kurnik and Reid 1982; Kwiatkowski, Lisicki, and Majewski 1984,), and *cosolvent* or *entrainer effects* (Brunner 1983; Van Alsten and Eckert 1993), respectively. Even though attempts to explain the existence of these solvation phenomena have been usually based on equations of state involving a variety of mixing rules (Brennecke and Eckert 1989), the earliest true molecular-based descriptions targeted a few fundamental issues concerning (a) the nature of the intermolecular asymmetry that originates the effect, (b) the link between this asymmetry and the system's microstructural manifestation, and (c) the translation of the microstructural evidence into meaningful macroscopic counterparts as a desirable target for any modeling effort (Chialvo 1993b; Jonah and Cochran 1994).

The microscopic interpretation of the expansion coefficients in Section 8.3.2 provides relevant insights into the suitability (or lack thereof) of a widely used set of mixing rules, in particular, for the description of the so-called mixed-solute and entrainer effects (Brennecke and Eckert 1989). In fact, as we discussed in Section 8.3.2 and elsewhere (Chialvo 1993b), the mixed-solute (and entrainer) system of interest is microscopically characterized by the radial distribution functions $g_{22}(r)$ and $g_{33}(r)$ decaying always to unity from above, (i.e., the slope $[dg_{ii}(r \rightarrow \text{large})/dr]_{i=2,3}$ < 0), while the solute–solvent $g_{12}(r)$ and cosolute–solvent $g_{13}(r)$ as well as the solute–cosolute $g_{23}(r)$ radial distribution functions might in principle decay to unity either from below, (i.e., $[dg_{ij}(r \rightarrow \text{large})/dr] > 0$) or from above, (i.e., $[dg_{ij}(r \rightarrow \text{large})/dr] < 0$) (Chialvo and Debenedetti 1992; Pfund and Cochran 1993).

The behavior of this set of radial distribution functions and, consequently, the thermophysical properties of the system, are not compatible with the ideas underlying the popular vdW-1f conformal mixing rules. In fact, vdW-1f are based on the conformity of the radial distribution functions, that is,

$$g_{ij}\left(r/\sigma_{ij}, k_B T/\varepsilon_{ij}, \rho\sigma_{ij}^3\right) = g_x\left(r/\sigma_x, k_B T/\varepsilon_x, \rho\sigma_x^3\right) \quad \text{for all } ij\text{-pairs} \qquad (8.64)$$

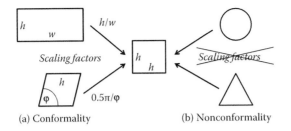

(a) Conformality   (b) Nonconformality

FIGURE 8.15   Schematic representation of conformal and nonconformal geometrical shapes.

where $\sigma_x^3 = \Sigma_{i,j} x_i x_j \sigma_{ij}^3$ and $\varepsilon_x = \Sigma_{i,j} x_i x_j \varepsilon_{ij} \sigma_{ij}^3 / \sigma_x^3$ represent the vdW-1f mixing rules (Henderson 1974), that is, the size and energy parameters defining the conformal intermolecular potential $u(r^{\circledR}) \equiv \varepsilon_{ij} \Im(r/\sigma_{ij}) = \varepsilon_x \Im(r/\sigma_x)$ of the hypothetical pure fluid whose isothermal compressibility is described by $\rho k_B T \kappa_T = 1 + 4\pi \rho \sigma_x^6 \int [g_x(r/\sigma_x, \rho \sigma_x^3, k_B T / \varepsilon_x) - 1] r^2 dr$. In order for the vdW-1f mixing rules to be compatible with the behavior of the radial distribution functions described above, that is, $[dg_{ii}(r \to \text{large})/dr]_{i=2,3} < 0$ and $[dg_{ij}(r \to \text{large})/dr] > 0$, the hypothetical fluid would have to describe simultaneously a positive (physical) and negative (unphysical) isothermal compressibility. The lack of conformality of the radial distribution functions exhibiting opposite limiting slopes can be depicted geometrically as illustrated in Figure 8.15, where we highlight the scaling factors that will allow the collapsing of the two conformal figures, case (a), and the lack of scaling factors for the nonconformal figures in case (b).

8.4   CONCLUSIONS

We have discussed some formal developments around the fluctuation theory of mixtures from Kirkwood and Buff, with special emphasis on the behavior of dilute species in highly compressible media. These formal results were then used to interpret a few specific solvation phenomena, to provide support to the selection of macroscopic quantities for the successful regression of experimental data, and to facilitate the development of truly molecular-based modeling of these challenging systems.

ACKNOWLEDGMENTS

Research sponsored by the Division of Chemical Sciences, Geosciences, and Biosciences, Office of Basic Energy Sciences, U.S. Department of Energy.

9 Molecular Thermodynamic Modeling of Fluctuation Solution Theory Properties

John P. O'Connell and Jens Abildskov

CONTENTS

Abstract: Fluctuation Solution Theory (FST) provides relationships between integrals of the molecular pair total and direct correlation functions and the pressure derivative of solution density, partial molar volumes, and composition derivatives of activity coefficients. For dense fluids, the integrals follow a relatively simple corresponding-states behavior even for complex systems, show well-defined relationships for infinite dilution properties in complex and near-critical systems, allow estimation of mixed-solvent solubilities of gases and pharmaceuticals, and can be expressed by simple perturbation models for densities and gas solubilities, including ionic liquids and complex mixtures such as coal liquids. The approach is especially useful in systems with strong nonidealities. This chapter describes successful application of such modeling to a wide variety of systems treated over several decades and suggests how to test equation of state (EOS) mixing rules.

9.1 INTRODUCTION

Among the many thermodynamic properties that can be treated by fluctuation solution theory, those for mixtures have had the longest history. Chapter 1 develops the relations from fluctuations in the grand ensemble for partial derivatives of the mole numbers with respect to chemical potentials at fixed temperature, volume, and other chemical potentials in terms of integrals of pair total correlation functions. Matrix inversions and thermodynamic manipulations connect these integrals to partial derivatives of the pressure with respect to mole numbers which are related to the partial molar volume, and derivatives with respect to density, that are related to the isothermal compressibility. Expansions of these properties about infinitely dilute solution were the focus of the seminal paper from Kirkwood and Buff (1951), and further works by Buff and coworkers where partial molar energies and heat capacities (Buff and Brout 1955), as well as perturbation theory for corresponding states (Buff and Schindler 1958) were developed. Pearson and Rushbrooke (1957) further extended the relations by involving the direct correlation function (DCF) to provide closure relations for correlation function integrals.

After some time, the basic formulation for activities and compressibilities was redone for the whole range of concentrations and for expressions that avoided matrix complexities by involving integrals of DCFs (O'Connell 1971b). It was noted much later that for molecular systems, these relations depended on the applicability of the "weak approximation" for angle-dependent intermolecular forces (Gray and Gubbins 1984; O'Connell 1994). In the absence of molecular simulations, this justified the successful applications of early works that used molecular theory (Gubbins and O'Connell 1974) and practical (Brelvi and O'Connell 1972, 1975a, 1975b, 1975c) perspectives, mainly via corresponding states. This FST approach was extended to correlate densities and activities of supercritical components in liquids and liquid mixtures (Mathias and O'Connell 1979, 1981; O'Connell 1981; Perry and O'Connell 1984) and of aqueous electrolyte solutions (Cabezas and O'Connell 1986; Cooney and O'Connell 1987; Perry, Cabezas, and O'Connell 1988; O'Connell, Hu, and Marshall 1999). Properties of near- and supercritical systems were also well described (Campanella, Mathias, and O'Connell 1987; O'Connell, Sharygin, and Wood 1996; Liu and O'Connell 1998; O'Connell and Liu 1998; Plyasunov, O'Connell, and Wood 2000; Plyasunov et al. 2000, 2001; Sedlbauer, O'Connell, and

Wood 2000; Plyasunov, Shock, and O'Connell 2006). In recent years, extensions have been made to ionic liquid systems and to solubilities of solids, such as pharmaceuticals, in mixed solvents (Abildskov, Ellegaard, and O'Connell 2009, 2010a, 2010b; Ellegaard, Abildskov, and O'Connell 2009; Ellegaard 2011).

The purpose of this chapter is to review these applications of FST methodology for correlating and predicting properties of a wide variety of systems and as a basis for validating EOS models. Some of the material is already in the previous review monograph (O'Connell 1990), though little of that discussion is repeated here. Since that time, advances have been made in several directions; these are the present focuses. Section 9.2 describes applications to pure component and mixture densities; Section 9.3 treats phase equilibria with a focus on dilute solutions; and Section 9.4 describes the use of FST formulations to test EOS models and mixing rules against data for binary total and direct correlation function integrals (TCFIs) and (DCFIs).

9.2 FLUCTUATION SOLUTION THEORY MODELING OF PURE COMPONENT AND MIXTURE DENSITY DEPENDENCES ON PRESSURE AND COMPOSITION

Relations between the TCFIs and derivatives of pressure with respect to molar density can be written for any number of components, as in Section 1.1.6 in Chapter 1. Matrix inversion techniques can provide expressions for all of the pair TCFIs of the system. Section 1.2 in Chapter 1 gives the full relations for applications to pure, binary, and ternary systems. As shown in Section 1.2.3 in Chapter 1, there is also a set of relations for the derivatives in terms of the DCFI, which are somewhat simpler and more direct. There are two modeling objectives with these relations. One is to obtain a solution density at elevated pressures; the other is to obtain the component partial molar volumes for the solution density variations with composition. The next section describes approaches that have been used for both objectives in a wide variety of pure and binary systems.

9.2.1 COMPRESSIBILITIES

FST gives relations for the reduced bulk modulus and the partial molar volume of component i to integrals of TCFIs and DCFIs (Section 1.2 in Chapter 1). The greatest use of this relationship has been for compressed liquids where the TCFI and DCFI show simple corresponding states dependence on density with weak temperature dependence (Brelvi and O'Connell 1972, 1975a, 1975b; Mathias and O'Connell 1979, 1981; Campanella, Mathias, and O'Connell 1987; Huang and O'Connell 1987; Abildskov, Ellegaard, and O'Connell 2010a). In terms of DCFIs, the expression is

$$\left(\frac{\partial \beta p}{\partial \rho}\right)_{T,\{x\}} = \sum_{i=1}^{n_c} x_i \sum_{j=1}^{n_c} x_j \left[1 - C_{ij}\left(T, \{\rho\}\right)\right] \tag{9.1}$$

The general expression for TCFIs involves matrix inverses. For pure components, Equation 9.1 becomes Equation 1.65 of Section 1.3.1 in Chapter 1, which is simple for TCFIs and DCFIs,

$$1/\rho\kappa_T k_B T = 1 - C = 1/(1 + \rho G) \tag{9.2}$$

Following the initial approach of Brelvi (Brelvi and O'Connell 1972) for $(1 - C)$, the most comprehensive corresponding states correlation was by Huang (Huang and O'Connell 1987) where C, scaled by a characteristic parameter C^*, is correlated by polynomials in ρ, scaled by V^*, and T, scaled by T^*,

$$C = C^* \sum_{i=0}^{3} \sum_{j=0}^{2} a_{ij} (\rho V^*)^i (\tau)^j \tag{9.3}$$

where $\tau = T/T^*$, which is always less than unity. The parameters C^*, V^*, and T^* are given for many substances by Huang (Huang and O'Connell 1987) and Poling, Praunitz, and O'Connell (2000) as well as for ionic liquids by Abildskov, Ellegaard, and O'Connell (2010a). The correlation has been applied to mixtures by employing mixing rules for V^* and T^*. An atomic group-contribution method was also established (Huang and O'Connell 1987). Figure 9.1 shows the behavior of the scaled DCFI. Note that there is essentially no temperature dependence of C/C^* when ρV^* is less than 1.0, about 2.5 times the critical density. Close inspection shows there is a crossover of the isotherms near $\rho V^* = 0.93$ (Huang and O'Connell 1987).

An alternative corresponding states model for DCFI formulates a perturbation about hard spheres (HS), with an added term linear in density. This approach has been used successfully for organic substances (Mathias and O'Connell 1981; Campanella, Mathias, and O'Connell 1987) and also for ionic liquids (Abildskov, Ellegaard, and O'Connell 2010a),

$$1 - C_{ij} = \left(1 - C_{ij}\right)^{HS} + \rho\left\{2V_{ij}^* \tilde{b}(T/T_{ij}^*) - \left[V_{ii}^* \tilde{b}^{HS}(T/T_{ii}^*) + V_{jj}^* \tilde{b}^{HS}(T/T_{jj})\right]\right\} \tag{9.4}$$

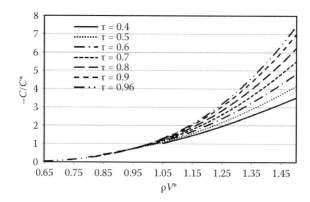

FIGURE 9.1   Scaled DCFI from Equation 9.3.

The reduced covolumes, $\tilde{b}$ and $\tilde{b}^{HS}$, are generalized functions of reduced temperature and multiplied by a characteristic volume, V^*. Expressions for the hard-sphere DCFI of Equation 9.4 are given by Mathias (O'Connell 1981) and Campanella, Mathias, and O'Connell (1987). Abildskov, Ellegaard, and O'Connell (2010a) give the most recent correlations for the hard-sphere diameters and linear-density coefficients as functions of reduced temperature. Ionic liquid (IL) characteristic parameters for Equations 9.3 and 9.4 are listed, along with group-contribution characteristics (Abildskov, Ellegaard, and O'Connell 2010a).

Comparisons of these correlations with experiment are given in the next section. Experimental compressibilities are typically values manipulated from densities measured at discrete pressures, so direct comparisons of densities of compressed fluids are more reliable than compressibilities.

9.2.2 Pure Component and Solution Densities

The pressure dependence of liquid densities can be obtained by isothermal integration of Equation 9.1 over a change in density from a reference state, ρ^0, p^0, T, and $\{x\}$, such as ambient or saturated, to the desired state, p, T, and $\{x\}$. In terms of the DCFIs, for which most applications have been made,

$$
\begin{aligned}
\frac{p(T,\rho,\{x\}) - p(T,\rho^0,\{x\})}{k_B T} &= \int_{x_i\rho^0}^{x_i\rho} \sum_{i=1}^{n_c} x_i \frac{\bar{V}_i}{k_B T \kappa_T} d\rho_i \\
&= \int_{x_i\rho^0}^{x_i\rho} \sum_{i=1}^{n_c} x_i \left(\sum_{j=1}^{n_c} x_j \left[1 - C_{ij}(T,\rho_i) \right] \right)_{T,\{x\}} d\rho_i
\end{aligned}
\tag{9.5}
$$

A general integration procedure for these relations is described by O'Connell (1981, 1994, 1995), though for models of the forms 9.3 and 9.4, analytic relations exist. The equations below are given by Abildskov, Ellegaard, and O'Connell (2009).

For Equation 9.3, the pressure difference can be solved iteratively to obtain ρ,

$$
\frac{p(T,\rho) - p^0(T,\rho^0)}{RT} = \left[1 - C^* b_1(\tilde{T}) \right](\rho - \rho^0)
\tag{9.6}
$$

$$
- C^* \left[b_2(\tilde{T}) V^* \frac{\rho^2 - (\rho^0)^2}{2} + b_3(\tilde{T}) V^{*2} \frac{\rho^3 - (\rho^0)^3}{3} + b_4(\tilde{T}) V^{*3} \frac{\rho^4 - (\rho^0)^4}{4} \right]
$$

where the coefficients $b_i(\tilde{T})$ are combinations of those in Equation 9.3 and are listed by Abildskov, Ellegaard, and O'Connell (2009). It should be noted that predictions of pressure/density differences, such as done here, are much more accurate than those that attempt to obtain high-pressure densities directly, since errors in estimation of the reference density, ρ^0, a quantity that is normally easy to measure, are eliminated.

A worked example of this correlation is described in Section 4-12 of Poling, Praunitz, and O'Connell (2000). Parameters for many substances are given by O'Connell (1990, 1995) and Huang (Huang and O'Connell 1987). For densities greater than twice the critical density, the accuracy is extremely good for a wide variety of substances.

Figure 9.2 shows pressures computed at measured densities at different temperatures for the ionic liquid $[C_1C_2Im][Tf_2N]$ using the correlation of Equation 9.6 (Abildskov, Ellegaard, and O'Connell 2010a). The agreement is normally better than for comparable methods (Paduszynski and Domanska 2012). Figure 9.2 shows that the measurements by different workers (Gardas et al. 2007; Jacquemin et al. 2007) are different for both the reference densities and compressibilities, since the low

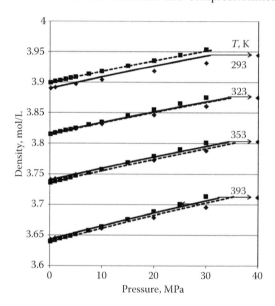

FIGURE 9.2 Density versus pressure for the IL $[C_1C_2Im][TfN]$. (Data (■) from R. L. Gardas, M. G. Freire, P. J. Carvalho, I. M. Marrucho, I. M. A. Fonseca, A. G. M. Ferreira, and J. A. P. Coutinho, 2007, P ρ T Measurements of Imidazolium-Based Ionic Liquids, *Journal of Chemical and Engineering Data*, 52, 1881 and (◆) from J. Jacquemin, P. Husson, V. Mayer, and I. Cibulka, 2007, High-Pressure Volumetric Properties of Imidazolium-Based Ionic Liquids: Effect of the Anion, *Journal of Chemical and Engineering Data*, 52, 2204.) Calculations of p using Equation 9.6 with p^0, ρ^0, and ρ. (From J. Jacquemin, P. Husson, V. Mayer, and I. Cibulka, 2007, High-Pressure Volumetric Properties of Imidazolium-Based Ionic Liquids: Effect of the Anion, *Journal of Chemical and Engineering Data*, 52, 2204 [—] and from R. L. Gardas, M. G. Freire, P. J. Carvalho, I. M. Marrucho, I. M. A. Fonseca, A. G. M. Ferreira, and J. A. P. Coutinho, 2007, P ρ T Measurements of Imidazolium-Based Ionic Liquids, *Journal of Chemical and Engineering Data*, 52, 1881 [- - -].) The arrows identify the data associated with the lines. (Modified from J. Abildskov, M. D. Ellegaard, and J. P. O'Connell, 2010a, Densities and Isothermal Compressibilities of Ionic Liquids—Modeling and Application, *Fluid Phase Equilibria*, 295, 215. With permission from Elsevier.)

pressure intercepts and the slopes of the lines are not the same. While the correlation cannot distinguish which reference density might be correct, the lines associated with the calculations suggest that the compression data of Gardas et al. (2007) may be more reliable at lower temperatures, though they are between the slopes from the data at higher T.

The relations based on Equation 9.4 from Abildskov, Ellegaard, and O'Connell (2010a) are

$$\frac{p - p^0}{k_B T} = \left(\frac{p^{HS}}{k_B T}\right) - \left(\frac{p^{HS}}{k_B T}\right)^0 + \sum_{i=1}^{n_c} \sum_{j=1}^{n_c} \left(x_j x_i \rho^2 - x_i^0 x_j^0 \left(\rho^0\right)^2\right)\left(b_{ij} - b_{ij}^{HS}\right) \quad (9.7)$$

For pure components, this becomes

$$\frac{p - p^0}{k_B T} = \left(\frac{p^{HS}}{k_B T}\right) - \left(\frac{p^{HS}}{k_B T}\right)^0 + \left(\rho^2 - \left(\rho^0\right)^2\right)\left(b_{ii} - b_{ii}^{HS}\right) \quad (9.8)$$

where the HS expressions are given below. The elements of the $n_c \times n_c$ matrix $\mathbf{b}$ are calculated using

$$b_{ij} = \tilde{b}_{ij} V_{ij}^* \quad (9.9)$$

with

$$\tilde{b}_{ij} = c_1 + \frac{c_2}{\tilde{T}_{ij}} + \frac{c_3}{\tilde{T}_{ij}^2} + \frac{c_4}{\tilde{T}_{ij}^3} + \frac{c_5}{\tilde{T}_{ij}^8} \quad (9.10)$$

and the set of $\{c\}$ values is

$$c^T = \left\{0.3625065 \quad -0.7140666 \quad -1.7543882 \quad 0.47075 \quad -0.0041793\right\} \quad (9.11)$$

The characteristic V_{ij}^* of Equation 9.9 is

$$V_{ij}^* = \frac{\left(\sqrt[3]{V_i^*} + \sqrt[3]{V_j^*}\right)^3}{8} \quad (9.12)$$

while the characteristic $\tilde{T}_{ii}$ for Equation 9.10 when $i \neq j$ is

$$\tilde{T}_{ij} = \frac{T}{(1 - k_{ij})\sqrt{T_{ii}^* T_{jj}^*}} \quad (9.13)$$

The hard-sphere terms are

$$\left(\frac{p^{\text{HS}}}{RT}\right) = \frac{6}{\pi}\left(\frac{\xi_0}{1-\xi_3} + \frac{3\xi_1\xi_2}{\left(1-\xi_3\right)^2} + \frac{\left(\xi_2\right)^3\left(3-\xi_3\right)}{\left(1-\xi_3\right)^3}\right) \tag{9.14}$$

The hard-sphere quantities are

$$\xi_q = \frac{\pi}{6}\sum_{j=1}^{n_C}\rho_j\sigma_j^q = \frac{\pi}{6}\rho\sum_{j=1}^{n_C}x_j\sigma_j^q \quad ; \quad q = 0,1,2,3$$

$$\xi_q^0 = \frac{\pi}{6}\sum_{j=1}^{n_C}\rho_j^0\sigma_j^q = \frac{\pi}{6}\rho^0\sum_{j=1}^{n_C}x_j^0\sigma_j^q \quad ; \quad q = 0,1,2,3 \tag{9.15}$$

where

$$\sigma_i = \sqrt[3]{\frac{3}{2\pi}b_{ii}^{\text{HS}}} \tag{9.16}$$

For a pure component, 9.14 becomes

$$\left(\frac{p^{\text{HS}}}{RT}\right) = \rho\left(\frac{\xi}{1-\xi} + \frac{3\xi}{\left(1-\xi\right)^2} + \frac{\left(\xi\right)^2\left(3-\xi\right)}{\left(1-\xi\right)^3}\right) \tag{9.17}$$

and Equation 9.15 plus Equation 9.16 give

$$\xi = \frac{1}{2}b_{ii}^{\text{HS}} \tag{9.18}$$

The elements of the $n_c \times n_c$ matrix $\mathbf{b}^{\text{HS}}$ in Equation 9.7 are calculated with the following equations,

$$b_{ij}^{\text{HS}} = \frac{b_{ii}^{\text{HS}} + b_{jj}^{\text{HS}}}{2} \tag{9.19}$$

$$b_{ii}^{\text{HS}} = V_i^*\tilde{b}_{ii}^{\text{HS}} \quad , \quad \tilde{b}_{ii}^{\text{HS}} = \begin{cases} \dfrac{\alpha_1}{\tilde{T}_i^{\alpha_2}} & \tilde{T}_i = \dfrac{T}{T_i^*} > 0.73 \\[4mm] \beta_1 e^{\beta_2\tilde{T}_i} & \tilde{T}_i = \dfrac{T}{T_i^*} < 0.73 \end{cases} \tag{9.20}$$

where the sets of constants $\{\alpha\}$ and $\{\beta\}$ are

$$\alpha = \begin{Bmatrix} 0.65386227 \\ 0.16067976 \end{Bmatrix} \quad , \quad \beta = \begin{Bmatrix} 0.807662393 \\ -0.22010926 \end{Bmatrix} \tag{9.21}$$

Though the forms of all the other equations are the same, the values used in Equation 9.11 and Equation 9.21 by Mathias and O'Connell (1981) and Campanella, Mathias, and O'Connell (1987) differ somewhat from those developed by Abildskov, Ellegaard, and O'Connell (2009) and Ellegaard, Abildskov, and O'Connell (2011), which were revised to more successfully correlate the properties of ionic liquids. Those given here are from Campanella (Campanella, Mathias, and O'Connell 1987). As in Equation 9.6, the pressure difference is analytic in reduced density and temperature, providing a solution for ρ at a specified p or for p at a specified ρ.

9.2.3 PARTIAL MOLAR VOLUMES

There is a direct FST connection of partial molar volumes of components in solution to DCFI,

$$\left(\frac{\partial \beta p V}{\partial N_i} \right)_{T,V,N_{j \neq i}} = \frac{\overline{V}_i}{k_B T \kappa_T} = \sum_{j=1}^{n_c} x_j \left[1 - C_{ij} \left(T, \{\rho\} \right) \right] \tag{9.22}$$

This relation has a number of aspects that lead to successful correlations, including those in the near-critical region where $\overline{V}_i^{\infty}$ may diverge (O'Connell 1994; O'Connell and Liu 1998; Plyasunov, O'Connell, and Wood 2000; Plyasunov et al. 2000, 2001; Sedlbauer, O'Connell, and Wood 2000; Plyasunov, Shock, and O'Connell 2006). It also gives a correction about a misinterpretation of the pressure dependence of gas solubility (Mathias and O'Connell 1979, 1981; O'Connell 1981), and notes errors in the values of partial molar volumes of gases from supercritical fluid chromatography (Liu and O'Connell 1998). The practical range of dilute solution compositions where a property is independent of the component's concentration, or is effectively at infinite dilution, depends on the property and the desired accuracy. Composition sensitivity is greater for activity coefficients than for partial molar volumes and enthalpies. This section describes several aspects of partial molar volume modeling, focusing on infinitely dilute solutions and the near-critical region.

9.2.3.1 Partial Molar Volumes at Infinite Dilution

For dilute solutions, the volume of a dilute binary solution might be approximated as

$$V_{\mathrm{m}}(T, p, x_1) \approx (1 - x_2) \overline{V}_1^{\circ} + x_2 \overline{V}_2^{\infty} \quad x_2 < 0.1 \tag{9.23}$$

Under this assumption, Equation 9.22 for a binary becomes

$$\lim_{N_2 \to 0} \left(\frac{\partial \beta p V}{\partial N_2} \right)_{T,V,N_1} = \frac{\bar{V}_2^\infty}{k_B T \kappa_{T,1}^\circ} \equiv A_{Kr} = 1 - C_{12}^\infty \left(T, \rho_1^\circ \right) \qquad (9.24)$$

The ratio of properties in Equation 9.24 is called the *Krichevskii function*, A_{Kr}, and identified at the solvent critical point as the *Krichevskii parameter* (Levelt Sengers 1991). The first FST correlation for partial molar volumes of gases in liquids was done by Brelvi (Brelvi and O'Connell 1975c), using characteristic properties for his correlation of the reduced bulk modulus (Brelvi and O'Connell 1972). Recent work (Ellegaard, Abildskov, and O'Connell 2011) has used the form of Equation 9.4 for partial molar volumes of gases in ILs. The comparisons with data for these systems seem not to be as successful as for their compressibilities and phase equilibria, for reasons that are not apparent.

For systems far from the critical point of component 1, the value of $\bar{V}_2^\infty$ is relatively small and positive. However, Cooney and O'Connell (1987) noted that the variation with pressure of $\bar{V}_i^\infty$ for aqueous salts, even at conditions approaching the critical where $\bar{V}_2^\infty$ is negative, could be successfully correlated with water density and requiring temperature dependence only below 100°C. This was followed by recognizing from data as well as theory, that A_{Kr} for aqueous nonelectrolytes is well behaved from ambient conditions through the critical region, even though as $T \to T_{c1}$, $\bar{V}_2^\infty$ diverges strongly (Levelt Sengers 1991). Typically, if $T_{c1} > T_{c2}$, such as for volatile nonelectrolytes (Hnědkovský, Majer, and Wood 1995), $\bar{V}_2^\infty \to \infty$, while for $T_{c1} < T_{c2}$, such as for inorganics (Hnědkovský, Wood, and Majer 1996) and salts (Sedlbauer, O'Connell, and Wood 2000), $\bar{V}_2^\infty \to -\infty$. The heat capacity at infinite dilution, $\bar{C}_p^\infty$, also diverges for all systems (see Figure 9.8, presented later in this chapter). However, since A_{Kr} is only very weakly divergent (Levelt Sengers 1991), $1 - C_{12}^\infty$ is well behaved over essentially all experimentally accessible conditions. In fact, data over wide ranges of density show that A_{Kr}, like pure component DCFI, is virtually independent of temperature for both nonelectrolytes and electrolytes (O'Connell 1995; O'Connell, Sharygin, and Wood 1996). For example, Figure 9.3 shows the variation of A_{Kr} from experiment for several small solutes in water for wide ranges of temperature and density along with the correlation,

$$A_{Kr} = 1 + a_1 \rho + a_2 \left[\exp(c\rho) - 1 \right] \qquad (9.25)$$

where a_1 and a_2 are solute-dependent parameters and c is a universal constant (O'Connell, Sharygin, and Wood 1996).

Further investigations have shown the generality of this behavior. Both Plyasunov (Plyasunov et al. 2001; Plyasunov 2011) and Majer, Sedlbauer, and Bergin (2008) have studied volumetric and phase behavior of many aqueous systems and developed group-contribution methods for a large variety of nonelectrolytes in water at extreme

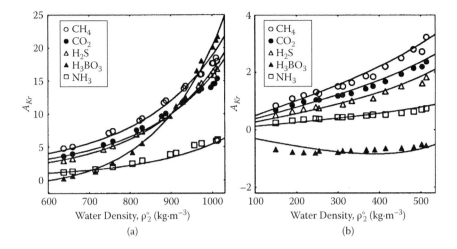

FIGURE 9.3 Variations of Krichevskii function, A_{Kr}, for various small solutes in water; (a) for high solvent densities, (b) for lower densities. Symbols for data, lines for Equation 9.25. (Reprinted with permission from J. P. O'Connell, A. V. Sharygin, and R. H. Wood, 1996, Infinite Dilution Partial Molar Volumes of Aqueous Solutes over Wide Ranges of Conditions, *Industrial and Engineering Chemistry Research*, 35, 2808. With permission from ACS Publishing.)

conditions. This behavior has led to extensive models for thermodynamic properties and phase behavior of dilute aqueous systems (Plyasunov, O'Connell, and Wood 2000; Plyasunov et al. 2000, 2001; Sedlbauer, O'Connell, and Wood 2000; Plyasunov, Shock, and O'Connell 2006; Majer, Sedlbauer, and Bergin 2008; Plyasunov 2011). Details of these models are described below.

9.2.4 Partial Molar Volumes at Finite Concentrations

One issue associated with the above systems is defining the mole fraction of solute where infinite dilution is no longer a satisfactory assumption. Liu (Liu and O'Connell 1998; O'Connell and Liu 1998) investigated this in the context of supercritical chromatography measurements for partial molar volumes. Expanding the DCFI in solute mole fraction about infinite dilution shows that first-order terms can be significant even at mole fractions of the order of 10^{-3}. This suggests that common conditions and literature treatments of data could be unreliable. The analysis was extended to solubilities in supercritical fluids, noting that at least one set of data did not display the required thermodynamic consistency for phase equilibria and volumetric behavior.

9.3 FLUCTUATION SOLUTION THEORY MODELING OF PHASE EQUILIBRIA INVOLVING DILUTE SOLUTIONS

Issues in dilute solution modeling have been discussed extensively (Cabezas and O'Connell 1993). The application of FST to these systems provides a powerful method for both solution nonidealities and densities as shown in the above sections, by integrating the FST partial derivatives of chemical potential from a reference state to a solution state. The most successful cases have been for solutions with strong nonidealities such as for supercritical gas components (Mathias and O'Connell 1981; Campanella, Mathias, and O'Connell 1987; O'Connell and Liu 1998; Plyasunov, O'Connell, and Wood 2000; Plyasunov et al. 2000, 2001; Sedlbauer, O'Connell, and Wood 2000; Plyasunov, Shock, and O'Connell 2006) and for solids in liquids (Ellegaard, Abildskov, and O'Connell 2009; Abildskov, Ellegaard, and O'Connell 2010a, 2010b; Ellegaard 2011). Also, FST formulations give the basis for correlations used for supercritical extraction of solids into near-critical solvents (Liu and O'Connell 1998; O'Connell and Liu 1998). The usual form is done with fugacities rather than chemical potentials since complexities of the latter at infinite dilution are eliminated (see Section 1.1.4 in Chapter 1).

9.3.1 Models Based on Activity Coefficients

The thermodynamic formulation of the fugacities in a binary appropriate for dilute solutions is

$$f_1 = (1 - x_2)\gamma_1 f_1^0$$
$$f_2 = x_2\gamma_2^* H_{21}$$

(9.26)

where

$$\lim_{x_1 \to 1} \gamma_1 = 1 \text{ and } \lim_{x_1 \to 1} \gamma_2^* = 1$$

and H_{21} is Henry's constant or the fugacity of the hypothetical pure component 2 in a state consistent with linear mole fraction dependence at infinite dilution, and f_1^0 is the fugacity of pure component 1 in its standard state. The property γ_2^* is the activity coefficient, related to the excess chemical potential, in the "unsymmetric convention" (Prausnitz, Lichtenthaler, and Gomes de Azevedo 1999). See below for dealing with mixed solvents.

Kirkwood and Buff (1951) included expressions for γ_2^* in single solvents from a Taylor's series expansion in solute mole fraction. The coefficients were collections of infinite-dilution (pure component 2) pair KBIs at the first order and pair and triplet KBIs at the second order. The form used for applications is

$$\ln\gamma_2^* = \Delta_{12}^0\left(-2x_2 + x_2^2\right) + \Delta_{112}^0 x_2^3 + \dots$$

(9.27)

The Gibbs–Duhem equation gives,

$$\ln\gamma_1 = \Delta_{12}^0 x_2^2 + \ldots \tag{9.28}$$

O'Connell (1971a) also gave these formulas in terms of direct correlation function integrals with

$$\Delta_{12}^0 = \frac{1}{2}\left[\left(1 - C_{11}^0\right) - \frac{\left(1 - C_{12}^0\right)^2}{\left(1 - C_{22}^0\right)}\right] \tag{9.29}$$

and an equivalent for Δ_{112}^0 in terms of pair and triplet DCFI. Expressions were also given for the partial molar volumes and the reduced bulk modulus to lowest order of the expansion in mole fraction. It is expected that this correlation, with an empirical value of Δ_{12}^0, would be adequate up to solute mole fractions of $x_2 \sim 0.1$. Brelvi (Brelvi and O'Connell 1975c) developed a correlation for Δ_{12}^0 in the spirit of his prior work (Brelvi and O'Connell 1972).

If models for TCFIs or DCFIs are available, the complete expression for both $\ln\gamma_1$ and $\ln\gamma_2^*$ can be obtained through appropriate integration. The expression is simpler for DCFIs,

$$\ln\gamma_i^* = \sum_{j=1}^{n_c} \int_{x_j^0\rho^0}^{x_j\rho} \frac{\left[1 - C_{ij}(T,\rho_i)\right]}{\rho}d\rho_j \tag{9.30}$$

If the solution state is specified by pressure rather than density, the same model for DCFI is used in both Equation 9.5 and Equation 9.30 to obtain ρ and γ_1^* for a specified state of T, p, and $\{x\}$ relative to the state T, p^0, ρ^0, and $\{x^0\}$ where γ_2^* is unity. The methodology for this approach is fully described by O'Connell (1981, 1994, 1995).

This method was used for gas solubility and solution densities of liquids by Mathias and O'Connell (Mathias and O'Connell 1979; O'Connell 1981) and expanded by Campanella, Mathias, and O'Connell (1987) for a wide variety of systems including hydrocarbons, organics, and aqueous solvents. The method has recently been extended by Ellegaard, Abildskov, and O'Connell (2009) and Abildskov, Ellegaard, and O'Connell (2010b) for use in IL systems. For a model of the form of Equation 9.4, this becomes

$$\ln\gamma_i\left(T,\rho,\{x\};\rho^0,\{x\}^0\right) = \ln\gamma_i^{HS}\left(T,\rho,\{x\};\rho^0,\{x\}^0\right)$$

$$+ 2\sum_{j=1}^{n_c}\left(x_j\rho - x_j^0\rho^0\right)\left(b_{ij} - b_{ij}^{HS}\right) \tag{9.31}$$

where

$$\ln \gamma_i^{HS}(T,\rho,\{x\};\rho^0,\{x\}^0) =$$

$$\left[\ln\rho - \left[\ln(1-\xi_3) \right] \left[1 - \left(\frac{\xi_2\sigma_i}{\xi_3} \right)^2 \left(3 - 2\frac{\xi_2\sigma_i}{\xi_3} \right) \right] \right.$$

$$\left. + \frac{3\left(\xi_2\sigma_i + \xi_1\sigma_i^2\right) - \left(\xi_2\sigma_i\right)^3 \left(2-\xi_3\right)/\xi_3^2}{1-\xi_3} + \frac{3}{\xi_3}\left(\frac{\xi_2\sigma_i}{1-\xi_3} \right)^2 + \frac{\pi}{6}\sigma_i^3 \frac{p^{HS}}{RT} \right] \qquad (9.32)$$

$$-\left[\ln\rho^0 - \left[\ln(1-\xi_3^0) \right] \left[1 - \left(\frac{\xi_2^0\sigma_i}{\xi_3^0} \right)^2 \left(3 - 2\frac{\xi_2^0\sigma_i}{\xi_3^0} \right) \right] \right.$$

$$\left. + \frac{3\left(\xi_2^0\sigma_i + \xi_1^0\sigma_i^2\right) - \left(\xi_2^0\sigma_i\right)^3 \left(2-\xi_3^0\right)/\left(\xi_3^0\right)^2}{1-\xi_3^0} + \frac{3}{\xi_3^0}\left(\frac{\xi_2^0\sigma_i}{1-\xi_3^0} \right)^2 + \frac{\pi}{6}\sigma_i^3 \left(\frac{p^{HS}}{RT} \right)^0 \right]$$

There are two aspects of this treatment that are significant. First, one or more binary constants are used in models such as Equation 9.4, for example, in Equation 9.13. Therefore, when Equation 9.30 is taken to the limit of pure solvent 3, there is a relation between the Henry's constants,

$$\lim_{x_3\to 1} \ln \gamma_2^* = \ln\left(\frac{H_{2R}}{H_{23}} \right) = \int_{\rho_{pure\,R}}^{0} \frac{\left[1 - C_{2R}(T,\{\rho\}) \right]}{\rho}d\rho_R + \int_{0}^{\rho_{pure\,3}} \frac{\left[1 - C_{23}(T,\{\rho\}) \right]}{\rho}d\rho_3 \qquad (9.33)$$

Therefore, if the binary constant, k_{2R} is set or determined from experiment, Equation 9.33 determines k_{23}. Second, this also means that one can predict solute solubility in a solvent i from that in a convenient reference solvent, R,

$$f_i = (1-x_2)\gamma_i f_i^0$$

$$f_2 = x_2\gamma_2^+ H_{2R} \qquad (9.34)$$

with

$$\lim_{x_R\to 1} \ln \gamma_2^+ = 1 \qquad (9.35)$$

Here,

$$\ln \gamma_2^+ = \int_{\rho_{pure\,R}}^{x_R\rho} \frac{\left[1 - C_{2R}(T,\{\rho\}) \right]}{\rho}d\rho_R + \int_{0}^{x_3\rho} \frac{\left[1 - C_{23}(T,\{\rho\}) \right]}{\rho}d\rho_3 \qquad (9.36)$$

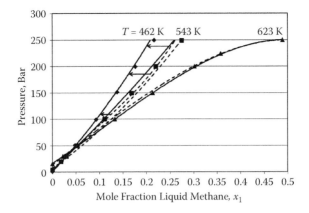

FIGURE 9.4 Pressure versus liquid mole fraction for isotherms of methane with m-cresol. Symbols for data from Simnick et al. (From J. J. Simnick, H. M. Sebastian, H. M. Lin, and K. C. Chao, 1979a, Gas-Liquid Equilibrium in Mixtures of Methane+m-Xylene, and Methane-Meta-Cresol, *Fluid Phase Equilibria*, 3, 145, With permission from Elsevier.) (—) Regression of temperature dependence of Henry's constant in Equation 9.26. (- - -) Prediction using Equation 9.34 with Henry's constant for methane in quinoline fitted to data of Simnick et al. (From J. J. Simnick, H. M. Sebastian, H. M. Lin, and K. C. Chao, 1979b, Vapor-Liquid-Equilibrium in Methane + Quinoline Mixtures at Elevated-Temperatures and Pressures, *Journal of Chemical and Engineering Data*, 24, 239.) Arrows indicate prediction at $T = 462$ K is not accurate.

Examples of using this with good success for hydrogen in different solvents, including for model coal liquids, based on a single reference solvent are given by Campanella, Mathias, and O'Connell (1987) and O'Connell (1995). Treatment of complex mixed solvents, including for actual coal liquids, is also described in those references. This concept also works quite well for ILs (Abildskov, Ellegaard, and O'Connell 2009; Ellegaard, Abildskov, and O'Connell 2010).

Figure 9.4 shows predictions of pressure versus liquid mole fraction for the methane (1) and m-cresol system where the Henry's constant of Equation 9.26 was obtained from fitting data directly (solid lines) or predicted with quinolone (dashed lines) as the reference solvent in Equation 9.34. As the arrows show, the prediction is not uniformly accurate in this case. Better results have been obtained with other solutes, especially hydrogen, nitrogen, and carbon monoxide, including mixed solvents, as shown by Campanella, Mathias, and O'Connell (1987). Figures 9.5 and 9.6 show similar results for a system containing hydrogen and an ionic liquid (Abildskov, Ellegaard, and O'Connell 2009).

9.3.2 MODELS BASED ON EQUATIONS OF STATE

For infinitely dilute solutions, the temperature independence of A_{Kr} at higher densities leads to the Henry's constant of Equation 9.26 by integrating Equation 9.24. The relations are

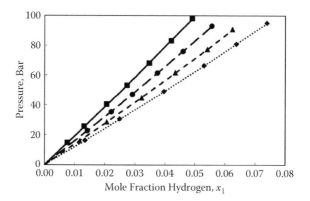

FIGURE 9.5 Pressure versus liquid mole fraction of hydrogen (1) in the IL [hmim][NTf2] calculated using Henry's constant of Equation 9.26 fitted to data of Kumelan et al. (From J. Kumelan, A. P. S. Kamps, D. Tuma, and G. Maurer, 2006, Solubility of H_2 in the Ionic Liquid [bmim][PF_6], *Journal of Chemical and Engineering Data*. 51, 11.) — ■ $T = 293.2$ K. —●— · $T = 333.2$ K. - - - ▲ $T = 373.2$ K. · · · ◆ $T = 413.2$ K. (Modified with permission from J. Abildskov, M. D. Ellegaard, and J. P. O'Connell, 2009, Correlation of Phase Equilibria and Liquid Densities for Gases with Ionic Liquids, *Fluid Phase Equilibria*, 286, 95, With permission from Elsevier.)

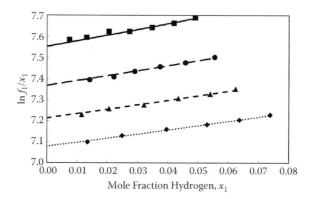

FIGURE 9.6 Log ratio of fugacity to liquid mole fraction of hydrogen (1) in the IL [hmim] [NTf2] calculated using Henry's constant of Equation 9.26 fitted to data of Kumelan et al. (From J. Kumelan, A. P. S. Kamps, D. Tuma, and G. Maurer, 2006, Solubility of H_2 in the Ionic Liquid [bmim][PF_6], *Journal of Chemical and Engineering Data*, 51, 11.) — ■ $T = 293.2$ K. —●— · $T = 333.2$ K. - - - ▲ $T = 373.2$ K. · · · ◆ $T = 413.2$ K. (Reprinted with permission from J. Abildskov, M. D. Ellegaard, and J. P. O'Connell, 2009, Correlation of Phase Equilibria and Liquid Densities for Gases with Ionic Liquids, *Fluid Phase Equilibria*, 286, 95, With permission from Elsevier.)

$$\ln\left(\frac{\beta\varphi_2^\infty p}{\rho_1^0}\right) = \int_0^{\rho_1^0}(A_{Kr}-1)\frac{d\rho}{\rho} \tag{9.37}$$

From this relation, a formulation can be made in terms of the infinite dilution properties of hydration for water as the solvent, such as the difference of Gibbs energies between the fluid at the designated reference pressure, p^0, and the infinitely dilute solute at p,

$$\Delta_h G_2^\infty(T,p) = RT\left[\ln\frac{\varphi_2^\infty p}{p^0}\right] \tag{9.38}$$

This yields Henry's constants via

$$H_{21}(T,p) = \exp\left[-\Delta_h G_2^\infty(T,p)/RT\right] \tag{9.39}$$

Standard thermodynamic manipulations yield the infinite dilution enthalpy, entropy, and isobaric heat capacity of hydration (Plyasunov, O'Connell, and Wood 2000; Sedlbauer, O'Connell, and Wood 2000). For example,

$$\Delta_h C_{p,2}^\infty = -T^2\left(\frac{d^2\,\Delta_h G_2^\infty/T}{dT^2}\right) \tag{9.40}$$

The issues are how to utilize known solution information at low temperatures and pressures that do depend on T and how to express the properties at conditions below the critical temperature where there is temperature dependence of A_{Kr}. This approach need not be limited to aqueous systems, but those are the only ones which have been treated.

The first approach (Plyasunov, O'Connell, and Wood 2000) was to augment the high-density function of Equation 9.25 with temperature-dependent correlations of cross (B_{12}) and pure solvent (B_{11}) second virial coefficients to include the low-density behavior for the integral of Equation 9.37. An analytically integrable form was adopted:

$$A_{Kr} = (1-\alpha_1)+\alpha_1\left[\left(1-C_{11}^0\right)\right]+2\rho\{B_{12}(T)-B_{11}(T)\}\exp(k_1\rho)$$
$$+\rho\left(\frac{\alpha_2}{T^5}+\alpha_3\right)\left[\exp(k_2\rho)-1\right] \tag{9.41}$$

where values of B_{11} and B_{12} are obtained from the square-well potential model, k_1 and k_2 are universal constants, and α_1, α_2, and α_3 are solute-dependent parameters. The full final relations of Equation 9.37 through Equation 9.40 are given by Plyasunov et al. (2000). These have been used to obtain infinite dilution thermodynamic properties of many organic substances in water (Plyasunov et al. 2001; Plyasunov,

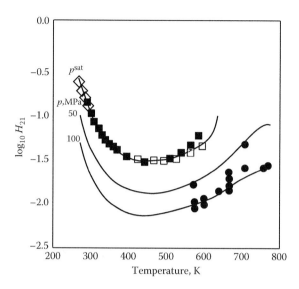

FIGURE 9.7 Experimental (symbols) and fitted (lines) results for Henry's constants (H_{21}) for Hydrogen sulfide (2) in water (1) from Equation 9.37 through Equation 9.41. (Reprinted with permission from A. Plyasunov, J. P. O'Connell, R. H. Wood, and E. L. Shock, 2000, Infinite Dilution Partial Molar Properties of Aqueous Solutions of Nonelectrolytes. II. Equations for the Standard Thermodynamic Functions of Hydration of Volatile Nonelectrolytes over Wide Ranges of Conditions Including Subcritical Temperatures, *Geochimica Et Cosmochimica Acta*, 64, 2779, With permission from Elsevier.)

Shock, and O'Connell 2006). Figure 9.7 shows results for H_2S in water at various pressures and temperatures.

The alternative route to aqueous solute properties based on FST was to use a finite pressure reference state where properties could be obtained, and compute the difference in Gibbs energy between the desired state and the reference state (Sedlbauer, O'Connell, and Wood 2000). This allows calculations for electrolytes, which would not have second virial coefficients that could be used at low densities as in Equation 9.41. The relation here for A_{Kr} is similar to Equation 9.41,

$$A_{Kr} = (1-\beta_1) + \beta_1\left[\left(1-C_{11}^0\right)\right] + \rho\left\{\beta_2 + \beta_3\exp(1500/T) + \beta_4\exp(c_1\rho) + \delta\left[\exp(c_2\rho)-1\right]\right\} \qquad (9.42)$$

where values of c_1 and c_2 are universal constants and β_1, β_2, β_3, and β_4 are solute-dependent parameters. The value of δ has a specific value for the class of solute: cation, anion, nonelectrolyte.

Again, analytic expressions for $\Delta_h\bar{G}_2^\infty$ and H_{21} are found using Equation 9.37 through Equation 9.39 (Sedlbauer, O'Connell, and Wood 2000), though slightly differently than by Plyasunov et al. (Plyasunov, O'Connell, and Wood 2000; Plyasunov et al. 2000). The modification to deal with temperatures below the critical is to add a temperature-dependent correction term to the heat capacity, which is different for nonelectrolytes and electrolytes,

$$\Delta_h \overline{C}_{p,2}^{\infty \, corr} = \beta_5 \frac{(T - T_c)^2}{(T - 228)} \quad T < T_{c1}$$

$$= 0 \qquad\qquad T \geq T_{c1}$$

aqueous nonelectrolytes,   (9.43)

$$\Delta_h \overline{C}_{p,2}^{\infty \, corr} = (T - T_c)\left[\frac{\beta_5}{(T - 228)} + \beta_6\right] \quad T < T_{c1}$$

$$= 0 \qquad\qquad\qquad T \geq T_{c1}$$

aqueous electrolytes. (9.44)

These analytically integrable expressions are included in the evaluations for $\Delta_h \overline{G}_2^{\infty}$ and for H_{21} (Sedlbauer, O'Connell, and Wood 2000). Figure 9.8 shows results for this model for aqueous NaCl with data for infinite dilution partial molar volumes

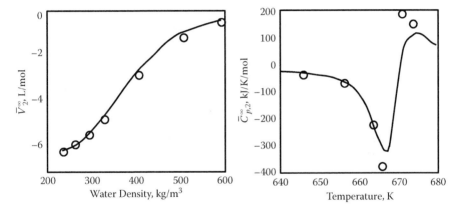

FIGURE 9.8 Left, infinite dilution partial molar volume of aqueous NaCl versus water density at $p = 38$ MPa over $600 < T < 725$ K. (—) from Equation 9.42. (○) Extrapolations by Sedlbauer et al., (From J. Sedlbauer, E. M. Yezdimer, and R. H. Wood, 1998, Partial Molar Volumes at Infinite Dilution in Aqueous Solutions of NaCl, LiCl, NaBr, and CsBr at Temperatures from 550 K to 725 K, *Journal of Chemical Thermodynamics*, 30, 3), of data from Majer et al., (From V. Majer, L. Hui, R. Crovetto, and R. H. Wood, 1991, Volumetric Properties of Aqueous 1-1 Electrolyte-Solutions near and above the Critical-Temperature of Water. 1. Densities and Apparent Molar Volumes of NaCl(*aq*) from 0.0025 mol·kg⁻¹ to 3.1 mol·kg⁻¹, 604.4-K to 725.5-K, and 18.5-MPa to 38.0-MPa, *Journal of Chemical Thermodynamics*, 23, 213). (Right) Infinite dilution heat capacity of aqueous NaCl at $p = 28$ MPa over $640 < T < 680$ K. (—) from Equation 9.37 through Equation 9.40, and Equation 9.42 through Equation 9.44. (○) Data from Carter. (R. W. Carter, 1992, Ph.D. Thesis, University of Delaware). (Modified from J. Sedlbauer, J. P. O'Connell, and R. H. Wood, 2000, A New Equation of State for Correlation and Prediction of Standard Molal Thermodynamic Properties of Aqueous Species at High Temperatures and Pressures, *Chemical Geology*, 163, 43, With permission from Elsevier.)

and heat capacities at elevated temperatures. Results for Henry's constants are also quite accurate.

9.3.3 SOLUBILITIES IN MIXED LIQUID SOLVENTS

While the above treatments are for solutes in single solvents, many applications of interest involve mixed solvents, especially for gases and pharmaceutical compounds. FST has been used to develop successful descriptions of such systems. Chapter 10 describes the approach of Shulgin and Ruckenstein. The focus here is on *excess solubility*, $\ln x^{Ex}$, the difference in solubility in a real solvent mixture and that for the solvents as an ideal solution (Prausnitz, Lichtenthaler, and Gomes de Azevedo 1999). For very dilute solutions, the solubility is the reciprocal of the Henry's constant, so typically the *excess Henry's constant*, H_2^{Ex}, is modeled

$$\ln H_2^{Ex} \equiv \ln H_{2m} - \sum_i^{n_S} x_i \ln H_{2i} \qquad (9.45)$$

where the sum is over all solvents. Most measurements and models have been for binary solvents where the variation of H_2^{Ex} can be positive or negative or both. Empirical models have been developed, especially because the variation of H_2^{Ex} with composition is similar for all solutes and is mostly determined by the substances of the solvent mixture. FST approaches have been among the most successful for both gaseous and solid solutes in mixed solvents as also described in Chapter 10.

9.3.3.1 Gas Solubility in Mixed Solvents

There have been two FST approaches to gas solubility in mixed solvents. The first (O'Connell 1971a) expressed H_2^{Ex} in terms of collections of DCFI with simple parameterization; details of the relations are given in Section 9.3.3.2. The second (Mathias and O'Connell 1979; Campanella, Mathias, and O'Connell 1987) used the DCFI model of Equation 9.4 by integrating the DCFI for the solute from one pure solvent, identified as reference solvent, $_R$, to the mixture composition, which for a binary is given by x_R. Then,

$$H_2^{Ex} \equiv (1 - x_R) \ln(H_{2R}/H_{23}) + \ln \gamma_2^* \qquad (9.46)$$

where $\ln \gamma_2^*$ is found from Equation 9.30 with ρ^0 being the pure reference solvent density, and the solution component densities being those for the mixed solvent, $\rho_i = x_i \rho_m$ for $i = R, 3$. For most systems, the mixture density used can be that for an ideal solution, though for aqueous systems, the excess volume is large enough that it must be taken into account (Campanella, Mathias, and O'Connell 1987; O'Connell 1995). Figure 9.9 shows results for H_2^{Ex} of ethylene in aqueous acetone and aqueous methanol. The complex behavior of the alcohol system is captured reasonably well, and quantitative agreement for the acetone system is obtained only if the solution excess

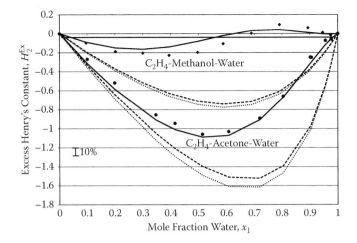

FIGURE 9.9   Excess Henry's constants of Equation 9.46 for ethylene in aqueous methanol and aqueous acetone solutions at 298.15 K. (Data of S. Zeck and H. Knapp, 1985, Solubilities of Ethylene, Ethane, and Carbon Dioxide in Mixed Solvents Consisting of Methanol, Acetone, and Water, *International Journal of Thermophysics*, 6, 643.) Calculations with Equation 9.46 using experimental densities (-), zero excess volumes (- - -) and with an excess Gibbs energy rule (· · ·). (From J. P. O'Connell and J. M. Prausnitz, 1964, Thermodynamics of Gas Solubility in Mixed Solvents, *Industrial and Engineering Chemistry Fundamentals*, 3, 347.)

volume is included. Without this contribution, the results are similar to those of the simplest model where H_2^{Ex} is set equal to the negative of the excess Gibbs energy of the solvent mixture (O'Connell and Prausnitz 1964). It should be noted that the method was applied to the solubility of ethane and ethylene in the ternary solvent water, methanol, and acetone with good success when the solvent excess volumes were estimated by binary additivity.

9.3.3.2   Solid Solubility in Mixed Solvents

The dissolution of solids into liquids is often a dilute solution situation and, as for gases, FST modeling following the approach of O'Connell (1971a) has been applied (Ellegaard, Abildskov, and O'Connell 2009, 2010). If the solid is pure and the solution is ideal, the solubility can be found from the properties of the pure solute. This is given in Ellegaard, Abildskov, and O'Connell (2010) as,

$$\ln x_2^{id} = -\frac{\Delta h_{m2}}{R}\left(\frac{1}{T}-\frac{1}{T_{m2}}\right)+\frac{\Delta C_{pm2}}{R}\left[\ln\left(\frac{T}{T_{m2}}\right)-1+\frac{T_{m2}}{T}\right] \qquad (9.47)$$

where T_{m2} is the melting temperature, Δh_{m2} is the enthalpy of melting, and ΔC_{pm2} is the difference in heat capacity of the hypothetical subcooled liquid and the solid from T_{m2} to T.

For mixed solvent systems, it is common to define an excess solubility,

$$x_2^{\text{Ex}} \equiv \ln x_2 - x_1 \ln x_{2,1} - (1 - x_1) \ln x_{2,3} \tag{9.48}$$

where x_{2_j} is the mole fraction of solute 2 in solvent j with $j = 1,2$. In the limit

$$\lim_{x_2 \to 0} x_2^{\text{Ex}} = -H_2^{\text{Ex}} \tag{9.49}$$

For estimations of dilute solid solubility, an expansion similar to Equation 9.27 is done for a ternary system and to the first order,

$$
H_2^{\text{Ex}} = \lim_{x_2 \to 0} \frac{x_3}{2} \left(\frac{\partial \ln \gamma_3}{\partial x_3} \right)_{T,p,N_1,N_2}
$$
$$
+ \frac{x_1}{2} \left(\frac{\Delta_{12}^{\text{m}}}{1 + x_1 x_3 \Delta_{13}^{\text{m}}} - \Delta_{12}^0 \right) + \frac{x_3}{2} \left(\frac{\Delta_{23}^{\text{m}}}{1 + x_1 x_3 \Delta_{13}^{\text{m}}} - \Delta_{23}^0 \right) \tag{9.50}
$$

where ^{m} indicates the solvent mixture and ^{0} indicates pure solvent. With suitable approximations (O'Connell 1971a; Ellegaard, Abildskov, and O'Connell 2009, 2010), in particular that $\Delta_{1i}^{\text{m}} = \Delta_{1i}^0$, the final form becomes

$$
H_2^{\text{Ex}} = \lim_{x_2 \to 0} \frac{x_3}{2} \left(\frac{\partial \ln \gamma_3}{\partial x_3} \right)_{T,p,N_1,N_2} \left(1 + x_1 \Delta_{12}^0 + x_3 \Delta_{23}^0 \right) \tag{9.51}
$$

The partial derivative is obtained from parameters found by regressing vapor–liquid equilibrium (VLE) data for the solvent mixture. The estimated solubilities are not very sensitive to these parameters. There are multiple ways to obtain the Δ_{1j}^0 values. One is to use data from solute solubilities in the pure solvents and regress the parameters Δ_{12}^0 and Δ_{23}^0 separately. Then,

$$
\Delta_{2i}^0 = 2 \frac{\ln x_{2i}^{\text{id}} - \ln x_{2i}}{\left(1 - x_{2i} \right)^2} \tag{9.52}
$$

This method requires values of the pure solute properties. Alternatively, if solute solubilities in mixed solvents are available, these data can be regressed for the parameters simultaneously and the pure solute properties are not needed. Note that the theory demands that parameters for a particular solute–solvent pair be the same for all systems involving the pair, regardless of the additional solvent. In Ellegaard, Abildskov, and O'Connell (2010), those regressed from ternary data

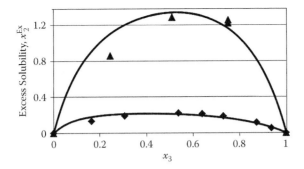

FIGURE 9.10 Calculated (lines) and experimental (symbols) excess solubilities, x_2^{Ex}, of solid solutes in mixed solvents: Lower curve, anthracene(2)/n-octane(1)/1-butanol(3), ◆ Data of J. M. Lepree, et al. (From J. M. Lepree, M. J. Mulski, and K. A. Connors, 1994, Solvent Effects on Chemical Processes. 6. The Phenomenological Model Applied to the Solubility of Naphthalene and 4-Nitroaniline in Binary Aqueous-Organic Solvent Mixtures, *Journal of the Chemical Society-Perkin Transactions, 2*, 1491.) Upper curve, β-carotene (2)/acetone (1)/n-hexane (3), ▲. (From A. Zvaigzne, I. Teng, E. Martinez, J. Trejo, and W. Acree, Jr., 1993, Solubility of Anthracene in Binary Alkane + 1-Propanol and Alkane + 1-Butanol Solvent Mixtures, *Journal of Chemical Engineering Data*, 38, 389.)

were assumed independent of temperature, though the obtained binary values did vary with T.

In Ellegaard, Abildskov, and O'Connell (2010), data were analyzed from nine pharmaceutical solutes in a total of 68 binary mixtures of 10 solvents, with some of the mixtures at different temperatures. The absolute average relative deviation of using parameters from binary data was 23% while that from correlation of ternary data was 11%. Additional study of this method is described in Ellegaard (2011) with more solutes and binary solvents, as well as in some ternary solvents. Figure 9.10 shows excess solubility results for representative systems.

9.4 FLUCTUATION SOLUTION THEORY PROPERTIES OF REACTIVE COMPONENTS AND STRONG ELECTROLYTE SOLUTIONS

The FST statistical mechanics described above are for nonelectrolyte systems. The treatment of solutions of ionizing salts, or in fact any system where speciation of the input components occurs, requires additional analysis. This section describes some elements of this area; other aspects are discussed in Chapter 8.

There are two general situations in which the particles in a solution do not have the same identity as those that are put into the system: when a component dissociates and when components can combine to form other species. The first case is commonly identified with salts forming ions with electrostatic charges. In both cases, the reactions need not be complete, so that the solution can have the original species (components) as well as the new species. Further, the speciation need not lead to detectable entities. Thus, the "chemical theory of solutions" and the "solution of groups" descriptions of solution nonideality (Prausnitz, Lichtenthaler, and Gomes de Azevedo 1999) involve virtual species considered to have thermodynamic properties. The general thermodynamics of such systems was established by Perry, Telotte, and O'Connell (1981) and their FST analysis was later articulated by Perry and O'Connell (1984).

The concept is that all species considered to have thermodynamic properties will also have pair correlation functions. However, because of stoichiometry, there are constraints on both the properties and the correlation function integrals. Further, though the collection of charges on ions requires electroneutrality only as a consequence of the stoichiometry, there are long-range electrostatic correlations of the ions that affect FST modeling of electrolytes. This aspect has also been treated (Cabezas and O'Connell 1986; Perry, Cabezas, and O'Connell 1988; O'Connell 1993). The purpose of this section is to briefly outline the basis and results available.

9.4.1 FLUCTUATION SOLUTION THEORY OF REACTIVE COMPONENTS

The principal effect of speciation is the expansion of the composition space associated with the input components to the larger composition space of the species. The goal is to express with models the properties of the species in the larger space and then project these properties into the smaller component space that has the proper number of independent variables and properties, that is, the number of components plus 2, to obtain measurable properties. The key mathematical entity is the projection operator, $\mathbf{W}$, a nonsquare ($n \times n_c$) matrix accounting for the stoichiometric connection between the n_c components and the n species. The full development, mathematical details, and specific examples are given by Perry (Perry, Telotte, and O'Connell 1981; Perry and O'Connell 1984). Applications to strong electrolytes are given by Perry, Cabezas, and O'Connell (1988), and O'Connell (O'Connell 1993; O'Connell, Hu, and Marshall 1999), as discussed in the next section. We provide a simple example here.

A traditional thermodynamics approach to reactions uses the set of the initial moles of species, $\{n_i^0\}$, the matrix ($n \times \mathbf{R}$) of reaction stoichiometric coefficients, $\mathbf{v}$, and the set of extents of the $\mathbf{R}$ independent reactions, $\{\xi\}$, to obtain the moles of species at some point in the reaction,

$$\{n_i\} = \{n_i^0\} + \mathbf{v}^{\mathrm{T}} \{\xi\} \tag{9.53}$$

An example is the reaction $A_1 + A_2 = 2A_3$ where the components are 1 and 2 and the species are 1, 2, and 3. Equation 9.53 would give

$$\begin{Bmatrix} n_1 \\ n_2 \\ n_3 \end{Bmatrix} = \begin{Bmatrix} n_1^0 \\ n_2^0 \\ n_3^0 \end{Bmatrix} + \begin{Bmatrix} -\xi \\ -\xi \\ 2\xi \end{Bmatrix} \tag{9.54}$$

An approach that allows description of the independent thermodynamic properties of reacting systems uses a slight variation,

$$\{n\} = \mathbf{W}\{n_o\} \tag{9.55}$$

where $\{n_o\}$ is the set of mole numbers of the input components. Then,

$$\mathbf{W} = \begin{bmatrix} 1-\xi & 0 \\ -\xi & 1 \\ 2\xi & 0 \end{bmatrix} \tag{9.56}$$

where the extent of reaction has been scaled to component 1. The importance of the latter form is that all independent partial molar properties can be related by

$$\mathbf{W}^T\{\bar{X}\} = \{\bar{X}_o\} \tag{9.57}$$

where X can be V, H, G, and so forth. In our example for partial molar volume,

$$\begin{bmatrix} \bar{V}_1 \\ \bar{V}_2 \\ \bar{V}_3 \end{bmatrix} = \begin{Bmatrix} \bar{V}_{1o} \\ \bar{V}_{2o} \\ \frac{1}{2}\left(\bar{V}_{1o} + \bar{V}_{2o}\right) \end{Bmatrix} \tag{9.58}$$

Similar connections can be made for properties of mixing, excess properties, and activity coefficients. The use of this approach on fluctuation properties is described by Perry and O'Connell (1984). The relations for TCFI among pairs of species are extremely complex with the extents of reactions embedded in the TCFI. However, those for DCFI are much more direct and the extents of reaction are contained in the projectors. Defining the desired matrix,

$$\left[\mathbf{A}^{-1}\right]_{ij} = \left[\frac{\partial\left(\mu_{oi}/RT\right)}{\partial n_{oj}}\right]_{T,V,n_{ok\neq j}} \tag{9.59}$$

The connection of the matrix of FST component derivatives, $\mathbf{A}^{-1}$, to the matrix of species DCFI, $\mathbf{C}$, is

$$\mathbf{A}^{-1} = \mathbf{K}^T\left(\mathbf{x}^{-1} - \mathbf{C}\right)\mathbf{K} \tag{9.60}$$

where the elements of the matrix $\mathbf{K}$ are

$$K_{ij} = W_{ij} + \left[v\xi'n_o \right]_{ij} \tag{9.61}$$

and, the elements of the matrix ξ' are

$$\left[\xi' \right]_{ij} = \left(\frac{\partial \xi_i}{\partial n_{oj}} \right)_{T,V,\mu_{ok \neq j}} \tag{9.62}$$

Ultimately obtaining the desired properties such as density differences and activity coefficients would be complicated, but feasible. This seems not to have been attempted. A much simpler case for complete reactions, such as dissociation of strong electrolytes, is shown in the next section.

9.4.2 FLUCTUATION SOLUTION THEORY OF STRONG ELECTROLYTE SOLUTIONS

Equations 9.60, 9.61, and 9.62 above can be used directly for FST properties of solutions with fully ionized salts (Perry and O'Connell 1984). In this case, the projection of the extent of reaction terms in Equation 9.62 is null and

$$\mathbf{A}^{-1} = \mathbf{W}^T \left(\mathbf{x}^{-1} - \mathbf{C} \right) \mathbf{W} \tag{9.63}$$

For dissociation of a salt $a_{v_a} b_{v_b}$ into ions, the FST derivatives are

$$\left[\frac{\partial \beta \mu_{o1}}{\partial n_{o1}} \right]_{T,V,n_{o2}} = x_1^{-1} - C_{11}$$

$$\left[\frac{\partial \beta \mu_{o1}}{\partial n_{o2}} \right]_{T,V,n_{o1}} = -v_a C_{1a} = -v_b C_b = \left[\frac{\partial \beta \mu_{o2}}{\partial n_{o1}} \right]_{T,V,n_{o2}} \tag{9.64}$$

$$\left[\frac{\partial \beta \mu_{o2}}{\partial n_{o2}} \right]_{T,V,n_{o1}} = \frac{v_a + v_b}{x_2} - \left[v_a^2 C_{aa} + 2 v_a v_b C_{ab} + v_b^2 C_{bb} \right]$$

where $x_1 = N_{o1}/[N_{o1} + (v_a + v_b)N_{o2}]$ and $x_2 = 1 - x_1$. Models for the species DCFI can be used in Equations 9.64 to obtain solution thermodynamic properties in the same manner as was done with nonelectrolytes above.

However, examination of the Debye–Hückel limiting law (Perry, Cabezas, and O'Connell 1988) shows a major complication in that the long-range electrostatic interactions lead to divergences of the ion–ion correlation functions, c_{aa}, c_{ab}, and c_{bb} in the limit of infinite dilution. Analysis of the long-range contributions (Stell, Patey, and Høye 1981) leads to rigorous formulas for the limiting law DCFIs, while more appropriate models for the extended limiting law DCFIs are found (Perry, Cabezas, and O'Connell 1988; O'Connell 1993). The short-range contributions can be modeled by 2nd virial terms similar to those of Pitzer (1995); full expressions are given by O'Connell, Hu, and Marshall (1999). The pressure and activity coefficient relations from Equation 9.5 and Equation 9.30 are, respectively,

$$
\frac{p-p^0}{k_BT} = \left(\rho-\rho^0\right) - \left\{\frac{1}{2}\left(\rho_{o1}^2 - \left(\rho_{o1}^0\right)^2\right)\Delta F_{11} + \rho_{o1}\rho_{o2}\Delta F_{12} + \rho_{o2}^2\Delta F_{22}\right\}
$$

$$
-\left\{\frac{1}{3}\left(\rho_{o1}^3 - \left(\rho_{o1}^0\right)^3\right)\Delta\Phi_{111} + \rho_{o1}^2\rho_{o2}\Delta\Phi_{112} + \rho_{o1}\rho_{o2}^2\Delta\Phi_{112} + \rho_{o2}^2\Delta\Phi_{222}\right\}
$$

$$
+\rho_{o1}S_\gamma I^{3/2}\left\{\frac{\partial\ln\varepsilon}{\partial\rho_{o1}}\right\}_{T,\rho_{o2}=0} - cI^{3/2} - \frac{1}{3}S_\gamma I^{3/2} - S_\gamma\left\{\frac{a}{b^3}\left[\frac{2}{9}\left(1+bI^{1/2}\right)^3\right.\right. \quad (9.65)
$$

$$
-\frac{1}{b^2}I + 2\left(1+bI^{1/2}\right) - \frac{2}{3}\ln\left(1+bI^{1/2}\right) - \frac{11}{9}\right]
$$

$$
+\frac{a}{3}I^{3/2}\ln\left(1+bI^{1/2}\right) + \beta\alpha^2 I^2 \exp\left(-\alpha I^{1/2}\right)\Big\}
$$

$$
\ln\gamma_2 = \ln\left(\frac{\rho}{\rho^0}\right) - \frac{1}{v_2}\left\{\left(\rho_{o1} - \rho_{o1}^0\right)\Delta F_{12} + \rho_{o1}\Delta F_{22}\right\}
$$

$$
-\frac{1}{v_2}\left\{\frac{1}{2}\left(\rho_{o1}^2 - \left(\rho_{o1}^0\right)^2\right)\Delta\Phi_{112} + \rho_{o1}\rho_{o2}\Delta\Phi_{122} + \rho_{o2}^2\Delta\Phi_{222}\right\} - \frac{\omega}{v_2}S_\gamma I^{1/2} \quad (9.66)
$$

$$
-\frac{\omega}{v_2}S_\gamma\left\{aI^{1/2}\ln\left(1+bI^{1/2}\right) + 2\beta\left[1 - \exp\left(-\alpha I^{1/2}\right)\left(1+\alpha I^{1/2} - \alpha^2/2I\right)\right]\right\}
$$

where I is the molar ionic strength and the limiting law property, S_γ is

$$
S_\gamma = \left[\frac{2\pi e^6 N_A}{\left(\varepsilon kT\right)^3}\right]^{\frac{1}{2}} \quad (9.67)
$$

The parameters are ΔF_{11}, ΔF_{12}, ΔF_{22}, $\Delta\Phi_{111}$, $\Delta\Phi_{112}$, $\Delta\Phi_{122}$, and $\Delta\Phi_{222}$. It was found (O'Connell, Hu, and Marshall 1999) that, except for ΔF_{22} and $\Delta\Phi_{222}$, these are ionically additive or follow simple combining rules. Using salt-specific parameters for ΔF_{22} and $\Delta\Phi_{222}$ gave excellent agreement in densities and activity coefficients for fifteen 1-1, 1-2, and 2-1 salts over the entire range of data available. Figure 9.11 shows mean ionic activity coefficients and densities for the systems $NaNO_3$ and $CaCl_2$. The sources of experimental information are given in Table 9.1. The agreement is quite good up to high salt concentrations.

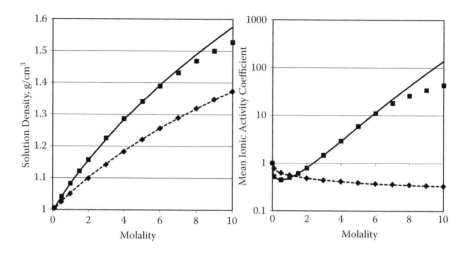

FIGURE 9.11   Densities and mean ionic activity coefficients for $NaNO_3$ (◆) and $CaCl_2$ (■) at 298.15 K. Data sources given in Table 9.1. Calculations (—) and (- - -) from Equation 9.65 through Equation 9.67.

TABLE 9.1
Sources of Data for Figure 9.11

Salt	Property	Reference
$NaNO_3$	Density	Janz et al. 1970
	Activity coefficient	Hamer and Wu 1972
$CaCl_2$	Density	Romankiw and Chou 1983
	Activity coefficient	Staples and Nuttall 1977

Work has not continued on this approach because of the complexity of the expressions and the lack of ionic additivity, combined with only limited improvement in accuracy compared to other models with the same number of parameters. However, the tabulations of salt DCFI from evaluated data (Hu 1997) can provide tests for such other models.

9.5   ANALYSIS OF EQUATION OF STATE MIXING AND COMBINING RULES USING FLUCTUATION SOLUTION THEORY

The simplest fundamental connection between equations of state, expressed as Helmholtz functions of T, v, and $\{x\}$, and correlation function integrals is via the second composition derivative of the residual Helmholtz energy and DCFI,

$$1 - C_{ij} = N \left[\frac{\partial^2 \beta A^r}{\partial N_i \partial N_j} \right]_{T,V,N_k} \tag{9.68}$$

As usual, the TCFI expressions are more complex because they involve matrix inverses of the second derivatives. Many applications use an EOS model that is cubic in molar volume, such as the generalized van der Waals form, with pure component parameters that depend upon composition according to mixing rules that employ combining rules with unlike interaction parameters of the solution components. An example is the original van der Waals EOS,

$$p = \frac{RT}{v - b(\{x\})} - \frac{a(\{x\})}{v^2} \tag{9.69}$$

The Helmholtz energy, when written in the form for taking the derivatives of Equation 9.68 is

$$\frac{A^r(T,V,\{N\})}{RT} = N \ln\left(\frac{N}{V - Nb(\{N\})}\right)^2 - \frac{N^2 a(\{N\})}{RTV} \tag{9.70}$$

The derivatives can be taken for any set of mixing and combining rules. The well-known 1-fluid mixing rules are

$$N^2 a(\{N\}) = \sum_{i=1}^{n_c} \sum_{j=1}^{n_c} N_i N_j a_{ij}$$

$$Nb(\{N\}) = \frac{\displaystyle\sum_{i=1}^{n_c} \sum_{j=1}^{n_c} N_i N_j b_{ij}}{\displaystyle\sum_{i=1}^{n_c} N_i} \tag{9.71}$$

and, among the combining rule options, a common choice is

$$a_{ij} = \left(a_{ii} a_{jj}\right)^{\frac{1}{2}} \left(1 - k_{ij}\right)$$

$$b_{ij} = \frac{1}{2}\left(b_{ii} + b_{jj}\right)\left(1 - l_{ij}\right) \tag{9.72}$$

where k_{ij} and l_{ij} are binary parameters which are zero for pure components $(i = j)$. For these choices, Equation 9.68 becomes

$$1 - C_{ii} = \left(\frac{v}{v - b_{ii}}\right)^2 - \frac{2a_{ii}}{k_B T v} \tag{9.73}$$

and

$$1 - C_{ij} = 1 + \frac{\left(b_{ii} + b_{jj}\right)\left(1 - l_{jj}\right)}{v - b\left(\{x\}\right)} + \frac{b_{ii}b_{jj}}{\left[v - b\left(\{x\}\right)\right]^2} - \frac{2\left(a_{ii}a_{jj}\right)^{\frac{1}{2}}\left(1 - k_{ij}\right)}{k_B T v} \tag{9.74}$$

Equation 9.74, and expressions from other EOS models can be used to test the accuracy of mixing and combining rules by comparisons with experimental DCFI values. The EOS studied here is the Peng and Robinson (1976) model,

$$\frac{p^{\text{res}}}{RT} = \frac{Nb\left(\{x\}\right)}{V - Nb\left(\{x\}\right)} - \frac{N^2 a\left(T,\{x\}\right)/RT}{V^2 + 2Nb\left(\{x\}\right)V + \left[Nb\left(\{x\}\right)\right]^2} \tag{9.75}$$

An initial effort in this direction was presented by O'Connell and Clairmont (2010). The databank of binary DCFI and TCFI values for a variety of subcritical systems given by Wooley and O'Connell (1991) was used for DCFI comparisons as well as for composition variations of excess volume. The EOS models were the cubic original van der Waals and Peng–Robinson models, and the quintic DeSantis model, with mixing rules of the original van der Waals, Huron and Vidal, and Wong–Sandler forms (Poling, Prausnitz, and O'Connell 2000). The last two rules use excess Gibbs energy results, usually from fitting VLE data, to establish mixture parameters that are more accurate for the composition dependence of liquid solutions. Figures 9.12 and 9.13 show DCFIs and V^E from this analysis for the binary CCl_4 (1) and methanol. The questions posed and conclusions were:

1. Do the EOS and mixing rules follow the composition variations for $1 - C_{11}$ and $1 - C_{22}$ from pure component to infinite dilution? The answer is sometimes, but not reliably.
2. Can binary parameters be found to describe the composition variations of $1 - C_{12}$ and V^E? The answer is generally yes, but not always.
3. Is there a difference in results for the van der Waals and G^E mixing rules? The answer is yes, but there is no preponderance of accuracy in one form or another. The van der Waals mixing rules, with two binary parameters, is quite satisfactory for nonassociating, nonsolvating systems. But no mixing rule is adequate for components with hydrogen bonds in dilute solution. The accuracy of the descriptions of excess volume was uneven, with no apparent pattern of success or failure.
4. If experimental liquid volumes were used in the EOS to compute total pressures, the results were generally poor. The sensitivity of p to density is too great to obtain accurate values from the EOS.

This work is far from complete because only a few of the systems in the database were tested and no contemporary EOS models, such as Statistical Associating Fluids Theory (SAFT) (Muller and Gubbins 2001; Economou 2002), Perturbed

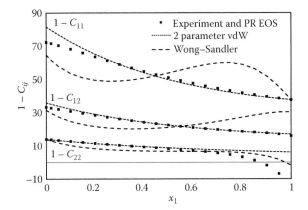

FIGURE 9.12 DCFIs for the CCl_4 – Methanol system at $T = 298.15$ K using Equation 9.74 with different mixing rules. EOS calculations from Peng–Robinson model, Equation 9.75, with the 2-parameter, 1-fluid, van der Waals mixing rule (- - -), Equation 9.74, and Wong–Sandler. (Data from D. S. H. Wong and S. I. Sandler, 1992, A Theoretically Correct Mixing Rule for Cubic Equations of State, *American Institute of Chemical Engineers Journal*, 38, 671.) G_m^E mixing rule (– – –). Data (■) from tabulations of Wooley and O'Connell. (From R. J. Wooley and J. P. O'Connell, 1991, A Database of Fluctuation Thermodynamic Properties and Molecular Correlation-Function Integrals for a Variety of Binary Liquids, *Fluid Phase Equilibria*, 66, 233.)

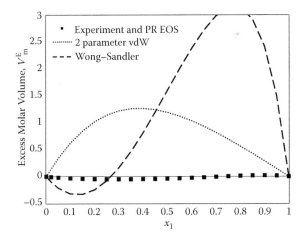

FIGURE 9.13 V_m^E for the CCl_4 – Methanol system at $T = 298.15$ K computed with different mixing rules. EOS calculations from Peng–Robinson model, Equation 9.75, with the 2-parameter, 1-fluid, van der Waals mixing rule (···) and Wong–Sandler. (Data from D. S. H. Wong and S. I. Sandler, 1992, A Theoretically Correct Mixing Rule for Cubic Equations of State, *American Institute of Chemical Engineers Journal*, 38, 671.) G^E mixing rule (– – –). Data (■) from tabulations of Wooley and O'Connell. (From R. J. Wooley and J. P. O'Connell, 1991, A Database of Fluctuation Thermodynamic Properties and Molecular Correlation-Function Integrals for a Variety of Binary Liquids, *Fluid Phase Equilibria*, 66, 233.)

Chain SAFT (PC-SAFT) (Gross and Sadowski 2001), and Cubic Plus Association (CPA) (Kontogeorgis et al. 1996), which have been reviewed by Kontogeorgis and Folas (Kontogeorgis et al. 2006a,b) and by Polishuk and Mulero (2011), were examined.

9.6 CONCLUSIONS

It has been shown that models for DCFIs can lead to successful descriptions of solution properties and phase equilibria, especially for strongly nonideal systems such as dilute solutions of gases and solids. Because FST formulations are for composition derivatives of pressure and chemical potential or fugacity, the evaluation can appear complex and requires property values at certain reference states. This may be the reason such models have not been implemented into process simulators. However, the reliability and accuracy of models based on perturbations from hard spheres, such as Equation 9.4, is quite good for very many systems, and the results can at least be used to generate local parameterizations and to validate EOS models. Ultimately, results from molecular simulations, as described in Chapter 6, may lead to new relations for the thermodynamic models for use in process simulators.

10 Solubilities of Various Solutes in Multiple Solvents
A Fluctuation Theory Approach

Ivan L. Shulgin and Eli Ruckenstein

CONTENTS

Abstract: As is well known, the methods of classical thermodynamics (activity coefficient or equation of state approaches) have frequently failed to correctly predict the solubilities of various solutes in pure or multiple solvents. This is true for the solubility of gases, large molecules (drugs, biomolecules), and very large molecules (polymers and proteins) in water and aqueous solutions. The main reason for this difficulty is that these systems are highly nonideal and the activity coefficients cannot be predicted using the traditional thermodynamic methods. Fortunately, the fluctuation theory of solutions (or the Kirkwood–Buff [KB] theory of solutions) has proven to be a very effective tool for predicting the solubility of various solutes, especially in mixed solvents. During the last decade, the KB theory was successfully applied to the solubility of solid solutes in supercritical solvents, to the solubility of gases in multiple solvents and polymer solutions, to the solubility of drugs and pollutants in aqueous mixtures and to the solubility of proteins in mixed solvents.

In this chapter, the KB theory of solutions is applied to the solubility in various systems, with the emphasis on the solubility in multicomponent (higher than binary) solvents. For such systems, only the fluctuation theory can provide useful results, and no traditional thermodynamic approach is available. In addition, the fluctuation theory is applied to the solubility of proteins and the salting-in or salting-out of proteins.

10.1 INTRODUCTION

According to IUPAC (McNaught and Wilkinson 1997), solubility is defined as the analytical composition of a saturated solution expressed as the proportion of a designated solute in a designated solvent. The solubility can be expressed as molar concentration, molality, mole fraction, mole ratio, and so forth. A solute and a solvent can be either individual substances or mixtures of substances. In addition, both solute and solvent can be in different aggregation states. Therefore, one can speak about the solubility of a gas in a liquid, of a solid in a liquid, of a gas in a solid, of a solid in a supercritical fluid, and so on. In an old book, Hildebrand wrote that: "The entire history of chemistry bears witness to the extraordinary importance of solubility. The somewhat mysterious nature of solution and recrystallization invited the speculations of ancient philosophers. The medieval alchemist took an interest in the *Alkahest*, or universal solvent, inferior only to his interest in gold and in eternal life" (Hildebrand 1924). At the present time, the importance of solubility in life, technology, and science is self-evident.

The solubility of gases in mixed solvents is important in many processes, such as biological and organic reactions, corrosion and oxidation of materials, aerobic fermentation, petroleum and natural gas exploitation, petroleum refining, coal gasification, gas antisolvent crystallization, formation of gas hydrates, and so forth (Wilhelm, Battino, and Wilcock 1977). The salting-out and salting-in effects are also relevant (Wilhelm, Battino, and Wilcock 1977). The solubility of naturally occurring and atmospheric gases in sodium chloride solutions, and its dependence on pressure, temperature, and salt concentration is of interest not only to the physical chemist, but also to the geochemist, because the aqueous salt solutions containing the gases can simulate the brine present in the Earth's crust. The information about solubility of atmospheric gases in water and seawater is also relevant to the understanding of the ecological balance between the fresh- and seawater systems (Wilhelm, Battino, and Wilcock 1977).

The solubility of solid substances in supercritical (SC) fluids is important in SC technologies such as the gas antisolvent technology, which allows one to separate a protein or a drug from a mixture, and in the cosolvent and entrainer processes, which enhance the solubility of a solute in a selective manner. The solubility of poorly soluble drugs in multicomponent solvents is important in the selection of appropriate solvents for drug formulations. The solubility of proteins and other biomolecules in mixed aqueous solvents, especially in various biological liquids such as blood, is important in the understanding of some biological processes in living organisms.

It is well known (Prausnitz, Lichtenthaler, and Gomes de Azevedo 1999) that the solubility can be calculated if the standard state properties and the activity coefficients of the solute(s) and solvent(s) are known. However, prediction of the activity coefficients of solutes (such as gases and large molecules of biomedical and environmental significance) in saturated solutions of multicomponent mixtures constitutes a major difficulty. Generally speaking, the activity coefficient of a solute in a saturated solution of a multicomponent mixture can be predicted using group-contribution methods, such as UNIFAC and ASOG (Derr and Deal 1969; Fredenslund, Jones, and Prausnitz 1975; Fredenslund, Gmehling, and Rasmussen 1977; Kojima and Tochigi 1979; Gmehling 1995; Prausnitz, Lichtenthaler, and Gomes de Azevedo 1999; Jakob et al. 2006; Schmid and Gmehling 2012). However, the above group-contribution methods cannot provide accurate results for the activity coefficients of large molecules, such as those typically required for the study of drug or environmental molecules in aqueous mixed solvents (Li, Doucette, and Andren 1994; Kan and Tomson 1996; Wienke and Gmehling 1998; Bouillot, Teychene, and Biscans 2011; Diedrichs and Gmehling 2011; Mota et al. 2012).

Another method for predicting the solubility of gases and large molecules in a mixed solvent involves the Kirkwood–Buff theory of solutions (Kirkwood and Buff 1951). This theory connects macroscopic properties of solutions, such as the isothermal compressibility, the derivatives of the chemical potentials with respect to concentration and the partial molar volumes, to their microscopic characteristics in the form of spatial integrals involving the radial distribution functions. It provides the opportunity to extract some microscopic characteristics of mixtures from measurable thermodynamic quantities. We employed the Kirkwood–Buff theory of solution to obtain expressions for the derivatives of the activity coefficients in ternary (Ruckenstein and Shulgin 2001b) and multicomponent (Ruckenstein and Shulgin 2003a) mixtures with respect to mole fractions. These expressions for the derivatives of the activity coefficients were used to calculate the solubilities of various solutes in aqueous mixed solvents, namely:

1. The solubilities of drugs and environmentally significant molecules in binary and multicomponent aqueous mixed solvents (Ruckenstein and Shulgin 2003b, 2003c, 2003d, 2004, 2005),
2. The solubilities of gases in binary and multicomponent aqueous mixed solvents (Ruckenstein and Shulgin 2002b; Shulgin and Ruckenstein 2002b; Ruckenstein and Shulgin 2003a; Shulgin and Ruckenstein 2003),
3. The solubilities of various proteins in aqueous solutions (Shulgin and Ruckenstein 2005b; Ruckenstein and Shulgin 2006; Shulgin and Ruckenstein 2006b).

Details regarding the applications of the KB theory of solutions to solubility can be found in a recently published book (Ruckenstein and Shulgin 2009). We also mention here the contributions of Mazo and Smith (Mazo 2006, 2007; Smith and Mazo 2008) to the application of the KB theory of solutions to the solubility in mixed solvents (see also Section 1.3.5 in Chapter 1 and Chapter 9).

In this chapter, the KB theory of solutions is applied to the study of solubility in various systems, the emphasis being on multicomponent (higher than binary) solvents. For these systems, only fluctuation theory can provide useful results, and no traditional thermodynamic approach is available. In addition, the fluctuation theory is applied to issues involving salting-in or salting-out effects.

10.2 DERIVATIVES OF ACTIVITY COEFFICIENTS VIA THE KIRKWOOD–BUFF THEORY OF SOLUTIONS

On the basis of fluctuation theory, the following expression for the derivative of the activity coefficient of a solute $(\gamma_{2,t})$ in a solvent (1)–solute (2)–cosolvent (3) mixture could be derived (Ruckenstein and Shulgin 2001b), and which is valid for any types of solutes and cosolvents,

$$
\left(\frac{\partial \ln \gamma_{2,t}}{\partial x_3^t}\right)_{T,p,x_2^t} =
$$

$$
-\frac{(c_1+c_2+c_3)\left(c_1\left[G_{11}+G_{23}-G_{12}-G_{13}\right]+c_3\left[-G_{12}-G_{33}+G_{13}+G_{23}\right]\right)}{c_1+c_2+c_3+c_1c_2\Delta_{12}+c_1c_3\Delta_{13}+c_2c_3\Delta_{23}+c_1c_2c_3\Delta_{123}} \tag{10.1}
$$

where x_i^t is the mole fraction of component i in the ternary mixture, c_k is the bulk molecular concentration of component k in the ternary mixture of 1–2–3, and $G_{\alpha\beta}$ is the usual Kirkwood–Buff integral (KBI) given by,

$$
G_{\alpha\beta} = 4\pi \int_0^\infty [g_{\alpha\beta}(r)-1]r^2 dr \tag{10.2}
$$

where $g_{\alpha\beta}$ is the radial distribution function between species α and β, r is the distance between the centers of molecules α and β, and $\Delta_{\alpha\beta}$ and Δ_{123} are defined as follows,

$$
\Delta_{\alpha\beta} = G_{\alpha\alpha} + G_{\beta\beta} - 2G_{\alpha\beta}, \quad \alpha \neq \beta \tag{10.3}
$$

and

$$
\Delta_{123} = G_{11}G_{22} + G_{11}G_{33} + G_{22}G_{33} + 2G_{12}G_{13} + 2G_{12}G_{23} + 2G_{13}G_{23} -
$$
$$
- G_{12}^2 - G_{13}^2 - G_{23}^2 - 2G_{11}G_{23} - 2G_{22}G_{13} - 2G_{33}G_{12} \tag{10.4}
$$

The factors in the square brackets in the numerator of Equation 10.1 and Δ_{123} can be expressed in terms of $\Delta_{\alpha\beta}$ as follows,

$$
G_{12} + G_{33} - G_{13} - G_{23} = \frac{\Delta_{13} + \Delta_{23} - \Delta_{12}}{2} \tag{10.5}
$$

$$G_{11} + G_{23} - G_{12} - G_{13} = \frac{\Delta_{12} + \Delta_{13} - \Delta_{23}}{2} \tag{10.6}$$

and

$$\Delta_{123} = -\frac{(\Delta_{12})^2 + (\Delta_{13})^2 + (\Delta_{23})^2 - 2\Delta_{12}\Delta_{13} - 2\Delta_{12}\Delta_{23} - 2\Delta_{13}\Delta_{23}}{4} \tag{10.7}$$

The insertion of Equation 10.5 through Equation 10.7 into Equation 10.1 provides expressions for the derivatives

$$\left(\frac{\partial \ln \gamma_{2,t}}{\partial x_3^t} \right)_{T,p,x_2^t}$$

in terms of $\Delta_{\alpha\beta}$ and concentrations.

It should be noted that Δ_{ij} is a measure of nonideality (Ben-Naim 1977) of the binary mixture α–β, because for an ideal mixture $\Delta_{\alpha\beta} = 0$. For a ternary mixture 1–2–3, Δ_{123} also provides a measure of nonideality. Indeed, inserting $G_{\alpha\beta}^{id}$ for an ideal mixture (Ruckenstein and Shulgin 2001a) into the expression of Δ_{123}, one obtains that for an ideal ternary mixture $\Delta_{123} = 0$.

At infinite dilution of component 2, Equation 10.1 becomes,

$$\lim_{x_2^t \to 0} \left(\frac{\partial \ln \gamma_{2,t}}{\partial x_3^t} \right)_{T,p,x_2^t} = -\frac{\left(c_1^0 + c_3^0\right)\left[\left(c_1^0 + c_3^0\right)\left(\Delta_{12} - \Delta_{23}\right)_{x_2^t=0} + \left(c_1^0 - c_3^0\right)\left(\Delta_{13}\right)_{x_2^t=0}\right]}{2\left[c_1^0 + c_3^0 + c_1^0 c_3^0 \left(\Delta_{13}\right)_{x_2^t=0}\right]} \tag{10.8}$$

where c_1^0 and c_3^0 represent the bulk molecular concentrations of components 1 and 3 in the solute-free binary solvent 1–3. In addition to Equation 10.8, one can write for the derivative of the activity coefficient in a binary mixture with respect to the mole fractions the expression,

$$\gamma_{11} \equiv \left(\frac{\partial \ln \gamma_1^0}{\partial x_1^0} \right)_{p,T} = -\frac{c_3^0 \Delta_{13}}{1 + c_1^0 x_3^0 \Delta_{13}} \tag{10.9}$$

where x_3^0 and γ_1^0 are the mole fraction of component 3 and the activity coefficient of component 1 in the solute-free binary solvent of 1–3.

By combining Equations 10.8 and 10.9, one obtains an expression for the derivative of the activity coefficient of an infinitely dilute solute with respect to the cosolvent mole fraction in terms of characteristics of the solute-free binary solvent (γ_{11}, c_1^0, and c_3^0) and the parameters Δ_{12} and Δ_{23}, which characterize the interactions of an infinitely dilute solute with the components of the mixed solvent. Even though Equation 10.8 constitutes a formal statistical thermodynamics relation in which all

Δ_{ij} are unknown, it could be successfully used to derive analytical expressions for the solubility of gases and large molecules of biomedical and environmental significance including proteins in aqueous media (Ruckenstein and Shulgin 2009). In addition, Equation 10.8 could be used to identify whether a cosolvent is a salting-out or salting-in agent (Ruckenstein and Shulgin 2002b).

10.3 SOLUBILITY OF GASES IN BINARY MIXED SOLVENTS

The solubility of gases (component 2) in the mixed solvent 1–3 can be expressed via the Henry constant H_{2t} (O'Connell 1971a). In order to obtain the composition dependence of the Henry constant in a binary solvent, one should consider either the derivative

$$\left(\frac{\partial \ln H_{2t}}{\partial x_3^t} \right)_{p,T,x_2^t=0}$$

or the derivative

$$\left(\frac{\partial \ln H_{2t}}{\partial x_1^t} \right)_{p,T,x_2^t=0}$$

which can be obtained starting from the following expression for the Henry constant in a binary solvent (O'Connell 1971a),

$$\ln H_{2,t} = \lim_{x_2^t \to 0} \ln \gamma_{2,t} + \ln f_2^0(p, T) \tag{10.10}$$

where $f_i^0(p, T)$ is the fugacity of component i (Prausnitz, Lichtenthaler, and Gomes de Azevedo 1999). The combination of Equations 10.8, 10.9, and 10.10 and the integration of the obtained equation provides the following relation for the composition dependence of the Henry constant in a binary solvent mixture, at constant temperature and pressure (Shulgin and Ruckenstein 2002b),

$$\ln H_{2t} = -\int \left(c_1^0 + c_3^0 \right) \frac{(\Delta_{12} - \Delta_{23})_{x_2^t=0}}{2} \left[1 + x_3^{b,1-3} \left(\frac{\partial \ln \gamma_3^{b,1-3}}{\partial x_3^{b,1-3}} \right)_{p,T} \right] dx_3^{b,1-3} +$$

$$+ \frac{1}{2} \left(\ln \gamma_1^{b,1-3} + \ln \gamma_3^{b,1-3} \right) + A \tag{10.11}$$

where $A(p,T)$ is a composition-independent integration constant and $x_i^{b,1-3}$ and $\gamma_i^{b,1-3}$ are the mole fraction and the activity coefficient of component i ($i = 1$ or 3) in the gas-free binary solvent 1–3. One can note that Equation 10.11 contains, with the exception of $(\Delta_{12} - \Delta_{23})_{x_2^1=0}$ at infinite dilution of solute (component 2), only solute-free variables.

The above equation can be applied to the solubility of a gas in various kinds of binary mixtures, a mixture of two solvents, or a mixture of a solvent and a solute (salt, polymer, or protein).

The oldest and simplest relationship between the Henry constant in binary solvents and those in individual solvents is that proposed by Krichevsky (Krichevskii 1937),

$$\ln H_{2,t} = x_1^{b,1-3} \ln H_{2,1} + x_3^{b,1-3} \ln H_{2,3} \qquad (10.12)$$

The Krichevsky equation (Equation 10.12) is valid when the binary mixtures 1–2 and 2–3 (gas solute/pure solvents) and the ternary mixture 1–2–3 are ideal (Shulgin and Ruckenstein 2002b). However, such conditions are often far from reality. Let us consider, for example, the solubility of a hydrocarbon in a water/alcohol solvent (for instance, water/methanol, water/ethanol, etc.). The activity coefficient of propane in water at infinite dilution is $\approx 4 \cdot 10^3$ (Kojima, Zhang, and Hiaki 1997), whereas the activity coefficients of alcohols and water in aqueous solutions of simple alcohols seldom exceed 10. It is therefore clear that the main contribution to the nonideality of the ternary gas/binary solvent mixture comes from the nonideality of the gas solute in the individual solvents, which are neglected in the Krichevsky equation.

For this reason, in a first step it will be assumed that only the binary solvent (1–3) behaves as an ideal mixture. One can therefore write that $\gamma_3^{b,1-3} = 1$ and

$$V_m = x_1^{b,1-3} V_1^0 + x_3^{b,1-3} V_3^0 \qquad (10.13)$$

where V_m is the molar volume of the binary mixture 1–3, and V_1^0 and V_3^0 are the molar volumes of the individual pure solvents 1 and 3. Under these conditions, Equation 10.11 becomes,

$$\ln H_{2t} = -\int \frac{(\Delta_{12} - \Delta_{23})_{x_2^1=0}}{2\left(x_1^{b,1-3} V_1^0 + x_3^{b,1-3} V_3^0\right)} dx_3^{b,1-3} + A \qquad (10.14)$$

Equation 10.14 can provide a number of analytical expressions for H_{2t} with various assumptions regarding the nonideality term $(\Delta_{12} - \Delta_{23})_{x_2^1=0}$. The simplest expression is obtained by assuming that $(\Delta_{12})_{x_2^1=0} = (G_{11} + G_{22} - 2G_{12})_{x_2^1=0}$ and $(\Delta_{23})_{x_2^1=0} = (G_{22} + G_{33} - 2G_{23})_{x_2^1=0}$ are independent of the composition of the solvent mixture. Then, Equation 10.14 becomes

$$\ln H_{2t} = A(p,T) - \frac{B(p,T)\ln\left(x_1^{b,1-3} V_1^0 + x_3^{b,1-3} V_3^0\right)}{V_3^0 - V_1^0} \qquad (10.15)$$

where $B(p,T) = (\Delta_{12} - \Delta_{23})_{x_2^1=0}/2$.

The constants $A(p,T)$ and $B(p,T)$ can be obtained using the following extreme expressions,

$$(\ln H_{2t})_{x_1^{b,1-3}=0} = \ln H_{2,3} \tag{10.16}$$

and

$$(\ln H_{2t})_{x_3^{b,1-3}=0} = \ln H_{2,1} \tag{10.17}$$

Combining Equation 10.15 through Equation 10.17 yields the final result,

$$\ln H_{2t} = \frac{\ln H_{2,1}\left(\ln V - \ln V_3^0\right) + \ln H_{2,3}\left(\ln V_1^0 - \ln V\right)}{\ln V_1^0 - \ln V_3^0} \tag{10.18}$$

Equation 10.18 provides the Henry constant for a binary solvent in terms of those for the individual solvents and their molar volumes. This simple equation was obtained using less restrictive approximations than those involved in the Krichevsky equation, but assuming that only the binary solvent 1–3 is an ideal mixture. Such an assumption is reasonable because the nonideality of the binary solvent is much lower than those of the solute gas and each of the constituents of the solvent.

Equation 10.11 can be, however, integrated using any of the analytical expressions available for the activity coefficient $\ln \gamma_3^{b,1-3}$, such as the Van Laar, Margules, Wilson, NRTL, and so forth. To take into account the nonideality of the molar volume, one can use the expression,

$$V_m = x_1^{b,1-3}V_1^0 + x_3^{b,1-3}V_3^0 + V_m^E \tag{10.19}$$

where the last term is the excess molar volume.

When the integration in Equation 10.11 cannot be performed analytically, one can first integrate it numerically between $0 < x_3^{b,1-3} < 1$ to obtain the expression

$$\ln H_{2,3} - \ln H_{2,1} = -B\int_0^1 \frac{1}{V}\left[1 + x_3^{b,1-3}\left(\frac{\partial \ln \gamma_3^{b,1-3}}{\partial x_3^{b,1-3}}\right)_{p,T}\right]dx_3^{b,1-3}$$
$$+ \frac{1}{2}\left(\ln \gamma_{1,3}^\infty - \ln \gamma_{3,1}^\infty\right) \tag{10.20}$$

where $\gamma_{1,3}^\infty = \lim_{x_1\to 0} \gamma_1^{b,1-3}$ and $\gamma_{3,1}^\infty = \lim_{x_3\to 0} \gamma_3^{b,1-3}$.

Equation 10.20 allows one to determine the constant B. Further, Equation 10.11 can be integrated between $0 < x_3^{b,1-3} < x$, to obtain the Henry constant for a mole fraction x in a binary solvent,

$$\ln H_{2,\mathrm{t}} = \ln H_{2,1} - B \int_0^x \frac{1}{V}\left[1 + x_3^{\mathrm{b},1-3}\left(\frac{\partial \ln \gamma_3^{\mathrm{b},1-3}}{\partial x_3^{\mathrm{b},1-3}}\right)_{p,T}\right] dx_3^{\mathrm{b},1-3} +$$

$$+ \frac{1}{2}\left(\ln \gamma_1^{\mathrm{b},1-3}(x) + \ln \gamma_3^{\mathrm{b},1-3}(x) - \ln \gamma_{3,1}^{\infty}\right)$$

(10.21)

This procedure allows one to account for the nonideality (activity coefficients and molar volume) of the binary solvent (Shulgin and Ruckenstein 2002b).

However, even the simple Equation 10.18 provides more accurate results for the Henry constant in binary solvents than the Krichevsky equation does (Shulgin and Ruckenstein 2002b), as indicated in Figure 10.1.

10.4 SOLUBILITY OF DRUGS AND HYDROPHOBIC ORGANIC POLLUTANTS IN BINARY MIXED SOLVENTS

Equation 10.11 can be used to predict the solubility of drugs and hydrophobic organic pollutants in binary mixed solvents. The majority of binary mixed solvents considered here are aqueous solvents.

The solubility of solid substances in pure and mixed solvents can be described by the usual solid–liquid equilibrium conditions (Acree Jr. 1984; Prausnitz, Lichtenthaler, and Gomes de Azevedo 1999). For the solubilities of a solid (solute, component 2) in water (component 3), cosolvent (component 1) and their mixture (mixed solvent 1–3), one can write the following equations:

$$f_2^{\mathrm{S}}/f_2^{\mathrm{L}}(T, p) = x_2^{\mathrm{b_1}}\gamma_2^{\mathrm{b_1}}\left(T, p, \{x\}\right)$$

(10.22)

$$f_2^{\mathrm{S}}/f_2^{\mathrm{L}}(T, p) = x_2^{\mathrm{b_3}}\gamma_2^{\mathrm{b_3}}\left(T, p, \{x\}\right)$$

(10.23)

$$f_2^{\mathrm{S}}/f_2^{\mathrm{L}}(T, p) = x_2^{\mathrm{t}}\gamma_2^{\mathrm{t}}\left(T, p, \{x\}\right)$$

(10.24)

In Equation 10.22 through Equation 10.24, $x_2^{\mathrm{b_1}}$, $x_2^{\mathrm{b_3}}$, and x_2^{t} are the solubilities (mole fractions) of the solid (component 2) in the cosolvent, water, and their mixture, respectively, $\gamma_2^{\mathrm{b_1}}$, $\gamma_2^{\mathrm{b_3}}$, and γ_2^{t} are the activity coefficients of the solid in its saturated solutions in cosolvent, water, and mixed solvent, $f_2^{\mathrm{L}}(T,p)$ is the hypothetical fugacity of a solid as a (subcooled) liquid at a given pressure and temperature, f_2^{S} is the fugacity of the pure solid (component 2), and $\{x\}$ designates that the activity coefficients of the solid depend on composition. If the solubilities of the pure and mixed solvents in the solid phase are negligible, then the left-hand sides of Equation 10.22 through Equation 10.24 depend only on the properties of the solute.

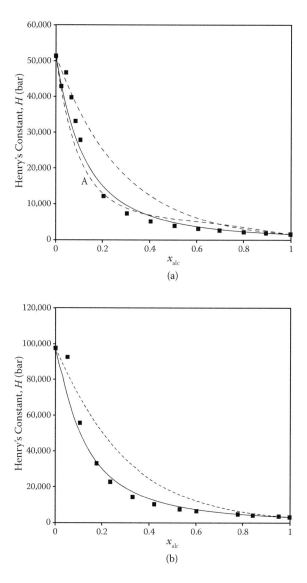

FIGURE 10.1 Henry constants of gases in binary solvent mixtures at 760 mmHg partial pressure: experimental (■) (From J. Tokunaga, 1975, Solubilities of Oxygen, Nitrogen, and Carbon-Dioxide in Aqueous Alcohol Solutions, *Journal of Chemical and Engineering Data*, 20, 41). Solid lines: calculated with Equation 10.18, dashed line: calculated with the Krichevsky equation (Equation 10.12), on Figure 10.1a, the dashed line (A) represents results obtained with Equation 10.11 when the activity coefficients of the mixed solvents are expressed through the two-suffix Margules equations: (a) oxygen (2) in 1-propanol (1) – water (3) at 40°C; (b) nitrogen (2) in 2-propanol (1) – water (3) at 40°C; (*continued on next page*)

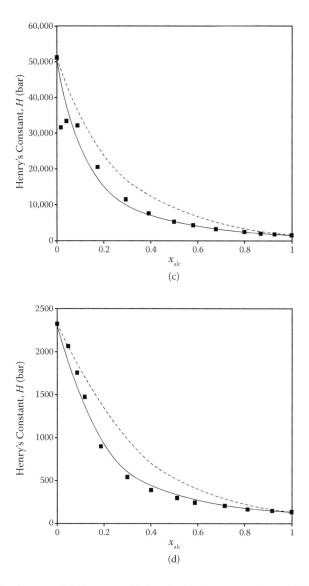

FIGURE 10.1 (continued). (c) oxygen (2) in ethanol (1) – water (3) at 40°C; and (d) carbon dioxide (2) in 1-propanol (1) – water (3) at 40°C. x_{alc} is the mole fraction of the alcohol in the gas-free mixture of solvents. (Adapted with permission from I. L. Shulgin and E. Ruckenstein, 2002b, Henry's Constant in Mixed Solvents from Binary Data, *Industrial and Engineering Chemistry Research*, 41, 1689 [Copyright 2002, American Chemical Society].)

Integration of Equation 10.8 combined with 10.9 provides the following expression for the activity coefficient ($\gamma_2^{t,\infty}$) of a solid solute in a binary mixed solvent at infinite dilution (Ruckenstein and Shulgin 2003b),

$$\ln \gamma_2^{t,\infty} = -\int \frac{B}{V} \left[1 + x_3^{b,1-3} \left(\frac{\partial \ln \gamma_3^{b,1-3}}{\partial x_3^{b,1-3}} \right)_{p,T} \right] dx_3^{b,1-3} +$$
$$+ \frac{1}{2} \int \frac{\left(x_1^{b,1-3} - x_3^{b,1-3} \right)}{x_1^{b,1-3}} \left(\frac{\partial \ln \gamma_3^{b,1-3}}{\partial x_3^{b,1-3}} \right)_{p,T} dx_3^{b,1-3} + A(p,T) \qquad (10.25)$$

where $x_i^{b,1-3}$ and $\gamma_i^{b,1-3}$ ($i = 1, 3$) are the mole fraction and the activity coefficient of component i in the binary solvent 1–3, V_m is the molar volume of the binary 1–3 solvent, $A(p,T)$ is a composition-independent constant of integration, and B is a function of the Kirkwood–Buff integrals (Ruckenstein and Shulgin 2003b). If B is considered independent of the composition of the binary mixed solvent 1–3, Equation 10.25 can be rewritten in the form,

$$\ln \gamma_2^{t,\infty} = -BI_1 + \frac{I_2}{2} + A(p,T) \qquad (10.26)$$

where

$$I_1 = \int \frac{\left[1 + x_3^{b,1-3} \left(\dfrac{\partial \ln \gamma_3^{b,1-3}}{\partial x_3^{b,1-3}} \right)_{p,T} \right]}{V_m} dx_3^{b,1-3} \qquad (10.27)$$

and

$$I_2 = \int \frac{\left(x_1^{b,1-3} - x_3^{b,1-3} \right)}{x_1^{b,1-3}} \left(\frac{\partial \ln \gamma_3^{b,1-3}}{\partial x_3^{b,1-3}} \right)_{p,T} dx_3^{b,1-3} \qquad (10.28)$$

Combining Equation 10.24 and Equation 10.26, one obtains the following expression for the solubility of a poorly soluble solid in a mixed binary solvent,

$$\ln x_2^l = BI_1 - \frac{I_2}{2} + \bar{A}(p,T) \qquad (10.29)$$

where $\bar{A}(p,T) = -A(p,T) + \ln[f_2^S/f_2^{l}(T,p)]$.

The integrals I_1 and I_2 can be calculated if the composition dependencies of the activity coefficients and of the molar volume V_m of the binary mixed solvent 1–3 are known. The solubilities of the solute in the individual solvents also are needed to calculate the composition-independent quantities $\bar{A}(p,T)$ and $B(p,T)$.

Equation 10.29 was used to derive a number of analytical expressions for the solubility of a poorly soluble solid such as drugs and hydrophobic organic pollutants in a mixed binary solvent. Two cases are relevant: (a) the mixed solvent (1–3) is an ideal solution, (b) the mixed solvent (1–3) is a nonideal solution.

In case (a), one can obtain a simple analytical equation for the solubility of a poorly soluble solid in a mixed binary solvent (Ruckenstein and Shulgin 2003b),

$$\ln x_2^l = \frac{\left(\ln V_m - \ln V_3^0\right)\ln x_2^{b1} + \left(\ln V_1^0 - \ln V_m\right)\ln x_2^{b3}}{\ln V_1^0 - \ln V_3^0} \tag{10.30}$$

where V_m is the molar volume of the binary mixture 1–3, and V_1^0 and V_3^0 are the molar volumes of the individual solvents 1 and 3.

However, Equation 10.30 cannot provide the maximum in the solubility versus mixed solvent composition, which was frequently observed in the solubility of drugs in aqueous mixed solvents (Jouyban-Gharamaleki et al. 1999, and references therein). To accommodate this feature of the solubility curve, the molar volume of the mixed solvent is replaced in Equation 10.30 by

$$V_m = x_1^{b,1-3}V_1^0 + x_3^{b,1-3}V_3^0 + ex_1^{b,1-3}x_3^{b,1-3} \tag{10.31}$$

where e is an empirical parameter which is evaluated from the solubility data in a mixed solvent. Equation 10.31 does not represent satisfactorily the molar volume of a mixed solvent. However, the insertion of Equation 10.31 into Equation 10.30 leads to a one-parameter semiempirical equation for the solubility of a solid in a mixed solvent, which exhibits a maximum in the solubility versus mixed solvent composition (Figures 10.2 and 10.3). A combination of Equations 10.30 and 10.31 provides the most accurate one-parameter equation, which is available for the solubility of a poorly soluble solid in a mixed binary solvent (Ruckenstein and Shulgin 2003b).

In case (b), various expressions for the activity coefficients in solute-free mixed solvent can be used (Ruckenstein and Shulgin 2003c, 2005; Shulgin and Ruckenstein 2003). The Margules, Flory, and Wilson equations were used to derive one- or two-parameter analytical equations for predicting the solubility of a poorly soluble solid in a mixed binary solvent (Ruckenstein and Shulgin 2003c, 2005; Shulgin and Ruckenstein 2003). The combination of Equation 10.29 with the Wilson equation provided the most accurate results of the two-parameter equations available in literature (Ruckenstein and Shulgin 2003c, 2005). One should point out that the solubilities of the solid in pure solvents are needed as well. Figures 10.4 and 10.5 illustrate the accuracy of Equation 10.29 combined with Flory and Wilson equations.

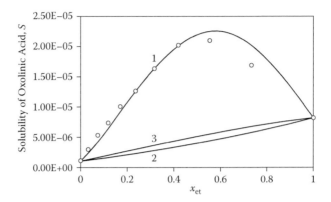

FIGURE 10.2 Comparison between experimental (o) (From A. Jouyban, S. Romero, H. K. Chan, B. J. Clark, and P. Bustamante, 2002, A Cosolvency Model to Predict Solubility of Drugs at Several Temperatures from a Limited Number of Solubility Measurements, *Chemical and Pharmaceutical Bulletin,* 50, 594) and predicted (solid lines) solubilities of oxolinic acid (*S* is the mole fraction of oxolinic acid) in the mixed solvent water/ethanol (x_{et} is the mole fraction of ethanol) at room temperature. 1-solubility calculated using Equations 10.30 and 10.31, 2-solubility calculated using Equation 10.30 and Equation 10.13, and 3-solubility calculated using log-linear model. (From S. H. Yalkowsky and T. J. Roseman, 1981, Solubilization of Drugs by Cosolvents, In *Techniques of Solubilization of Drugs,* edited by S. H. Yalkowsky, New York: Marcel Dekker.) (Reprinted from E. Ruckenstein and I. L. Shulgin, 2003b, Solubility of Drugs in Aqueous Solutions. Part 1: Ideal Mixed Solvent Approximation, *International Journal of Pharmaceutics,* 258, 193, With permission from Elsevier.)

10.5 SOLUBILITY OF SOLUTES IN MULTICOMPONENT MIXED SOLVENTS

Binary aqueous mixed solvents usually increase the solubility of a poorly soluble drug compared to that in pure water; however, they can also increase the risk of toxicity. The right selection of a ternary or a multicomponent aqueous mixed solvent can improve the solubility of the drug with minimal toxic effects. Pharmaceutical practice has shown that many marketed liquid formulations involve multiple solvents. However, the experimental determinations of the solubilities in multicomponent solutions are time consuming because of the large number of compositions needed to cover the concentration ranges of interest and can be very expensive because of the high prices of some modern drugs. For this reason, it is important to provide a reliable method for predicting the solubility of drugs in multicomponent mixed solvents from available experimental solubilities in subsystems such as pure solvents, binary mixed solvents, and so forth.

The KB theory can be successfully used to predict the solubility of poorly soluble solute in multicomponent mixed solvents from available experimental solubilities in subsystems such as pure solvents, binary mixed solvents, and so forth. Such a method can be used for preselecting a multicomponent mixed solvent with the use of

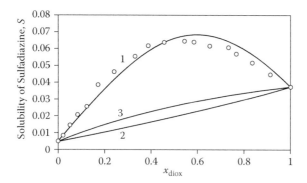

FIGURE 10.3 Comparison between experimental (o) (From P. Bustamante, B. Escalera, A. Martin, and E. Selles, 1993, A Modification of the Extended Hildebrand Approach to Predict the Solubility of Structurally Related Drugs in Solvent Mixtures, *Journal of Pharmacy and Pharmacology*, 45, 253) and predicted (solid lines) solubilities of sulfadiazine (S is the mole fraction of sulfadiazine) in the mixed solvent water/dioxane (x_{diox} is the mole fraction of dioxane) at room temperature. 1-solubility calculated using Equations 10.30 and 10.31, 2-solubility calculated using Equations 10.30 and 10.13, and 3-solubility calculated using log-linear model. (From S. H. Yalkowsky and T. J. Roseman, 1981, Solubilization of Drugs by Cosolvents, In *Techniques of Solubilization of Drugs*, edited by S. H. Yalkowsky, New York: Marcel Dekker.) (Reprinted from E. Ruckenstein and I. L. Shulgin, 2003b, Solubility of Drugs in Aqueous Solutions. Part 1: Ideal Mixed Solvent Approximation, *International Journal of Pharmaceutics*, 258, 193, With permission from Elsevier.)

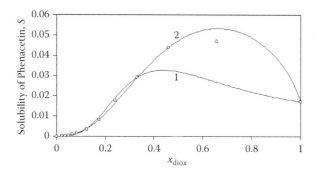

FIGURE 10.4 Comparison between experimental (o) and calculated (solid lines) solubilities of phenacetin (S is the mole fraction of phenacetin) in the mixed solvent water/dioxane (x_{diox} is the mole fraction of dioxane) at room temperature. The solubility was calculated using Equation 10.29. 1-activity coefficients expressed via the Flory–Huggins equation, 2-activity coefficients expressed via the Wilson equation. (From C. Bustamante, and P. Bustamante, 1996, Nonlinear Enthalpy-Entropy Compensation for the Solubility of Phenacetin in Dioxane-Water Solvent Mixtures, *Journal of Pharmaceutical Sciences*, 85, 1109. Reprinted from E. Ruckenstein, and I. L. Shulgin, 2003c, Solubility of Drugs in Aqueous Solutions. Part 2: Binary Nonideal Mixed Solvent, *International Journal of Pharmaceutics*, 260, 283, With permission from Elsevier.)

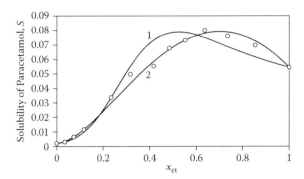

FIGURE 10.5 Comparison between experimental (o) and calculated (solid lines) solubilities of paracetamol (S is the mole fraction of paracetamol) in the mixed solvent water/ethanol (x_{et} is the mole fraction of ethanol) at room temperature. The solubility was calculated using Equation 10.29. 1-activity coefficients expressed via the Flory–Huggins equation, 2-activity coefficients expressed via the Wilson equation. (From S. Romero, A. Reillo, B. Escalera, and P. Bustamante, 1996, The Behavior of Paracetamol in Mixtures of Amphiprotic and Amphiprotic-Aprotic Solvents. Relationship of Solubility Curves to Specific and Nonspecific Interactions, *Chemical and Pharmaceutical Bulletin*, 44, 1061. Reprinted from E. Ruckenstein, and I. L. Shulgin, 2003c, Solubility of Drugs in Aqueous Solutions. Part 2: Binary Nonideal Mixed Solvent, *International Journal of Pharmaceutics*, 260, 283, With permission from Elsevier.)

a minimal number of experimental data. However, our further considerations involve aqueous multicomponent solvents, because almost all solubility data from literature (which are used in the examples) are based on aqueous multicomponent solvents.

Let us consider an n_c-component mixture containing a solute (component 2), water, and (n_c-2) organic cosolvents. If one considers the (n_c-1) mixed solvent as an ideal mixture, the following expression for the activity coefficient of a solid solute at infinite dilution in a multicomponent (ternary and higher) solvent can be obtained (Ruckenstein and Shulgin 2003a, 2004, 2005),

$$(\ln \gamma_2^{n_c,\infty})_{x_{i \neq 1,3}^{n_c}} = -\left(\frac{B \ln W}{(V_3^0 - V_1^0)} \right)_{x_{i \neq 1,3}^{n_c}} + A \tag{10.32}$$

where $\gamma_2^{n_c,\infty}$ is the activity coefficient of the solid solute at infinite dilution in an n_c-component mixture (solute+(n_c-1) component solvent), W is the molar volume of an ideal (n_c-1)-components solvent, V_i^0 is the molar volume of the individual cosolvent i, $x_i^{n_c}$ is the mole fraction of component i in the n mixture, and A and B are composition-independent constants. The constants A and B can be determined from the activity coefficients of the solid solute in two (n_c-1)-component mixtures with the mole fraction of component 1 equal to zero in one of them and the mole fraction of component 3 equal to zero in the other one. It should be noted that Expression 10.32 is valid on the line on which the sum of the mole fractions of components 1 and 3 is constant. Of course, a similar expression can be written for any pair of components of the mixed solvent.

In order to apply Equation 10.32 to the solubility of a solid solute in a $(n_c\text{-}1)$-component solvent, one must calculate the constants A and B. As already noted, Equation 10.32 is valid along the line for which

$$x_1^{n_c} + x_3^{n_c} = \text{const} \tag{10.33}$$

In particular, it is valid in the two limiting cases for which (I) $x_3^{n_c} = 0$ and (II) $x_1^{n_c} = 0$. These two limiting cases represent two different $(n_c\text{-}2)$-components mixed solvents with the following compositions: In the first limiting case (I), the mole fractions are $y_1^{n_c-1} = x_1^{n_c} + x_3^{n_c}$, $y_3^{n_c-1} = 0$, $y_4^{n_c-1} = x_4^{n_c}$, ..., $y_n^{n_c-1} = x_n^{n_c}$ with $y_1^{n_c-1} + \sum_{i=3}^{n_c} y_i^{n_c-1} = 1$ and the mole fraction of the solute is $y_2^{n_c-1}$. In the second limiting case (II), the mole fractions are $z_1^{n_c-1} = 0$, $z_3^{n_c-1} = x_1^{n_c} + x_3^{n_c}$, $z_4^{n_c-1} = x_4^{n_c}$, ..., $z_n^{n_c-1} = x_n^{n_c}$ with $\sum_{i=3}^{n_c} z_i^{n_c-1} = 1$ and the mole fraction of the solute is $z_2^{n_c-1}$.

In the limiting cases I and II, Equation 10.32 has the forms,

$$\ln \gamma_2^{n_c-1(\text{I}),\infty} = -\frac{B \ln V^{(\text{I})}}{(V_3^0 - V_1^0)} + A \tag{10.34}$$

$$\ln \gamma_2^{n_c-1(\text{II}),\infty} = -\frac{B \ln V^{(\text{II})}}{(V_3^0 - V_1^0)} + A \tag{10.35}$$

where $\gamma_2^{n_c-1(\text{I}),\infty}$ and $\gamma_2^{n_c-1(\text{II}),\infty}$ are the activity coefficients of the solid solute at infinite dilution in the $(n_c\text{-}2)$-component solvents I and II, respectively; $V^{(\text{I})}$ and $V^{(\text{II})}$ are the molar volumes of the mixtures composed of $(n_c\text{-}2)$-component solvents I and II and solid solute, respectively. Furthermore, for a poorly soluble solid, the molar volumes of the mixtures can be taken equal to the molar volumes of the solvents.

The combination of Equation 10.32 with Equations 10.34 and 10.35 provides the following expression for the activity coefficient of a poorly soluble solid at infinite dilution in an ideal $(n_c\text{-}1)$-component solvent mixture,

$$(\ln \gamma_2^{n_c,\infty})_{x_{i\neq1,3}^{n_c}} = \frac{(\ln W - \ln V^{(\text{II})}) \ln \gamma_2^{n_c-1(\text{I}),\infty} + (\ln V^{(\text{I})} - \ln W) \ln \gamma_2^{n_c-1(\text{II}),\infty}}{\ln V^{(\text{I})} - \ln V^{(\text{II})}} \tag{10.36}$$

Equation 10.36 relates the activity coefficient of a poorly soluble solid at infinite dilution in an ideal $(n_c\text{-}1)$-component mixed solvent to the molar volume W and the activity coefficients at infinite dilution in the two limiting cases I and II and their molar volumes. The same procedure can be applied to the activity coefficient of a poorly soluble solid at infinite dilution in two ideal $(n_c\text{-}2)$-component solvents I and II and so on. Ultimately, $\gamma_2^{n_c,\infty}$ in Equation 10.32 can be predicted from the activity coefficients of a poorly soluble solid at infinite dilution in binary or even in individual constituents of the solvent and their molar volumes.

By inserting into Equation 10.36 the expressions of $\gamma_2^{n_c,\infty}$, $\gamma_2^{n_c-1(I),\infty}$, and $\gamma_2^{n_c-1(II),\infty}$ from equations (which are similar to Equation 10.24 for binary mixed solvents) written for the solubilities of a solid in an ideal $(n_c\text{-}1)$-component solvent mixture and in two ideal $(n_c\text{-}2)$-component solvents I and II, one can obtain the following final expression for the solubility of a poorly soluble solid in an ideal $(n_c\text{-}1)$-component solvent in terms of its solubilities in ideal $(n_c\text{-}2)$-component mixed solvents I and II and their molar volumes,

$$\ln x_2^{n_c} = \frac{(\ln W - \ln V_m^{(II)})\ln y_2^{n_c-1} + (\ln V_m^{(I)} - \ln W)\ln z_2^{n_c-1}}{\ln V_m^{(I)} - \ln V_m^{(II)}} \qquad (10.37)$$

Furthermore, the solubilities of a poorly soluble solid in an ideal $(n_c\text{-}2)$-component mixed solvents I and II can be expressed through those in ideal $(n_c\text{-}3)$-component mixed solvents, and so on. Therefore, the suggested procedure allows one to predict the solubility of a poorly soluble solid in an ideal $(n_c\text{-}1)$-component mixed solvent from solubilities in binary constituents of solvents or even from those in individual constituents of the solvent.

The solubilities of naphthalene and anthracene in the ternary aqueous mixed solvents (see Table 10.1) were calculated using Equation 10.37. The prediction of the solubility of naphthalene in water/methanol/1-propanol mixed solvent (system 1 of Table 10.1) is provided as an example.

TABLE 10.1

Information about the Experimental Data Regarding the Solubilities of Naphthalene and Anthracene in Ternary, Quaternary, and Quinary Aqueous Mixed Solvents Used in the Calculations

Number	Solute	Multicomponent Solvent	Reference
		Ternary Mixed Solvent	
1	Naphthalene	water/methanol/1-propanol	Dickhut et al. 1989
2	Naphthalene	water/methanol/1-butanol	Dickhut et al. 1989
3	Anthracene	water/methanol/acetone	Morris 1988
4	Naphthalene	water/methanol/acetonitrile	Morris 1988
		Quaternary Mixed Solvent	
5	Naphthalene	water/methanol/1-propanol/1-butanol	Dickhut et al. 1989
		Quinary Mixed Solvent	
6	Naphthalene	water/methanol/ethanol/isopropanol/acetone	Morris 1988
7	Anthracene	water/methanol/ethanol/isopropanol/acetone	Morris 1988

Source: Adapted with permission from E. Ruckenstein and I. L. Shulgin, 2005, Solubility of Hydrophobic Organic Pollutants in Binary and Multicomponent Aqueous Solvents, *Environmental Science and Technology*, 39, 1623. (Copyright 2005, American Chemical Society.)

As mentioned above, the solubility can be calculated in two different ways:

1. From solubilities in individual constituents of the solvent and their molar volumes. The molar volume of a ternary mixed solvent (W) in Equation 10.37 can be obtained from the molar volumes of water, methanol, and 1-propanol as an ideal molar volume,

$$W = x_{H_2O}^t V_{H_2O}^0 + x_{MeOH}^t V_{MeOH}^0 + x_{PrOH}^t V_{PrOH}^0 \tag{10.38}$$

where $x_{H_2O}^t$, x_{MeOH}^t, and x_{PrOH}^t are the mole fractions of water, methanol, and 1-propanol in the mixed solvent and $V_{H_2O}^0$, V_{MeOH}^0, and V_{PrOH}^0 are the molar volumes of the pure water, methanol, and 1-propanol. There are three options for selecting the binary subsystems I and II: (1) water/methanol and water/1-propanol, (2) water/methanol and methanol/1-propanol, or (3) water/1-propanol and methanol/1-propanol. One can show that all these selections lead to the same result. For our calculations, we will select the aqueous binary subsystems (water/methanol and water/1-propanol), because the solubilities of naphthalene in binary aqueous mixed solvents are available in the same references as those for ternary data (Morris 1988; Dickhut, Andren, and Armstrong 1989). Hence, the binary subsystem I is the binary system water/1-propanol with the mole fractions of components: $y_{PrOH}^{b,I} = x_{MeOH}^t + x_{PrOH}^t$ and $y_{H_2O}^{b,I} = 1 - y_{PrOH}^{b,I}$ and molar volume,

$$V_m^{(I)} = y_{H_2O}^{b,I} V_{H_2O}^0 + y_{PrOH}^{b,I} V_{PrOH}^0 \tag{10.39}$$

The binary subsystem II is water/methanol with the mole fractions of components: $z_{MeOH}^{b,II} = x_{MeOH}^t + x_{PrOH}^t$ and $z_{H_2O}^{b,II} = 1 - y_{MeOH}^{b,II}$ and molar volume,

$$V_m^{(II)} = z_{H_2O}^{b,II} V_{H_2O}^0 + z_{MeOH}^{b,II} V_{MeOH}^0 \tag{10.40}$$

The solubilities of naphthalene in the binary subsystems I and II can be calculated using Equation 10.37 for binary mixed solvents, which requires only solubilities in the individual constituents of the solvent.

2. From the experimental solubilities in binary solvents and molar volumes of the individual constituents, the calculation procedure is the same as above, with the exception that the solubilities of naphthalene in the binary subsystems I and II are obtained from experiment. The calculation of the solubilities in binary aqueous solvents was described in previous sections. It should be noted that the water/methanol/1-butanol mixed solvent and the binary subsystem water/1-butanol are not completely miscible. Only the homogeneous regions of mixed solvents are considered here.

The solubilities of naphthalene and anthracene in ternary aqueous mixed solvents are predicted and compared with experiment in Table 10.2.

TABLE 10.2

Comparison between the Experimental Solubilities of Naphthalene and Anthracene in Multicomponent (Ternary, Quaternary, and Quinary) Aqueous Mixed Solvents and the Solubilities Predicted by Equation 10.37

| Solute | Multicomponent Mixed Solvent | Deviation (%) from Experimental Data | | | |
| | | Using Solubilities in Individual Solvents | | Using Solubilities in Binaries | |
		MPD$_1$[a]	MPD$_2$[b]	MPD$_1$[a]	MPD$_2$[b]
	Ternary Mixed Solvent				
Naphthalene	water/methanol/1-propanol	7.8	0.7	3.3	0.3
Naphthalene	water/methanol/1-butanol	11.9	1.1	4.8	0.4
Anthracene	water/methanol/acetone	76.4	16.0	36.1	4.8
Naphthalene	water/methanol/acetonitrile	62.2	18.9	14.8	2.6
	Quaternary Mixed Solvent				
Naphthalene	water/methanol/1-propanol/1-butanol	7.7	0.7	2.6	0.2
	Quinary Mixed Solvent				
Naphthalene	water/methanol/ethanol/isopropanol/ acetone	54.3	14.7	28.8	3.8
Anthracene	water/methanol/ethanol/isopropanol/ acetone	76.1	14.7	30.6	2.7

Source: Adapted with permission from E. Ruckenstein and I. L. Shulgin, 2005, Solubility of Hydrophobic Organic Pollutants in Binary and Multicomponent Aqueous Solvents, *Environmental Science and Technology*, 39, 1623. (Copyright 2005, American Chemical Society.)

[a] Deviation from experimental data calculated as MPD (%) (the mean percentage deviation) defined as

$$100 \sum_{i=1}^{N_j} \frac{\left| x_i^{exp} - x_i^{calc} \right|}{x_i^{exp}} \Big/ N_j$$

where x_i^{exp} and x_i^{calc} are experimental and calculated solubilities, and N_j is the number of experimental points in the dataset j.

[b] Deviation from experimental data calculated as

$$100 \sum_{i=1}^{N_j} \frac{\left| \ln(x_i^{exp}) - \ln(x_i^{calc}) \right|}{\ln(x_i^{exp})} \Big/ N_j$$

where x_i^{exp} and x_i^{calc} are experimental and calculated solubilities, and N_j is the number of experimental points in the dataset j.

10.5.1 QUATERNARY MIXED SOLVENT

The solubilities of naphthalene in the water/methanol/1-propanol/1-butanol solvent mixture are predicted using Equation 10.37. The molar volume of the ideal quaternary mixed solvent was calculated as

$$W = x^q_{H_2O}V^0_{H_2O} + x^q_{MeOH}V^0_{MeOH} + x^q_{PrOH}V^0_{PrOH} + x^q_{BuOH}V^0_{BuOH} \qquad (10.41)$$

where x^q_i is the molar fraction of component i in the quaternary mixed solvent. The two selected ternary subsystems I and II were water/methanol/1-propanol and water/methanol/1-butanol with the compositions:

I $(y^{t,I}_{PrOH} = x^q_{BuOH} + x^q_{PrOH},\ y^{t,I}_{H_2O} = x^q_{H_2O}$ and $y^{t,I}_{MeOH} = x^q_{MeOH})$ and

II $(z^{t,II}_{BuOH} = x^q_{BuOH} + x^q_{PrOH},\ z^{t,II}_{H_2O} = x^q_{H_2O}$ and $z^{t,II}_{MeOH} = x^q_{MeOH}).$

The solubilities of naphthalene in the ternary subsystems I and II were calculated as developed earlier in ternary mixed solvents. It should be noted that the water/methanol/1-propanol/1-butanol solvent is not completely miscible. Only the homogeneous regions of mixed solvents are considered here. The solubility predictions for the quaternary mixed solvent are listed and compared with experiment in Table 10.2.

10.5.2 QUINARY MIXED SOLVENT

The solubilities of naphthalene and anthracene in the quinary mixed solvents (water/methanol/ethanol/isopropanol/acetone) were predicted using Equation 10.37. The molar volume of an ideal quinary mixed solvent was calculated as

$$W = x^{qu}_{H_2O}V^0_{H_2O} + x^{qu}_{MeOH}V^0_{MeOH} + x^{qu}_{EtOH}V^0_{EtOH} + x^{qu}_{i-PrOH}V^0_{i-PrOH} + x^{qu}_{Acet}V^0_{Acet} \qquad (10.42)$$

where x^{qu}_i is the molar fraction of component i in the quinary mixed solvent. The two selected quaternary subsystems I and II were water/methanol/ethanol/isopropanol and water/methanol/ethanol/acetone with the compositions:

I $(y^{q,I}_{i-PrOH} = x^{qu}_{Acet} + x^{qu}_{i-PrOH},\ y^{q,I}_{H_2O} = x^{qu}_{H_2O},\ y^{q,I}_{MeOH} = x^{qu}_{MeOH}$ and $y^{q,I}_{EtOH} = x^{qu}_{EtOH})$ and

II $(z^{q,I}_{Acet} = x^{qu}_{Acet} + x^{qu}_{i-PrOH},\ z^{q,I}_{H_2O} = x^{qu}_{H_2O},\ z^{q,I}_{MeOH} = x^{qu}_{MeOH}$ and $z^{q,I}_{EtOH} = x^{qu}_{EtOH}).$

The solubilities of naphthalene and anthracene in the quaternary subsystems I and II were calculated as described in Section 10.5.1 "Quaternary Mixed Solvent." The results of the solubility prediction for quinary mixed solvent are listed and compared with experiment in Table 10.2.

10.6 SOLUBILITY OF PROTEIN IN MIXED SOLVENTS

The solvation behavior of a macromolecule such as a protein in a binary aqueous solvent is important in the understanding of such solutions (Schachman and Lauffer 1949; Casassa and Eisenberg 1964; Wyman 1964; Kuntz Jr. and Kauzmann 1974; Timasheff 1993). A macromolecule can be preferentially hydrated when the concentration of water in the vicinity of the macromolecule (local concentration of water) is higher than in bulk. The macromolecule can be preferentially solvated when the concentration of the cosolvent in the vicinity of the macromolecule is higher than in bulk. A measure of the solvation (or hydration) is provided by the preferential binding parameter (Casassa and Eisenberg 1964; Wyman 1964; Kuntz Jr. and Kauzmann 1974; Timasheff 1992, 1993), which can be defined using various concentration scales (component 1 is water, component 2 is a protein, and component 3 is a cosolvent) in various ensembles.

1. In molal concentrations,

$$\Gamma_{23}^{(m)} \equiv \lim_{m_2 \to 0} \left(\partial m_3 / \partial m_2 \right)_{T,p,\mu_3} \tag{10.43}$$

where $\Gamma_{23}^{(m)}$ is the preferential binding parameter defined in molal concentrations, m_i is the molality of component i, and μ_i is the chemical potential of component i. Only isothermal–isobaric conditions are considered.

2. In molar concentrations,

$$\Gamma_{23}^{(c)} \equiv \lim_{c_2 \to 0} \left(\partial c_3 / \partial c_2 \right)_{T,p,\mu_3} \tag{10.44}$$

where $\Gamma_{23}^{(c)}$ is the preferential binding parameter defined in molar concentrations, c_i is the molar concentration of component i. It should be noted that $\Gamma_{23}^{(m)}$ and $\Gamma_{23}^{(c)}$ are defined at infinite dilution of the protein.

Many characteristics of a protein in aqueous solvents are connected to its preferential solvation (or preferential hydration). The protein stability is a well-known example. Indeed, the addition of certain compounds (such as urea) can cause protein denaturation, whereas the addition of other cosolvents, such as glycerol, sucrose, and so forth, can stabilize at high concentrations the protein structure and preserve its enzymatic activity (Kuntz Jr. and Kauzmann 1974; Timasheff 1992, 1993, 1998a). Analysis of literature data demonstrates that as a rule $\Gamma_{23}^{(m)} > 0$ for the former and $\Gamma_{23}^{(m)} < 0$ for the latter compounds. Recently, we showed how the excess (or deficit) number of water (or cosolvent) molecules in the vicinity of a protein molecule can be calculated in terms of $\Gamma_{23}^{(m)}$, the molar volume of the protein at infinite dilution, and the properties of the protein-free mixed solvent (Shulgin and Ruckenstein 2005a). The protein solubility in an aqueous mixed solvent is another important quantity which can be connected to the preferential solvation (or hydration) (Arakawa and Timasheff 1984, 1985b, 1987; Timasheff and Arakawa 1988; Arakawa, Bhat, and Timasheff 1990) and can help to understand the protein behavior (Cohn and Edsall 1943; Arakawa and Timasheff 1984, 1985b, 1987; Timasheff and Arakawa 1988; Arakawa, Bhat, and Timasheff 1990; Qu, Bolen, and Bolen 1998; Bolen 2004; Pace et al. 2004).

The goal of the present part is to establish a relation between: (1) the preferential solvation (or hydration) of a protein, and (2) the protein solubility in an aqueous mixed solvent. The relation obtained will be used to predict the protein solubility in an aqueous solvent in terms of the preferential binding parameter.

The preferential binding parameter $\Gamma_{23}^{(m)}$ can be measured experimentally using various methods, such as sedimentation (Kuntz Jr. and Kauzmann 1974), dialysis equilibrium (Timasheff 1998a), vapor pressure osmometry (Cohn and Edsall 1943), and so forth, and was determined for numerous systems (Casassa and Eisenberg 1964; Wyman 1964; Kuntz Jr. and Kauzmann 1974; Arakawa and Timasheff 1984, 1985b, 1987; Timasheff and Arakawa 1988; Arakawa, Bhat, and Timasheff 1990; Timasheff 1992, 1993, 1998a; Zhang et al. 1996; Record, Zhang, and Anderson 1998; Courtenay, Capp, and Record 2001; Anderson, Courtenay, and Record 2002; Felitsky and Record 2004).

The results obtained will be presented as follows: (1) a relation between the protein solubility and the preferential binding parameter in a binary solvent will be established; (2) the established relation will be used to derive criteria for the effect of cosolvents (salting-in or salting-out), and (3) the experimental data for the preferential binding parameter $\Gamma_{23}^{(m)}$ will be used to predict the protein solubility and the results obtained will be compared with available experimental data.

Previously (Shulgin and Ruckenstein 2005a), the following expression for the preferential binding parameter $\Gamma_{23}^{(c)}$ was derived on the basis of the Kirkwood–Buff theory of ternary solutions,

$$\Gamma_{23}^{(c)} = \frac{c_1 c_3 \left(J_{21} \overline{V}_1 - J_{11} \overline{V}_2^{\infty} \right)}{\left(c_1 + c_1 J_{11} + c_3 \right)} + \frac{c_3 \left(c_1 + c_3 \right) \left(\overline{V}_1 - \overline{V}_2^{\infty} \right)}{\left(c_1 + c_1 J_{11} + c_3 \right)} \tag{10.45}$$

where $\overline{V}_i^{\infty}$ is the partial molar volume of component i, $\overline{V}_2^{\infty}$ is the partial molar volume of a protein at infinite dilution in a mixed solvent,

$$J_{11} = \lim_{x_2 \to 0} \left(\frac{\partial \ln \gamma_1}{\partial x_1} \right)_{x_2}$$

$$J_{21} = \lim_{x_2 \to 0} \left(\frac{\partial \ln \gamma_2}{\partial x_1} \right)_{x_2}$$

x_i is the mole fraction of component i, and γ_i is the activity coefficient of component i at a mole fraction scale. It should be noted that the quantities $\Gamma_{23}^{(c)}$, $\overline{V}_2^{\infty}$, and J_{21} of Equation 10.45 depend on the nature of the protein, while all the other ones are related to the properties of the protein-free mixed solvent.

Equation 10.45 can be rewritten as

$$J_{21} = \left(\frac{\partial \ln \gamma_2}{\partial x_1}\right)_{x_2=0} = -\frac{c_3(c_1+c_3)\overline{V}_1 - \left(\Gamma_{23}^{(c)} + c_3\overline{V}_2^{\infty}\right)(c_1+c_1J_{11}+c_3)}{c_1c_3\overline{V}_1} \quad (10.46)$$

Because (Casassa and Eisenberg 1964; Eisenberg 1976; Shulgin and Ruckenstein 2005a)

$$\Gamma_{23}^{(c)} = \left(1 - c_3\overline{V}_3\right)\Gamma_{23}^{(m)} - c_3\overline{V}_2^{\infty} \quad (10.47)$$

and experiment provides $\Gamma_{23}^{(m)}$, Equation 10.46 can be recast in the form,

$$\left(\frac{\partial \ln \gamma_2}{\partial x_1}\right)_{x_2=0} = -\frac{c_3(c_1+c_3)\overline{V}_1 - \Gamma_{23}^{(m)}\left(1-c_3\overline{V}_3\right)(c_1+c_1J_{11}+c_3)}{c_1c_3\overline{V}_1} \quad (10.48)$$

For poorly soluble solids, such as the proteins, one can use the infinite dilution approximation and consider that the activity coefficient of the protein in a mixed solvent is equal to that at infinite dilution. Therefore, for the solubility y_2 of a protein (solute, component 2) in a mixed solvent 1–3, one can write an equation similar to Equation 10.24,

$$f_2^S / f_2^l (T, p) = y_2\gamma_2^{\infty} \quad (10.49)$$

where γ_2^{∞} is the activity coefficient of a protein in a mixed solvent at infinite dilution, $f_2^l(T,p)$ is a hypothetical fugacity of a solid as a (subcooled) liquid at a given pressure and temperature, and f_2^S is the fugacity of the pure solid component 2. If the solubility of the mixed solvent in the solid phase is negligible, then the left-hand side of Equation 10.49 depends only on the properties of the solute.

Combining Equation 10.48 and Equation 10.49 yields the following relation for the solubility of a protein in a mixed solvent,

$$\left(\frac{\partial \ln y_2}{\partial x_1}\right) = \frac{c_3(c_1+c_3)\overline{V}_1 - \Gamma_{23}^{(m)}\left(1-c_3\overline{V}_3\right)(c_1+c_1J_{11}+c_3)}{c_1c_3\overline{V}_1} \quad (10.50)$$

Equation 10.50 allows one to derive a criterion for salting-in or salting-out for small cosolvent concentrations. Using the Gibbs–Duhem equation for a binary mixture, one can conclude that

$$\lim_{x_3 \to 0} J_{11} = 0 \quad (10.51)$$

Equation 10.50 can be therefore rewritten for $c_3 \to 0$ in the form,

$$\left(\frac{\partial \ln y_2}{\partial x_3}\right) = -\left(\frac{\partial \ln y_2}{\partial x_1}\right) = \frac{\alpha}{V_1^0} - 1 \quad (10.52)$$

where $\alpha = \lim\limits_{c_3 \to 0} \dfrac{\Gamma_{23}^{(m)}}{c_3}$ and V_1^0 is the molar volume of pure water.

Salting-in occurs when

$$\left(\frac{\partial \ln y_2}{\partial x_3}\right) > 0, \quad \text{hence when } \alpha > V_1^0 \qquad (10.53)$$

and salting-out occurs when

$$\left(\frac{\partial \ln y_2}{\partial x_3}\right) < 0, \quad \text{hence when } \alpha < V_1^0. \qquad (10.54)$$

It is well known (Record, Zhang, and Anderson 1998; Courtenay et al. 2000; Baynes and Trout 2003; Shulgin and Ruckenstein 2005a) that the preferential binding parameter $\Gamma_{23}^{(m)}$ is proportional to the concentration of the cosolvent, at least at low concentrations. Consequently, the salting-in or salting-out depends on the slope of the curve $\Gamma_{23}^{(m)}$ versus concentration for small c_3. The application of the established criteria to salting-in or salting-out to real systems is illustrated in Table 10.3.

The above criteria (Equations 10.53 and 10.54) are valid:

1. For $c_3 \to 0$, hence when a small amount of cosolvent is added to pure water;
2. For ternary mixtures [water (1)–protein (2)–cosolvent (3)] (the experimental results regarding the preferential binding parameter $\Gamma_{23}^{(m)}$ and the solubilities were obtained for mixtures which involve, in addition, a buffer, whose effect is taken into account only indirectly via the preferential binding parameter $\Gamma_{23}^{(m)}$);
3. For infinite dilution [this means that the protein solubility is supposed to be small enough to satisfy the infinite dilution approximation ($\gamma_2 \cong \gamma_2^\infty$)];
4. For experimental preferential binding parameters $\Gamma_{23}^{(m)}$ and solubilities determined at low cosolvent concentrations. It should be noted that the preferential binding parameter $\Gamma_{23}^{(m)}$ and the solubilities were usually determined for molalities larger than 0.5, and those values had to be used for the cases listed in Table 10.3 because no other experimental data were available.

The combination of Equation 10.46 and Equation 10.49 leads to the following expression for the solubility of a protein in a mixed solvent,

$$\left(\frac{\partial \ln y_2}{\partial x_1}\right) = \frac{c_3(c_1 + c_3)\overline{V}_1 - \left(\Gamma_{23}^{(c)} + c_3\overline{V}_2^\infty\right)(c_1 + c_1 J_{11} + c_3)}{c_1 c_3 \overline{V}_1} \qquad (10.55)$$

The integration of Equation 10.55 yields, for the solubility y_2 of a protein in a mixed solvent at a water mole fraction x_1, the expression,

$$\ln \frac{y_2}{y_2^w} = \int_1^{x_1} \frac{\left(\overline{V}_1 - \overline{V}_2^\infty - \Gamma_{23}^{(c)}/c_3\right)dx_1}{x_1 \overline{V}_1} - \int_1^{x_1} \frac{\left(\Gamma_{23}^{(c)}/c_3 + \overline{V}_2^\infty\right)J_{11}}{\overline{V}_1}dx_1 \qquad (10.56)$$

TABLE 10.3
Application of Criteria (Equations 10.53 and 10.54) for Salting Proteins Into or Out of Aqueous Solutions

| | | Experimental Data Used | | |
| | | Solubility (Salting-In or Salting-Out, Conditions, | Preferential Binding Parameter $\Gamma_{23}^{(m)}$ (Conditions, | Do the Criteria (Equations 10.53 |
Protein	Cosolvent[a]	References)	References)	and 10.54) work?
Lysozyme	NaCl	Salting-out, T = 0–40°C, pH = 3–10 (Ries-Kautt and Ducruix 1989; Retailleau et al. 1997; Retailleau et al. 2002; Guo et al. 1999; Forsythe et al. 1999)	pH = 4.5 (Arakawa and Timasheff 1982), pH = 3–7 (Arakawa and Timasheff 1987)	Yes
Lysozyme	MgCl$_2$	Salting-out, T = 18°C, pH = 4.5 (Ries-Kautt and Ducruix 1989)	pH = 3.0, 4.5 (Arakawa, Bhat, and Timasheff 1990)	Yes
Lysozyme	NaAcO	Salting-out, T = 18°C, pH = 4.5, 8.3 (Ries-Kautt and Ducruix 1989)	pH = 4.5–4.71 (Arakawa and Timasheff 1982)	Yes
Ribonuclease Sa	Urea	Salting-in, T = 25°C, pH = 3.5, 4.0 (Pace et al. 2004)	pH = 2.0, 4.0, 5.8 (Lin and Timasheff 1994)[b]	Yes
Lysozyme	Glycerol	Salting-in, T = 25°C, pH = 4.6 (Kulkarni and Zukoski 2002)	pH = 2.0, 5.8 (Gekko and Timasheff 1981)	No
β-Lactoglobulin	NaCl	Salting-in, T = 25°C, pH = 5.15–5.3 (Treece et al. 1964)	pH = 1.55–10 (Arakawa and Timasheff 1987)	No

Source: Reprinted from I. L. Shulgin and E. Ruckenstein, 2005b, Relationship between Preferential Interaction of a Protein in an Aqueous Mixed Solvent and Its Solubility, *Biophysical Chemistry*, 118, 128. (With permission from Elsevier.)

[a] The term *cosolvent* is also used here for electrolytes.

[b] The preferential binding parameters were determined for ribonuclease A in 30% glycerol solution.

where y_2^w is the protein solubility in cosolvent-free water plus buffer. Equation 10.56 allows one to calculate the protein solubility if the composition dependencies of J_{11}, $\Gamma_{23}^{(c)}$ (or $\Gamma_{23}^{(m)}$), and partial molar volumes are available.

Equation 10.56 can be simplified if one takes into account that at least at low cosolvent concentrations, $\Gamma_{23}^{(c)}$ is proportional to the concentration c_3 ($\Gamma_{23}^{(c)} = \beta c_3$) (Record,

Zhang, and Anderson 1998; Courtenay et al. 2000; Baynes and Trout 2003; Shulgin and Ruckenstein 2005a) and by assuming, in addition, that the partial molar volumes of the solute and the primary solvent are composition independent. With these two approximations, Equation 10.56 becomes

$$\ln \frac{y_2}{y_2^w} = \frac{\left(\bar{V}_1 - \bar{V}_2^\infty - \beta\right)}{\bar{V}_1} \ln x_1 - \frac{\left(\beta + \bar{V}_2^\infty\right)}{\bar{V}_1} \left(\ln \gamma_1\right)_{x_2=0} \tag{10.57}$$

and, hence,

$$\ln \frac{y_2}{y_2^w} = -\frac{\left(\bar{V}_2^\infty + \beta\right) \ln a_w}{\bar{V}_1} + \ln x_1 \tag{10.58}$$

where a_w is the water activity in the protein-free mixed solvent.

Taking into account Equation 10.47 and the relation

$$\frac{\Gamma_{23}^{(m)}}{c_3} = \alpha$$

Equation 10.58 can be recast as

$$\ln \frac{y_2}{y_2^w} = -\frac{\left(\alpha - \bar{V}_3 \Gamma_{23}^{(m)}\right) \ln a_w}{\bar{V}_1} + \ln x_1 \approx -\frac{\left(\alpha - \bar{V}_3 \Gamma_{23}^{(m)}\right) \ln a_w}{\bar{V}_1} \tag{10.59}$$

Because, as noted a long time ago (Kuntz Jr. and Kauzmann 1974), $\bar{V}_3 \Gamma_{23}^{(m)}$ "is two order of magnitude smaller" than α, $\alpha \gg \bar{V}_3 \Gamma_{23}^{(m)}$, and Equation 10.59 can be further simplified to

$$\ln \frac{y_2}{y_2^w} = -\frac{\alpha \ln a_w}{\bar{V}_1} \tag{10.60}$$

Equations 10.56 and 10.60 provide interrelations between the preferential binding parameter $\Gamma_{23}^{(c)}$ (or $\Gamma_{23}^{(m)}$) and the protein solubility in a mixed solvent. Recently, Equation 10.60 was used to examine the synergistic effect of two cosolvents (additives) on the aqueous protein solubility (Shukla and Trout 2011).

Application of Equation 10.60 to protein solubility in real systems is presented in Figures 10.6 and 10.7 (Shulgin and Ruckenstein 2005b). One can see that Equation 10.60 provides accurate predictions regarding protein solubility.

In conclusion, a relationship between the derivative of the activity coefficient of the protein with respect to the mole fraction of water at infinite dilution of protein and the preferential binding parameter was used to connect the solubility of the protein in an aqueous mixed solvent to the preferential binding parameter. This relation

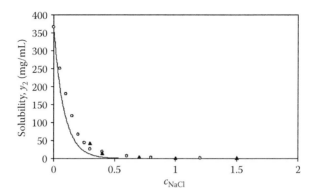

FIGURE 10.6 Lysozyme solubility in aqueous solutions of sodium chloride at pH = 4.5. The solid line represents the prediction based on Equation 10.60, (o) and (▲) are experimental data. (From M. M. Ries-Kautt and A. F. Ducruix, 1989, Relative Effectiveness of Various Ions on the Solubility and Crystal-Growth of Lysozyme, *Journal of Biological Chemistry*, 264, 745, and from P. Retailleau, M. Ries-Kautt, and A. Ducruix, 1997, No Salting-In of Lysozyme Chloride Observed at Low Ionic Strength over a Large Range of pH, *Biophysical Journal*, 73, 2156.) (Reprinted from I. L. Shulgin and E. Ruckenstein, 2005b, Relationship between Preferential Interaction of a Protein in an Aqueous Mixed Solvent and Its Solubility, *Biophysical Chemistry*, 118, 128, With permission from Elsevier.)

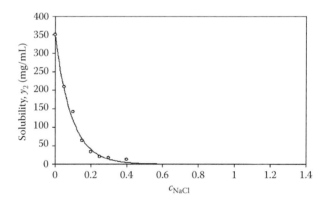

FIGURE 10.7 Lysozyme solubility in aqueous solutions of sodium chloride at pH = 6.5. The solid line represents predictions based on Equation 10.60, (o) are experimental data. (From P. Retailleau, A. Ducruix, and M. Ries-Kautt, 2002, Importance of the Nature of Anions in Lysozyme Crystallisation Correlated with Protein Net Charge Variation, *Acta Crystallographica Section D-Biological Crystallography*, 58, 1576.) (Reprinted from I. L. Shulgin and E. Ruckenstein, 2005b, Relationship between Preferential Interaction of a Protein in an Aqueous Mixed Solvent and Its Solubility, *Biophysical Chemistry*, 118, 128, With permission from Elsevier.)

was used to examine the salting-in and salting-out effect of various cosolvents on the protein solubility in water and to predict the protein solubility.

10.7   CONCLUSIONS

In this chapter, the application of the KB theory to the solubility in various systems is examined. The KB theory of solutions is a suitable tool to examine the solubility of a solute in a multicomponent solvent. The original Kirkwood and Buff equation for the composition derivative of the chemical potential (and respectively activity coefficient) of component i in an n_c mixture (Kirkwood and Buff 1951) provides the theoretical basis for such an analysis. Such an expression has advantages when compared to classical thermodynamics, which cannot provide information about the activity coefficients of the components of multicomponent (even binary) mixtures. In addition, the KB theory of solutions is valid for all systems under any conditions. For example, it can be applied to supercritical mixtures.

In the last one and a half to two decades, the KB theory of solutions was successfully used to predict the solubility in numerous systems. Indeed, it is quite impressive that one method was successful for such different systems as those examined in this chapter: the solubility of gases and drugs and hydrophobic organic pollutants in binary and multicomponent mixed solvents and the solubility of protein. In addition, this method was successfully applied to the solubility of solids in supercritical fluids, to the solubility of mixed gases in individual solvents, to the explanation of salting-out phenomena (Ruckenstein and Shulgin 2009). In some cases (multicomponent mixtures) the KB theory of solutions constitutes the most powerful theoretical method for investigating the thermodynamic properties.

11 Why Is Fluctuation Solution Theory Indispensable for the Study of Biomolecules?

Seishi Shimizu

CONTENTS

Abstract: How does urea denature proteins, whereas sugars, polyols, and crowders stabilize proteins? This question had been answered by the use of simple thermodynamic models assuming competitive solvent binding on protein surfaces. Yet the treatment of solvation in the language of stoichiometric binding has caused a number of difficulties and controversies. On the contrary, fluctuation theory has not only cleared up such confusions, but also has provided a unifying picture of the cosolvent effect, spanning from urea denaturation, through crowding, to the Hofmeister series.

11.1 INTRODUCTION AND SCOPE

Biological macromolecules, such as proteins and DNAs, are responsible for a number of biological functions necessary for the maintenance of life. Proteins exist in an environment consisting mainly of water, in which many of them fold into compact, well-defined folded structures, in order to perform their functions.

Solvent molecules play an important role in protein folding, as well as modifying the interaction (binding) between protein molecules. Not only are water molecules in the cell, but also there are other solvent molecules, which affect protein folding and binding. They include: (i) other biological macromolecules, which exist in the same environment, irrelevant to the specific function of the biomolecule in question—the effects due to the presence of such irrelevant molecules are called *crowding* (Davis-Searles et al. 2001); (ii) small organic molecules, such as sugars and amino acid derivatives, which prevent the protein molecules from unfolding (denaturation)—such molecules are called *stabilizers* or *osmolytes*, and are particularly important in the organisms living under extreme conditions (Timasheff 2002b); (iii) various inorganic salts, some of which stabilize the proteins, others denature the proteins—there is an empirical ranking of salts in terms of their effect on protein stability, which is known as the *Hofmeister series* (Baldwin 1996); and (iv) small organic molecules, such as urea, which unfold (denature) proteins—commonly referred to as *denaturants* (Timasheff 2002b). Protein stability and binding are thus modulated by cosolvents in a varied manner.

Cosolvents are also used in biochemical laboratories not only for the control of protein stability, but also in protein crystallization (Timasheff 2002b). Even artificial crowders (such as polyethylene glycol, PEG) are used for the stabilization (Davis-Searles et al. 2001). The understanding of cosolvent action at the molecular level is crucial in the designing of biochemical experiments.

Osmolytes have also been employed to modulate the behavior of water molecules around the biomolecules in question, in order to understand protein hydration; namely, the interaction between water and protein molecules (Parsegian, Rand, and Rau 1995). This is the area that was particularly controversial; this has initially necessitated a rigorous molecular understanding of protein solvation in mixed solution. Therefore, I shall start this chapter from this topic, in order to show why fluctuation solution theory (FST) is indispensable for the understanding of biomolecular solvation.

11.2 WHY IS FLUCTUATION SOLUTION THEORY INDISPENSABLE?

Had FST been applied to biomolecules much earlier, the following controversy might not have taken place. Yet, what has come to pass is an excellent illustration as to why FST is so important and indispensable for the understanding of biomolecular processes in aqueous solutions (Shimizu 2004).

The controversy concerned the estimation of the number of water molecules adsorbed or released from biomolecules as a result of a biochemical process (Timasheff 1998b; Parsegian, Rand, and Rau 2000). For example, when a ligand molecule binds to a protein, the water molecules hydrating the binding site have to be

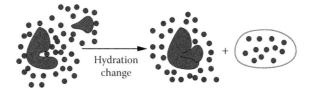

FIGURE 11.1 Changes of hydration accompany many biomolecular processes and reactions. The enumeration of the number of water molecules released (or adsorbed) for such reactions was a matter of debate. **(See color insert.)**

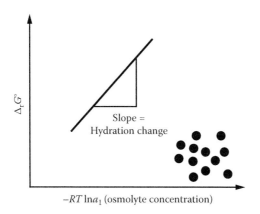

FIGURE 11.2 The principle of osmotic stress analysis (OSA), whose aim is to estimate the number of water molecules released during biomolecular processes and reactions by the use of osmolytes.

removed; this can be a bottleneck to ligand binding (Figure 11.1). The enumeration of the change in the number of water molecules is thus crucial for the understanding of the biomolecular processes.

Osmotic stress analysis (OSA) aims to estimate the number of water molecules adsorbed (or released) as a result of biomolecular processes purely from thermodynamic measurements (Parsegian, Rand, and Rau 1995). The experimental procedure is: (i) add an osmolyte (such as glycerol or PEG—known as *protein stabilizers*) to the system; (ii) measure the Gibbs energy change, $\Delta_r G^\circ$, that accompanies the biomolecular process as a function of osmolyte concentration; and (iii) infer the change of hydration number from the slope of the resulting plot (Figure 11.2). OSA is thus a simple and elegant method to obtain crucial information concerning hydration changes (Parsegian, Rand, and Rau 1995).

The elegance, robustness, and simplicity of OSA have led to many striking discoveries on the role of water in a number of biomolecular processes. The first success was hemoglobin's taut (T) to relaxed (R) transition: as many as 65 water molecules were inferred to be adsorbed upon this transition. This number obtained by OSA was claimed to be comparable to an independent estimation based upon the change of buried surface area. Encouraged by this consistency, OSA has been applied to a wide variety of biomolecular processes, including ion channels, DNA–protein, and carbohydrate–protein interactions, which yielded many surprising and insightful results (Parsegian, Rand, and Rau 1995).

How, then, does OSA work? It is based upon a long-established observation that protein-stabilizing osmolytes are preferentially excluded from protein surfaces (Parsegian, Rand, and Rau 1995, 2000). Consequently, the osmolytes cannot have any access to cavities, grooves, channels, or pockets formed by biomolecules, from whence the water molecules are released through the biomolecular processes in question. These regions are not subject to osmolyte binding, but to osmotic stress (Parsegian, Rand, and Rau 1995, 2000). This osmotic stress and the accompanied change of water activity modulate the equilibrium of the process, and the number of waters which are adsorbed upon the reaction in the absence of osmolytes are enumerated by measuring the change of the equilibrium constant with respect to osmotic pressure (Parsegian, Rand, and Rau 1995, 2000). Indeed, intuitively speaking, the osmotic pressure dependence of the reaction Gibbs energy change $\Delta_r G^\circ$ is the volume as provided by

$$\Delta_r V^\circ = \left(\frac{\partial \Delta_r G^\circ}{\partial \pi} \right)_{T,p} \tag{11.1}$$

It is this volume that corresponds to the released (or adsorbed) hydration water (Parsegian, Rand, and Rau 1995).

To summarize, the basis of the OSA is that the osmolytes are *inert*—they neither interact nor act directly on macromolecules because they are excluded (Parsegian, Rand, and Rau 1995, 2000)—and that intuitive thermodynamic argument can be used to evaluate the hydration changes.

In spite of its successes, simplicity, and elegance, the theoretical foundation of OSA has attracted much controversy. Timasheff pointed out that a cosolvent cannot be both excluded and inert at the same time, because exclusion requires a positive free-energy change (Timasheff 1998b). He emphasized that this free energy of exclusion (which is related to the experimentally measurable preferential hydration parameter) is the origin of the osmolyte-induced equilibrium shift. Timasheff also argued that this equilibrium shift is totally unrelated to the osmotic stress, because osmotic pressure is by definition a colligative property (Timasheff 1998b). In reply to this criticism, Parsegian, Rand, and Rau (2000) demonstrated the equivalence between the equations used in OSA and in the preferential hydration analysis, which will be discussed fully later. This demonstration of equivalence, however, was later questioned by Timasheff (2002a); he suggested that OSA is based on a misinterpretation of the equation of preferential hydration. He has shown further that the number of water molecules enumerated by OSA often depends on the choice of osmolytes, thereby questioning the fundamental assumption of the OSA: osmolytes do not always behave as inert and excluded. Furthermore, in spite of the demonstrated consistency between OSA's estimation and the buried surface area for the hemoglobin, cases have been reported in which OSA may underestimate the number of water molecules released upon reaction; the reason for this is that OSA ignores the osmolytes present with hydrated water (Timasheff 1998b, 2002a). The accuracy, applicability, and validity of OSA have been thus questioned.

The controversy necessitated a more rigorous theoretical foundation than Equation 11.1 for biomolecular processes in the presence of osmolytes. Parsegian, Rand, and Rau (2000) have indeed provided this foundation based upon the Gibbs–Duhem equation,

$$-\left(\frac{\partial \Delta_r G^\circ}{\partial \mu_1}\right)_{T,p}^{\infty} = \Delta N_{21} - \frac{c_1}{c_3}\Delta N_{23} \tag{11.2}$$

where ΔN_{21} and ΔN_{23} are the number of water (species 1) and osmolytes (species 3) bound to the dilute protein (species 2). This equation is identical in form to the one that was derived under the assumptions that: (i) there are binding sites for the solvent molecules on the surface of the protein molecule; and that (ii) water and osmolytes competitively bind to each of the binding sites (Schellman 1987).

In contrast to the binding model approaches, Parsegian and coworkers have been able to derive Equation 11.2 purely thermodynamically from the Gibbs–Duhem equation. Furthermore, according to the main assumption of OSA, $\Delta N_{23} = 0$, which leads to $-\beta \partial \Delta_r G^\circ / \partial \ln a_1 = \Delta N_{21}$. OSA has found a theoretical foundation in a solvent-binding model and the Gibbs–Duhem equation (Parsegian, Rand, and Rau 2000).

The debate, however, went on further: Is Equation 11.2 valid? Does it describe solvation accurately? Timasheff emphasized that ΔN_{21} and ΔN_{23} (although Timasheff used a different notation) are purely phenomenological parameters representing *site occupancy*, which have *no real physical meaning* (Timasheff 2002a, 2002b). He has also argued that these parameters are indeterminates, because they are coupled to conform to Equation 11.2, and that Equation 11.2 is the only relationship that is known to connect them. Are these statements valid? It is FST that answers this question, through which all the confusion regarding the foundation, interpretation, and validity of OSA can be resolved (Shimizu 2004). The following represent FST's answers to all the main theoretical points of the controversy.

QUESTION 1

Is there really any physical meaning to ΔN_{21} and ΔN_{23}?

Answer 1:

Yes (Shimizu 2004): (i) Equation 11.2 has been derived rigorously from FST (Chitra and Smith 2001b; Shimizu 2004); and (ii) FST shows that N_{21} and N_{23} (for each of the conformations) are defined microscopically through the solute–solvent and solute–cosolvent radial distribution function $g_{2i}(r)$,

$$N_{2i} = 4\pi c_i \int_0^\infty \left[g_{2i}(r) - 1\right] r^2 dr \tag{11.3}$$

QUESTION 2:

Are ΔN_{21} and ΔN_{23} indeterminates?

Answer 2:

No (Shimizu 2004). It has long been known that the partial molar volume change upon reaction ΔV_2 can be expressed in terms of ΔN_{21} and ΔN_{23} in the following form (Ben-Naim 1992),

$$\Delta V_2 = -\overline{V_1}\Delta N_{21} - \overline{V_3}\Delta N_{23} \qquad (11.4)$$

Thus, an independent relationship exists which connects ΔN_{21} and ΔN_{23} (Shimizu 2004). Therefore, ΔN_{21} and ΔN_{23} can be determined solely from the experimental data by combining osmotic stress and partial molar volume measurements (Shimizu 2004). Smith had reached the same simultaneous equations around the same time (Smith 2004).

QUESTION 3:

Is *exclusion* equivalent to *zero binding*?

Answer 3:

No (Shimizu 2004). The more excluded the osmolytes are, the larger the region in which $g_{23}(r) = 0$. This leads to a negative N_{23}, as one can see from Equation 11.3 and Figure 11.3. Therefore, cosolvent exclusion does not mean $N_{23} = 0$; N_{23} is actually large and negative!

All the above answers have significant meaning (Shimizu 2004). Answer 1 has provided a long-awaited modernization of the field, away from stoichiometric binding models and other similar approaches. Answer 2 has provided a breakthrough in our ability to probe solute–solvent interactions from thermodynamic measurements, via the combination of two thermodynamic measurements hitherto interpreted independently. Now the excess coordination numbers can be obtained directly from

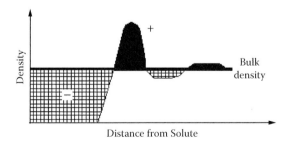

FIGURE 11.3   Exclusion does indicate an absence of binding. It does lead to a negative contribution to the excess coordination number, as shown here, as well as in Equation 11.3.

experiments. Answer 3 has shown that the basic hypothesis of the OSA is flawed. It has also clarified what it means to be excluded using a rigorous approach based upon radial distribution functions (Shimizu 2004).

The subsequent resolution of the debate regarding OSA has since brought a wealth of breakthroughs, which will be elaborated upon one by one in the remainder of this chapter.

11.3 PROTEIN HYDRATION AND ITS MICROSCOPIC MEANING

11.3.1 MOLECULAR CROWDING: THE DOMINANCE OF OSMOLYTE EXCLUSION IN OSMOTIC STRESS

We have seen that FST has finally brought a molecular understanding of biomolecular solvation in the presence of cosolvents. It has paved a way toward an experimental determination of the excess coordination numbers. Now, the practical question is: how can we estimate the change of hydration that accompanies a biomolecular process? FST provides a clear answer to this question through partial molar volume measurements, commonly called *volumetric analysis* (Shimizu 2004).

FST provides a clear theoretical framework for the solvation of biomolecules in pure water. In the absence of osmolytes, the change in partial molar volume of a protein, provided by Equation 11.4, reduces to (Ben-Naim 1992; Shimizu 2004; Shulgin and Ruckenstein 2005a; Pierce et al. 2008)

$$\Delta V_2 = - \overline{V_1}\Delta N_{21} \tag{11.5}$$

This means that the excess coordination number can directly be obtained from the partial molar volume measurements.

The excess coordination number obtained via FST consists of two contributions (Shimizu 2004; Chalikian 2011). The first contribution is the excluded volume, namely, the contribution from the region into which the solvent molecules cannot penetrate. The second is the contribution due to the change of water distribution in the solvation shell. If one is interested in the latter, then the excluded volume contribution should be subtracted out. Admittedly, the estimation of the excluded volume at present is a rough estimation at best; it nevertheless has provided some useful insight into biomolecular hydration (Chalikian 2003, 2011). Therefore, I employ the method proposed by Chalikian and coworkers in the estimation of the excluded volume V_E. This requires the calculation of the following two factors: the first is the inaccessibility of solvent molecules to biomolecules to intrinsic (core) volume V_I; and the second is thermal volume (volume inaccessible due to thermal motion) V_T. The former is calculated from the van der Waals volumes of the molecule, the second is estimated from the solvent-accessible surface area. The contribution of the above two to the excess coordination number is $-c_i V_E = -c_i (V_I + V_T)$. Thus, we obtain the solvation-shell contribution to the excess hydration number,

$$\Delta N'_{21} = \Delta N_{21} + c_1 (\Delta V_I + \Delta V_T) \tag{11.6}$$

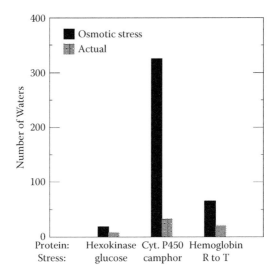

FIGURE 11.4 A comparison of the estimated number of released waters provided by OSA (black) and FST (grey). OSA grossly overestimates the number of waters. (From S. Shimizu, 2004, Estimating Hydration Changes upon Biomolecular Reactions from Osmotic Stress, High Pressure, and Preferential Hydration Experiments, *Proceedings of the National Academy of Sciences of the United States of America*, 101, 1195.)

A comparison of the estimated hydration numbers between the osmotic stress and FST (Figure 11.4) shows clearly that OSA overestimates the number of water molecules taken up during the reaction (Shimizu 2004). What is the reason for this large discrepancy?

The fact that OSA overestimates the excess coordination number can be explained from the experimental data. This is illustrated by comparing the osmolyte-induced equilibrium shift $-\partial\Delta_r G^{\circ}/\partial\mu_1$ with the pressure-induced shift, $\Delta V_2/\bar{V}_1$. The former is provided by the left-hand side of Equation 11.2, while the latter is obtained from Equation 11.5 and is simply $-\Delta N_{21}$. For a number of biomolecular processes extrapolated to the zero osmolyte concentration limit, Table 11.1 shows that the latter contribution is negligibly small. Therefore, the following relationship holds true,

$$-\left(\frac{\partial\Delta_r G^{\circ}}{\partial\mu_1}\right)_{T,p}^{\infty} \cong -\frac{c_1}{c_3}\Delta N_{23} \qquad (11.7)$$

This means that the contribution from the excess coordination number, ΔN_{21}, is negligibly small, in stark contrast to the supposition of OSA (Shimizu and Boon 2004; Gee and Smith 2009).

Equation 11.7 drastically simplifies the interpretation of cosolvent action. All we have to do is to consider the distribution of cosolvent molecules. The traditional molecular crowding approach (Davis-Searles et al. 2001) has indeed done this already, because it considers the large exclusion of cosolvents from proteins as a

TABLE 11.1

A Comparison of the Number of Water Molecules Released for a Series of Processes[a]

Reaction	Cosolvent	$-\partial\Delta_r G^\circ/\partial\mu_1$	$\Delta V_2/\overline{V}_1$
Hexokinase-glucose dissociation	PEGs	326	−4.2
Cytochrome P450-camphor dissociation	Osmolytes	19	−1.6
Hemoglobin T to R transition	PEGs and osmolytes	65	0
Lysozyme denaturation	GuHCl	−233	−3.0
Tendamistat denaturation	GuHCl	−99.5	−2.3

[a] Obtained using Equation 11.2 and Equation 11.7: The dominance of the former guarantees that the molecular crowding approximation (Equation 11.7) is an excellent approximation.

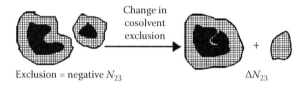

Change in cosolvent exclusion

Exclusion = negative N_{23} ΔN_{23}

FIGURE 11.5 Schematic diagram showing what the osmotic stress analysis actually measures: not the hydration change, but the change of cosolvent exclusion. The meshed area represents the region into which the osmolytes cannot penetrate. This provides a negative contribution to the excess coordination number.

dominant contribution to the cosolvent-induced equilibrium shift. FST thus provides a clear endorsement to the molecular crowding picture (Figure 11.5).

The crowding approach, which has been based upon McMillan–Mayer solution theory, has employed Equation 11.7 as a starting point (Davis-Searles et al. 2001; Shimizu and Boon 2004). This is based upon a second virial approximation. FST can even provide the condition upon which this approximation is accurate. Since Equation 11.7 holds under the condition that ΔN_{21} is negligibly small, and that this quantity is related to the partial molar volume via Equation 11.5, the proposed condition is

$$\left|\frac{\partial\Delta_r G^\circ}{\partial\mu_1}\right| \gg \left|c_1^0\Delta V_2\right| \tag{11.8}$$

This means that the osmolyte-induced equilibrium shift is much more significant than the pressure-induced one. This is satisfied indeed by biomolecular processes (see Table 11.1).

FST has thus established a clear relationship between OSA and the crowding approaches, which had been a matter of serious debate in the literature.

11.3.2 RELATIONSHIP TO THE VOLUMETRIC ANALYSIS

Before concluding our long story of controversy concerning biomolecular hydration, one must make a final mention of yet another facet of the same debate: volumetric analysis versus osmotic stress analysis, which initially led to a quest for a sound theoretical basis for studying biomolecular hydration.

Volumetric analysis has been established and formalized by Chalikian (2003, 2011), and is capable of estimating the number of water molecules involved in reaction by measuring the changes in partial molar volumes of biomolecules. Partial molar volumes are obtained from the change of equilibrium when hydrostatic pressure is applied, in contrast to OSA's "volume of water" determined by the application of osmotic pressure (Shimizu 2004). Partial molar volumes can be measured from high-pressure experiments, as well as densitometry (Chalikian 2003, 2011). Here, the numbers of waters estimated from OSA and from volumetric analyses were often inconsistent. For instance, a fourfold difference was seen in the equilibrium dissociation of human IFN-γ, and up to a threefold difference in the camphor binding of cytochrome P-450 (Shimizu 2004).

Before the introduction of FST, the reason for these discrepancies was not understood, although speculations were made concerning the different aspects of biomoleculer hydration that osmotic and hydrostatic pressures modulate (Robinson et al. 1995). Now we know that: (i) osmotic stress assumes exclusion as zero binding, instead of negative excess coordination numbers; and (ii) volumetric analysis aims at estimating the coordination numbers themself, instead of the excess coordination numbers, based upon the assumption of a two-state model for water (Chalikian 2003, 2011), or through the estimated water density increment in the vicinity of biomolecules (Kornblatt and Kornblatt 2002). However, the precise relationship between FST and the approach of Chalikian has yet to be determined.

11.3.3 SUMMARY

Prior to the application of FST, the lack of a rigorous theoretical framework had been the major hindrance in clarifying how hydration changes could be estimated from thermodynamic measurements. The FST provided a rigorous foundation, which is capable of relating the effect of osmolytes and hydrostatic pressure to the excess coordination numbers of the solvents surrounding the biomolecules. It has also established a rigorous theoretical foundation for the molecular crowding picture.

11.4 COSOLVENT-INDUCED DENATURATION AND STABILIZATION

11.4.1 HYPOTHESES CONCERNING THE ORIGIN OF COSOLVENT DENATURATION

We have seen that FST has the power of clarifying the behavior of solvent molecules underlying traditional thermodynamic measurements. It has also provided a unified framework for the interpretation of experimental data under a single aegis of the excess coordination numbers. Here, I apply FST to the interpretation of protein folding and stability, namely, to the conformational changes induced by the presence of cosolvents.

Protein molecules are denatured when denaturants, such as urea or guanidine hydrochloride, are added to the solution. Even though such an effect of urea has long been observed, reported, and exploited, a crucial question still remains unanswered: how do denaturants affect the stability of proteins? Numerous papers (focusing mainly on urea) have been published to address this question (Timasheff 2002a). They are based upon two different hypotheses as summarized below.

The first hypothesis states that the breaking of "water structure" by denaturants is the cause of protein denaturation (Frank and Franks 1968): urea has been assumed to play a role in reducing "the ability of the hydrocarbon solute to 'make structure' of water surrounding it." When the water structure is broken, the hydrophobic effect is weakened, thereby leading to protein denaturation (Frank and Franks 1968).

The second hypothesis claims that the denaturants preferentially bind to the surface of the proteins (Timasheff 2002a): the larger the surface area, the more denaturant molecules are bound to each protein; the denatured state therefore becomes more stable than the native state. Both of these proposals have been founded upon primitive and antiquated models of solutions; the lattice theory of solution is the foundation of the water structure breaker hypothesis (Frank and Franks 1968), whereas the stoichiometric binding model of solvation is the basis of the preferential binding hypothesis (Schellman 1987; Timasheff 2002a). Consequently, the weak theoretical foundation had prompted much debate, not only over the validity of these hypotheses, but also over the true meaning of these hypotheses at a molecular level.

Now the introduction of FST to the study of biomolecules has made it possible to formulate this problem microscopically and to quantify how the binding of denaturants can lead to denaturation (Shimizu 2004; Shulgin and Ruckenstein 2005a, 2006b; Smith 2006a; Pierce et al. 2008; Jiao and Smith 2011). The theoretical formulation has already been laid in the previous section. The goal here is to determine the change in excess coordination numbers and their solvation shell contributions. What we need for the calculation is the data on denaturant-induced equilibrium shifts and partial molar volume changes. They can again be obtained from experimental data, as will be discussed in the following sections.

11.4.2 *m*-VALUE ANALYSIS AND THE FLUCTUATION SOLUTION THEORY

The effectiveness of protein denaturants has long been quantified via the *m*-value, which is defined as the proportionality constant relating the molar denaturant concentration to the change in free energy upon protein denaturation. In short, the larger the *m*-value, the more susceptible the protein is to denaturation (Myers, Pace, and Scholtz 1995). How then can the *m*-values be interpreted in terms of interactions between the protein and solvent molecules? Myers et al. demonstrated that the *m*-values are proportional to the change of solvent-accessible surface area (SASA), which accompanies protein denaturation (Myers, Pace, and Scholtz 1995), namely, the change from its native structure to the fully extended unfolded structures. This means that the *m*-value reflects the degree of expansion of a protein upon denaturation.

This traditional interpretation of the m-values, however, is based upon an unrealistic assumption. Experiments suggest that denatured states make compact structures, even in the presence of highly concentrated denaturants (Klein-Seetharaman et al. 2002). Therefore, the fully extended unfolded structures, which have been assumed in the previous models do not reflect reality. In addition, size-exclusion chromatography has cast doubts upon the correlation between the m-value and the expansion in size of proteins upon denaturation (Wrabl and Shortle 1999). This suggests that the SASA proportionality of the m-values may not be an accurate assumption. We therefore need a better theory. FST can fill this gap (Shimizu and Boon 2004).

FST analysis of protein denaturation requires two sets of input data: the cosolvent-induced equilibrium shift and the change of partial molar volume. The former can be calculated from the m-value by (Shimizu and Boon 2004),

$$m = -\left(\frac{\partial \Delta_r G^\circ}{\partial c_3}\right)_{T,p,m_3 \to 0} = \frac{c_3}{c_1}\left(\frac{\partial \mu_3}{\partial c_3}\right)_{T,p,m_3 \to 0}\left(\frac{\partial \Delta_r G^\circ}{\partial \mu_1}\right)_{T,p,m_3 \to 0} \tag{11.9}$$

for infinitely dilute solutes. This converts the experimental m-values so they conform to the language of FST. Equation 11.2 and Equation 11.4 can now be solved to yield the excess coordination numbers.

Although m-values have been measured extensively for GuHCl and urea, the partial molar volume change accompanying the denaturation process has rarely been measured. GuHCl denaturation of hen egg lysozyme ($m = 10.445$ kJ/mol/M and $\Delta V_2 = -55$ ml/mol) and tendamistat ($m = 4.62$ kJ/mol/M and $\Delta V_2 = -41.2$ ml/mol) are the rare data available for the monomeric systems with two-state folding equilibria, which report ΔV_2 over a wide denaturant concentration range (Sasahara, Sakurai, and Nitta 1999, 2001; Pappenberger et al. 2000). The concentration dependence of ΔV_2 in both cases has been reported to be negligible.

Traditionally, biophysicists have always been interested in obtaining the contribution from the first solvation shell. This requires the estimation of ΔV_E for protein denaturation. Unfortunately, it is difficult to determine the excluded volume for the denatured state structures with sufficient accuracy. This is because the denatured state consists of an enormous number of possible configurations. Therefore, an accurate evaluation of excluded volume for such an ensemble is prohibitive at this stage (Shimizu and Boon 2004). Therefore, we are forced to make a bold approximation by using an empirical model proposed by Chalikian and Filfil (2003) for this purpose. Their model estimates the V_E for the denatured states from the molecular weight of the protein and the degree of unfolding ($\alpha = 0$ corresponds to the native state, and $\alpha = 1$ to the fully extended structure). The value of α for lysozyme has been estimated to be similar to acid denaturation, namely $\alpha = 0.7$–0.8. Therefore, we use $\alpha = 0.75 \pm 0.05$ for lysozyme (Chalikian and Filfil 2003; Shimizu and Boon 2004).

Figure 11.6 shows the FST-based analysis of GuHCl-denaturation of lysozyme. At lower GuHCl concentration, the denaturant-induced equilibrium shift $-\partial \Delta_r G^\circ/\partial \mu_3$ is determined predominantly by ΔN_{23}. As the concentration increases, the contribution from the protein–water interaction becomes more important. Combining this with Equation 11.2 and Equation 11.9, one can show that ΔN_{21} must have the opposite sign

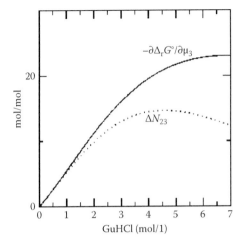

FIGURE 11.6   A comparison between $-\partial \Delta_r G^\circ / \partial \mu_3$ (solid) and ΔN_{23} (dashed) for the GuHCl-induced denaturation of bovine serum albumin (BSA). At low concentration, ΔN_{23} dominates. (From S. Shimizu and C. L. Boon, 2004, The Kirkwood–Buff Theory and the Effect of Cosolvents on Biochemical Reactions, *Journal of Chemical Physics*, 121, 9147.)

to ΔN_{23}, suggesting some kind of solvent-exchange mechanism consistent with the previous thermodynamic approaches (Timasheff 2002a; Shimizu and Boon 2004).

11.5   COSOLVENT-INDUCED MODULATION OF THERMAL DENATURATION

The previous section has clarified the mechanism of denaturation induced by the addition of cosolvents. Here, we discuss how the cosolvents can enhance or protect proteins from denaturation, which can be caused by the change in the environment (temperature, pressure, etc.). Some organisms, living under extreme conditions, exploit various cosolvents to protect their precious proteins from denaturation. Here, I focus particularly upon cosolvent effects on thermal denaturation.

A protein goes through heat denaturation when the temperature is raised. The melting temperature T_m (i.e., the temperature at which half of the proteins in solution are unfolded) increases when protein stabilizers—such as polyols, sugars, and amino-acid derivatives—are present (Timasheff 2002b). Protein stabilizers are used in cells and in laboratories to protect proteins from heat denaturation. T_m, on the other hand, is reduced when denaturants—such as urea and guanidine hydrochloride—are present in the solution (Timasheff 2002b). Cosolvents can thus enhance or prevent thermal denaturation.

Even though such modulation of heat denaturation by cosolvents has long been observed, reported, and exploited, a crucial question remains unanswered: how do cosolvents affect the heat denaturation of proteins? Numerous papers have been published to address this question (Timasheff 2002b; Shimizu 2011). However, as in the case of the cosolvent-induced denaturation, the mechanism was not clearly understood before FST was applied to this question. Indeed, just like the mechanism of cosolvent-induced denaturation, two hypotheses have been proposed. The first involves the cosolvent-induced promotion or destruction of the "water structure" (Frank and Franks 1968), the second is due to cosolvent–protein binding (Timasheff 2002b). The latter presupposes a correlation between chemical denaturation (denaturation of protein by adding cosolvents without changing the temperature) and the lowering of the melting temperature.

What remains to be done is to elucidate directly the mechanism upon which cosolvents work to alter the thermal denaturation of proteins, rather than to rely indirectly on the correlation with the chemical denaturation. Here, I apply FST to show that it is the binding or exclusion of the cosolvents that modulates thermal denaturation (Shimizu 2011).

The analysis can be made even simpler than the case of cosolvent-induced denaturation described in the previous section by focusing exclusively on the low-concentration limit of the cosolvent effect. Equation 11.2 and Equation 11.5 then simplify to

$$\left(\frac{\partial \ln K}{\partial \ln a_1} \right)_{T,p,m_2 \to 0} = c_1 (\Delta G_{21}^{\infty} - \Delta G_{23}^{\infty}) \tag{11.10}$$

$$\Delta V_2^{\infty} = -\Delta G_{21}^{\infty} \tag{11.11}$$

where the Kirkwood–Buff parameter, $G_{2i} = c_i^{-1} N_{2i}$, has been introduced to facilitate the discussion.

By a combination of experimentally determined cosolvent-induced equilibrium shifts and partial molar volume changes, one can obtain the information regarding the interaction between the protein and the cosolvents. A systematic set of data by Miyawaki, describing how K and the melting temperature T_m of thermal unfolding are affected by the presence of the cosolvent molecules, will be used. The values of $\ln K$ at a fixed constant temperature (333 K) for a number of cosolvents (guanidine hydrochloride, urea, ribose, glucose, sucrose, and trehalose) have been reported against $\ln a_1$ ranging from 0.95 and 1.0 (Miyawaki 2009).

Figure 11.7 shows that ΔG_{23} is positive for denaturants, negative for stabilizers. This suggests that more denaturants bind to the unfolded state than to the folded state, which is in agreement with the previous discussion. The negative ΔG_{23} is attributed to the increased depletion of stabilizers when the protein denatures. Indeed, as will be seen in the next section, the value of G_{23} for the protein's native state is also negative for osmolytes, due to the stabilizers' exclusion from protein surfaces (Shimizu and Matubayasi 2006). Consequently, the negative ΔG_{23} should mean that the cosolvents are more excluded from the denatured state than from the native state. This view is consistent with the earlier speculation derived from a molecular crowding perspective (Davis-Searles et al. 2001).

Figure 11.7 also indicates that the change of the cosolvent-induced equilibrium shift (left-hand side of Equation 11.10) is predominantly determined by

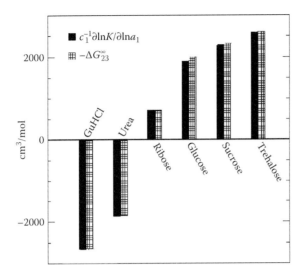

FIGURE 11.7 Cosolvent-induced equilibrium shift in the heat denaturation. A comparison between $c_1^{-1}\partial \ln K/\partial \ln a_1$ (black) and $-\Delta G_{23}^{\infty}$ (meshed) of Equation 11.10. The contribution from the protein–cosolvent interaction changes (meshed) dominate. (Data from S. Shimizu, 2011, Molecular Origin of the Cosolvent-Induced Changes in the Thermal Stability of Proteins, *Chemical Physics Letters*, 514, 156.)

the protein–cosolvent Kirkwood–Buff (KB) integral change, ΔG_{23}, and that the contribution from protein hydration, ΔG_{21}, is negligible. The only exception is α-chymotripsinogen A in the presence of ribose, which is a weak osmolyte. In all the other cases, ΔG_{21} is negligibly small compared to ΔG_{23} for both denaturants and stabilizers alike. This, again, simplifies Equation 11.10 to Equation 11.7; the latter is the fundamental equation for the molecular crowding theory, which has been derived originally via a second virial approximation (Davis-Searles et al. 2001). Indeed, the molecular crowding theory is a special case of FST, as has repeatedly been demonstrated (Shimizu and Boon 2004; Shimizu and Matubayasi 2006; Shimizu 2011).

What is even more powerful is that the molecular crowding approximation (Equation 11.7) is even reasonably valid for chemical denaturation by urea and guanidinium ions. The protein–cosolvent interaction thus seems to be a driving force for a wider range of biochemical processes, including protein–ligand interaction, cosolvent-induced denaturation, and the cosolvent-induced modulation of thermal denaturation.

11.6 PROTEIN–WATER AND PROTEIN–COSOLVENT INTERACTIONS

11.6.1 THEORY

The focus of our discussion so far has been the following: how do cosolvent molecules modulate biochemical reactions? Any biochemical reaction, by definition, involves more than two states—such as free and bound states, or unfolded and folded states. This chapter so far has focused upon the differences in solute–solvent interactions between the two states. How, then, can we determine, from experimental data, the interaction of a biomolecule in a single state with the surrounding solvent molecules?

The physical quantity responsible for answering the question is the preferential hydration parameter (Timasheff 2002b). It is this parameter, which has first shown that protein stabilizers, such as glycerol and trehalose, tend to be preferentially excluded from the protein surface, whereas denaturants, such as urea, are usually preferentially bound to protein surfaces (Timasheff 2002b). This parameter, however, cannot, by definition, give any information beyond preferential binding and exclusion. Understanding how each individual component (i.e., water or cosolvent) is solvating or distributing around the protein cannot be understood by preferential hydration parameters alone (Timasheff 2002a). To go beyond this difficulty, FST will again be indispensable. The theme here is again the FST-based reinterpretation of well-established thermodynamic measurements.

The preferential hydration parameter, ν_{21}, defined below, can be expressed by the excess coordination numbers through FST (Shulgin and Ruckenstein 2005a),

$$\nu_{21} = \left(\frac{\partial c_1}{\partial c_2}\right)_{T,p,\mu_1} = c_1\left(G_{21}^{\infty} - G_{13}\right) \tag{11.12}$$

Equation 11.12, solved in combination to the following relationship (namely, single-conformation equivalent of Equation 11.4) yields the microscopic information regarding protein–solvent interactions expressed through the KB integrals,

$$\overline{V}_2^{\infty} = -\overline{V}_1 N_{21}^{\infty} - \overline{V}_3 N_{23}^{\infty} + k_B T \kappa_T \tag{11.13}$$

where κ_T, the isothermal compressibility of the solution, is negligibly small.

The difference between Equation 11.12 and the following equation, which is the single-conformation equivalent of the cosolvent-induced equilibrium shift (which has been the focus of all the preceding discussions) should be noted with caution (Shulgin and Ruckenstein 2005a; Shimizu and Matubayasi 2006; Smith 2006a),

$$-\left(\frac{\partial \mu_2}{\partial \mu_1}\right)_{T,p,c_2 \to 0} = c_1\left(G_{21}^{\infty} - G_{23}^{\infty}\right) \tag{11.14}$$

We note here that prior to the introduction of FST, Timasheff and Xie have estimated the parameters equivalent to G_{21} and G_{23} based upon an intuitive concept of total site occupancy (Timasheff and Xie 2003). This has been a matter of debate, as seen in Section 11.1, and it is only FST, which has brought a clearly defined relationship (Equation 11.13) that is independent of Equation 11.14.

11.6.2　Urea and Trehalose

In order to understand the difference of action between protein denaturants and stabilizers, FST-based analysis of experimental data has been performed for ribonuclease in aqueous urea and trehalose solutions. Urea and trehalose are chosen

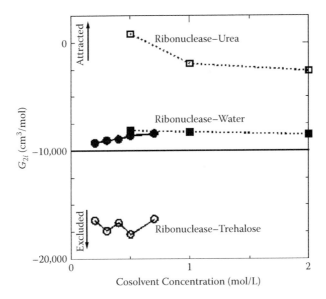

FIGURE 11.8 Kirkwood-Buff integrals for ribonuclease in aqueous urea (square) and tre-halose (circle) solutions. Filled symbols, G_{21}; open symbols, G_{23}. The line (approximately located at -10000 cm^3/mol) denotes the estimated excluded volume of the protein. (From S. Shimizu and N. Matubayasi, 2006, Preferential Hydration of Proteins: A Kirkwood–Buff Approach, *Chemical Physics Letters, 420*, 518.)

here as representative cosolvents of each class, and because they are among the most well-known chemicals. The experimental data has been taken from Timasheff (2002b). An inspection of Figure 11.8 shows that v_{21} for urea and trehalose have dif-ferent signs, clearly distinguishing the behavior of protein denaturants and stabiliz-ers. Since, as in Equation 11.14, and because μ_1 decreases upon the introduction of urea, the negative sign of v_{21} for urea shows that the chemical potential of a protein decreases upon the introduction of urea. This means that the solvation of proteins becomes more favorable on the addition of urea, consistent with the previous prefer-ential hydration analysis (Timasheff 2002b). Trehalose, on the other hand, behaves in the opposite manner: the positive sign of v_{21} suggests that the solvation process becomes unfavorable upon the addition of trehalose, consistent again with previous work (Timasheff 2002b).

What is the mechanism underlying the different signs of v_{21}? Again, the KB parameters provide an indispensable insight (Shulgin and Ruckenstein 2005a; Shimizu and Matubayasi 2006; Smith 2006a). For urea, Figure 11.8 shows that G_{23} is less negative than G_{21}. This, according to the considerations given in the previous section, suggests that urea is accumulated around protein molecules. For trehalose, G_{23} is more negative than G_{21}, suggesting that trehalose is excluded from the protein surface. Therefore, a comparison of KB parameters reveals the opposite behavior of urea and trehalose.

To help characterize this opposite behavior in more detail, it is instructive to com-pare the calculated G_{21} and G_{23} values with the excluded volume V_E for the protein in water estimated using Chalikian's method (Chalikian 2003). It is noteworthy (see Figure 11.8) that the excluded volume contribution to G_{21} as $c_3 \to 0$ is $-V_E$ (Shimizu 2004; Shimizu and Matubayasi 2006). Therefore, the difference between G_{21} and V_E as $c_3 \to 0$ provides an estimate for the contribution from the first solvation shell. The difference between G_{23} and V_E is also informative in characterizing the distribution of a cosolvent around a protein molecule relative to that of water, as will be discussed below (it is worth noting here that the excluded volume of a protein is larger with urea or trehalose than with water). Figure 11.8 indicates that G_{21} is more positive than V_E, suggesting an increase in the water density in the first solvation shell. (As a first approximation, the c_3 dependence of V_E has not been considered here.) G_{23} is even more positive than $-V_E$, implying, in spite of the increased size of urea compared to water (which makes the volume that is inaccessible to urea larger than V_E), urea molecules do accumulate around the protein. On the other hand, Figure 11.8 shows that G_{23} for trehalose is more negative than $-V_E$, showing a depletion in the trehalose density in the solvation shell and/or the effect of the larger size of trehalose. In any case, the relative depletion of trehalose in comparison with the case of water is evident from G_{21} and G_{23}. The opposite behavior of urea and trehalose in the solvation shell has been clarified by the FST (Shulgin and Ruckenstein 2005a; Shimizu and Matubayasi 2006; Smith 2006a).

11.6.3 MOLECULAR CROWDING

Large inert cosolvents, such as PEG, tend to stabilize the native structure of proteins (Davis-Searles et al. 2001). This effect is called *molecular crowding* and is considered to be an important difference between biochemical reactions *in vivo* and *in vitro*. Here, we present a FST analysis of the preferential hydration and volumetric data to help understand the molecular mechanism of crowding (Shulgin and Ruckenstein 2005a; Shimizu and Matubayasi 2006; Shulgin and Ruckenstein 2006b; Smith 2006a; Gee and Smith 2009).

Figure 11.9 shows the preferential hydration parameter v_{21} and $-c_3 G_{23}$ for BSA in aqueous PEG solutions of various molecular weights. It is observed that the larger the molecular weight of PEG, the more dominant the relative contribution of $-c_3 G_{23}$ to v_{21} becomes. However, for smaller PEGs and trehalose (data not shown), the contribution from hydration, G_{21}, is not negligible (Shimizu and Matubayasi 2006).

What is the mechanism of protein stabilization by PEG? We have seen from Figure 11.9 that $-c_3 G_{23}$ is the dominant contribution for large PEGs, which is large and positive. It follows that G_{23} is large and negative. The large and negative G_{23} values show that PEG molecules are strongly excluded from protein surfaces (Equation 11.3). This can indeed be understood more strikingly by comparing $-c_3 G_{23}$ to $c_1 V_E$, the contribution to the former due to the exclusion of PEGs from the excluded volume of proteins in water. For both BSA and ribonuclease, the value of $-c_1 V_E$ is much smaller than $-c_3 G_{23}$ for large PEGs (Shimizu and Matubayasi 2006). This suggests that the volume from which PEGs are excluded is much larger than V_E (i.e., the volume from which water molecules are excluded).

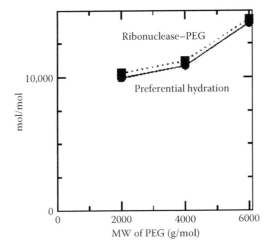

FIGURE 11.9 Preferential hydration parameter (solid) and the excess coordination number of PEG (dotted) around ribonuclease. Consequently, the contribution from protein–water interaction is negligibly small.

Thus, the KB analysis shows, at least for large PEGs, that the strong exclusion of PEGs from protein surfaces is the dominant cause of protein stabilization. This supports the previous view proposed by the molecular crowding approach.

11.6.4 THE HOFMEISTER EFFECT

It has long been known that cations and anions affect biomolecular processes with widely varied effectiveness (Baldwin 1996). Both cations and anions have been ranked separately in terms of their capacity to precipitate proteins. This ranking is called the *Hofmeister series* (Baldwin 1996). The same ranking applies to the stabilization of proteins. Since its initial discovery through the study of protein precipitation over a century ago, the Hofmeister series has been observed in a wide range of phenomena not only in protein chemistry but also in colloid, surface, membrane, and polymer chemistry (Baldwin 1996).

What, at a molecular level, causes the Hofmeister series? Again, an FST-based analysis of the experimental data provides useful insights (Shimizu, McLaren, and Matubayasi 2006). Figure 11.10 shows the G_{21} and G_{23} of bovine serum albumin in the presence of a number of Hofmeister salts as well as PEGs. The following observations are particularly noteworthy:

- A linear correlation between preferential hydration parameter and G_{23}, but not with G_{21}. This suggests that the protein–cosolvent interaction is responsible for the variation of preferential hydration phenomenon.
- G_{23} values for the Hofmeister salts and PEGs fall onto the same line, suggesting a unifying principle underlying both the Hofmeister effect and molecular crowding: namely, the variation of the protein–cosolvent interaction.

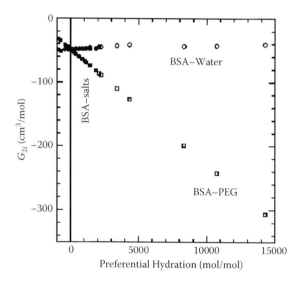

FIGURE 11.10 Kirkwood-Buff integrals, G_{21} (diamond) and G_{23} (square), of bovine serum albumin in the presence of the Hofmeister salts (filled) and PEGs (open). (Adapted from S. Shimizu and N. Matubayasi, 2006, Preferential Hydration of Proteins: A Kirkwood–Buff Approach, *Chemical Physics Letters*, 420, 518.)

Here, again, the protein–cosolvent interaction has been shown to be a crucial driving force describing the observed diversity of the cosolvent effect (Shimizu, McLaren, and Matubayasi 2006).

11.7 CONCLUSIONS

I have demonstrated with examples that FST is indispensable in modern biochemistry and biophysics. Not only has it provided a theoretical framework for the interaction between biomolecules and the surrounding solvent molecules, but also a powerful method for interpreting traditional and well-established thermodynamic data at a molecular basis. This is in stark contrast with the pre-FST approaches, whose theoretical foundations were at best simple solvent binding and stoichiometric binding models, which have been subject to unrealistic assumptions regarding the nature of solvation. An FST-based reinterpretation of the classical experiments has confirmed that it is the protein–cosolvent interactions, as well as its chemical variation, that gives rise to a wide range of cosolvent actions (Figure 11.11). This picture has unified the phenomena that have hitherto been classified under a number of different headlines—crowding, denaturation, Hofmeister effect, *m*-values, osmolyte effects, to name but a few. FST-based analyses have also shown that the contribution from the change in protein–water interactions is secondary. The precise molecular mechanism that determines protein–cosolvent interactions in various solutions, of course, still awaits fuller elucidation. More complete, molecular-based models, as well as the incorporation of more solvent components, are amongst the most fruitful potential developments (Pierce et al. 2008; Ploetz and Smith 2011a).

Stabilizers (sugars, polyols...)

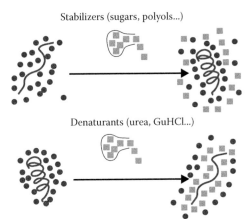

Denaturants (urea, GuHCl...)

FIGURE 11.11 A schematic diagram summarizing the difference between protein stabilizers and denaturants in terms of their interactions with a protein. **(See color insert.)**

12 Osmophobics and Hydrophobics
The Changing Landscape of Protein Folding

Matthew Auton and B. Montgomery Pettitt

CONTENTS

Abstract: We discuss recent experiments and theories concerning protein collapse and folding. Experiments using multicomponent solutions have revealed much about the mechanism of folding. Simulation and theory have been used to interpret thermodynamic and fluorescence correlation spectroscopy experimental results. We consider the theoretical arguments using variations of the free energy with respect to fluctuations in number and composition to consider recent experiments. We find new measures of protein stability tendencies offer a different view than the often poorly defined hydrophobic effect.

The underlying questions governing the transition of proteins from their unfolded state to their native state and recognition between proteins are as relevant now as they were 50 years ago (Anfinsen 1973). Misfolded (Weinreb et al. 1996) and unstructured domains (Dyson and Wright 2002, 2005) represent interesting examples of states where the understanding of recognition, and the self-recognition or folding process, has many biological implications (Weinreb et al. 1996; Wright and Dyson 1999; Uversky, Oldfield, and Dunker 2005; Chiti and Dobson 2006; Herczenik and Gebbink 2008).

We can probe the thermodynamics of the transitions of proteins in folding and recognition via several thermodynamic variables and their fluctuating conjugates including temperature, pressure, or concentration, especially of cosolvents. A specific class of cosolvents known as *osmolytes* are a particularly interesting biological perspective because of their evolutionarily selected use to regulate cell volume and stabilize proteins against environmental osmotic stress conditions to which cells must adapt (Yancey et al. 1982). Osmolytes have been utilized to regulate intracellular pressure in all kingdoms of life, providing a standard example of convergent evolution. There are two classes of osmolytes, protecting and denaturing osmolytes. Protecting osmolytes stabilize proteins thermodynamically by maintaining the native structure and conformational interactions. Denaturing osmolytes shift the native structure toward the denatured or unfolded state by disrupting native contacts and conformations. Osmolytes also affect the solution phase stability of proteins and thus have a direct effect on their solubility.

12.1 THERMODYNAMIC BACKGROUND

Considerable insight into the mechanisms of both solution stability and protein stability has been gained by studying the interactions between osmolytes and proteins (Auton and Bolen 2004, 2005, 2007). This has provided new models of the protein folding landscape (Rose et al. 2006). Thermodynamic measurements carried out on proteins and osmolyte aqueous solutions have determined that an important driving force responsible for osmolyte activity is the thermodynamic interactions of the osmolyte with the peptide backbone and thus the solution density distributions, fluctuations, and correlations (Liu and Bolen 1995; Auton and Bolen 2004).

Classically we think of osmolyte-induced stability as brought about by a thermodynamically unfavorable interaction with the protein (Lee and Timasheff 1981; Arakawa and Timasheff 1985a). We understand that if we decompose a protein into side chains and backbone that stability induced by such an osmolyte results in correlations or solution fluctuations, which thermodynamically yield a response less favorable than that with water (the osmophobic effect) (Rösgen, Pettitt, and Bolen 2004). Similarly, we consider solution stabilization or protein denaturation as having a substantial component that occurs by favorable interactions with the osmolyte (Pace 1986; Bolen and Baskakov 2001). More recently, it has become clear that the thermodynamics is not overwhelmingly dominated by the interactions with the side chains, but rather the backbone makes a major, perhaps dominant contribution in many cases (Auton, Holthauzen, and Bolen 2007; Auton et al. 2011). The precise nature of the balance is dependent on the concentration of osmolyte as recently shown (Auton et al. 2011) Some less concentrated solutions of osmolyte solutions do not show the backbone contributions as purely dominating the side chains. However, the backbone shows a classically surprising contribution to stability in all cases; the side chains are not responsible alone.

Bearing this in mind, the Tanford model of protein stability must be largely amended because of its relationship to the hydrophobic interactions between osmolytes and the amino acid side chains. The data for the systems with the strongest osmolyte effects clearly points to a de-emphasis of the role of the amino acid side chains (Auton et al. 2011). That data also show a sliding scale. The group transfer free energies of Tanford were apparent free energies and not corrected for activity and solubility (Nozaki and Tanford 1963) and as a result the importance of the protein backbone for protein folding, as envisioned at the onset of protein structural biophysics (Pauling, Corey, and Branson 1951), is now emphasized.

Recent thermodynamic (Auton, Holthauzen, and Bolen 2007; Auton, Bolen, and Rösgen 2008) and spectroscopic experiments (Teufel et al. 2011) that investigate osmolyte–protein interactions have challenged the prevailing paradigm of the protein folding mechanism (Jha et al. 2005; Rose et al. 2006). Those measurements carried out on proteins in aqueous solutions containing cosolvents have made clear that for osmolyte solutions, the interactions of the osmolyte with the peptide backbone is a driving force behind folding, which is augmented by the interactions with side chains (Liu and Bolen 1995; Auton and Bolen 2004; Jha et al. 2005).

The primary result of recent experimental evaluations of the nonideal corrections to Tanford's data on the changes in solubility of the component moieties of proteins challenges the dominant role of the side chains' hydrophobicity in the folding mechanism and sharpens the importance of the protein backbone for protein folding (Pauling, Corey, and Branson 1951). Model simulations have been equivocal on the subject to date (Hu et al. 2010b; Kokubo, Hu, and Pettitt 2011) and have not made the most stringent tests. While the importance of the backbone is clear, the relative weight of backbone versus side chain is more rich and varied than previously supposed. Side chains and backbone both play roles to select or capture a particular fold as well as display differential solvation free energies. The initial polymer collapse from an extended state is a free energy/solubility driven event. What drives the backbone solubility? Experimentally, the longer the backbone (Gly_n), the less soluble it becomes as the solutions become turbid (Auton et al. 2004; Teufel et al. 2011). Simulation work suggests the solvent effect on the initial collapse of proteins toward folding (or the reverse) does not even require side chains (Tran, Mao, and Pappu 2008; Hu et al. 2010b).

Recent progress in experimental physical chemistry aimed at quantifying transfer free energies of peptide backbone and side chain moieties from water to one molar osmolyte solutions created substantial cracks in the critical barrier to understanding protein solvent interactions (Auton and Bolen 2005). For any given osmolyte, knowledge of side chain and backbone transfer free energy contributions allows accurate prediction of the shift in free energy of the native to denatured equilibrium brought about by the osmolyte (Auton and Bolen 2007). This free energy shift is simply the sum of the transfer free energy contributions (also known as the *m*-value) of those parts of the protein that become occluded from solvent upon folding. Clearly, the important thermodynamic information for osmolyte-induced protein folding/unfolding quantitatively lies in the solvation differences of the relevant backbone and side chain moieties in water and in osmolyte solution. In Figure 12.1 we display data from Wayne Bolen's laboratory for the free energy of transfer from water to a urea solution for a small protein broken up into contributions of side chain and backbone, scaled by revealed accessible surface area in the ideal denatured state. The substantial contribution for urea denaturation plotted this way is in the backbone. The model of the denatured state accessible surface can be used to modulate the relative weights to some degree, but do not change the fact that the backbone has considerable importance (Canchi and Garcia 2011). The effect with the model above may be used to predict free energy differences in solution for a wide variety of proteins in several osmolyte solutions. In addition, those free energy contributions have been used to analyze folding versus aggregation in Huntington polyQ proteins (Friedman 1986).

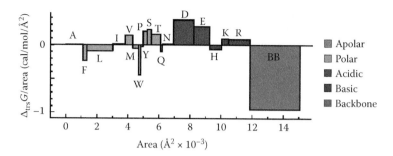

FIGURE 12.1 Free energy of transfer of the Nank4–7 denatured state plotted for each of the side chains and backbone contributions as a function of change in the accessible surface area from native (water) to denatured state (1 M urea). **(See color insert.)**

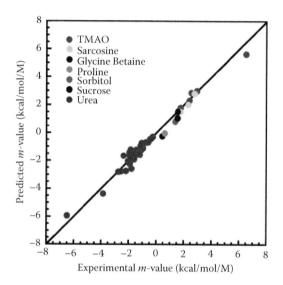

FIGURE 12.2 Predicted group transfer model m-values (related to the slope of the free energy change with respect to osmolyte concentration) versus that measured experimentally. **(See color insert.)**

The data in Figure 12.2 shows 40 proteins of different sizes that are either forced to unfold by the osmolyte, urea, or forced to fold by the six protecting osmolytes listed. The huge m-value range (~12 kcal/mol/M) of free energies predicted and the range of osmolytes used provides a rich array of experimental quantities. The transfer free energies of side chains and backbone and their use in this application provides an ontology for protein solvent interaction in conjunction with recent advances in the theoretical determination of multicomponent solution free energies (Kokubo and Pettitt 2007; Kokubo et al. 2007).

Recognition (or aggregation) between proteins/peptides and the protein folding process may be viewed as two facets of the same problem. Whether the contacts are between two individual peptides or between lengths of peptide covalently connected, the recognition and resulting structures are driven by similar interactions and (de) solvation phenomena.

Appreciating the roughness of the landscape for protein folding (Dill 1990; Bryngelson et al. 1995) has helped the computational folding field measurably (Cho, Levy, and Wolynes 2009). The interplay between theory, experiment, and computation has provided many fascinating results related to the characterization of the thermodynamic landscapes for folding and recognition. Yet more critically, we lack a simple theoretical mechanistic understanding of solution thermodynamics that is quantitative. Consider the plethora of hydrophobicity scales. The above measurements on activities and solubilities of proteins and their chemical constituents have called into question some of our more basic mechanistic tenets (Auton and Bolen 2004; Auton, Holthauzen, and Bolen 2007).

Progress in understanding the relation of protein sequence to folding/collapse and structure historically has come in bursts. In our current picture, in order to accurately describe protein folding in the true native environment, the protein and cellular milieu must be modeled as a thermodynamically nonideal system as opposed to dilute ideal solutions (Kokubo et al. 2007). By recognizing the utility of the influence of cosolvents on protein stability, different insights into protein folding have been obtained (Bolen and Baskakov 2001; Celinski and Scholtz 2002). Recent computational and theoretical descriptions of the nonideal behavior have begun to provide accurate descriptions of key experiments (Rösgen, Pettitt, and Bolen 2004; Smith 2004; Kokubo et al. 2007).

There has been considerable recent debate over the theoretical mechanism of action of various osmolytes (Zou et al. 2002; Bennion and Daggett 2003; Paul and Patey 2007; Hua et al. 2008; O'Brien et al. 2008; Rossky 2008; Hu, Pettitt, and Rösgen 2009; Canchi and Garcia 2011). The long-held concepts concerning the solution structure around proteins in stabilizing or destabilizing cosolvents, respectively, as either preferentially excluded or included in the first solvation shell, is a picture which has been tested and clearly works for most compounds (Timasheff and Arakawa 1989; Timasheff 1993; Marlow and Pettitt 1994). Yet, this picture lacks detail and structural specificity (Lim, Rösgen, and Englander 2009). Neither the suggestion that nonionic denaturants work by direct hydrogen bonding (Lim, Rösgen, and Englander 2009) nor the idea that there is a water structure change disrupting the hydrophobic effect, (Rossky 2008) are now well supported. Thus, such simple rules may be seen as the result of often but not universally competing effects. It is important to have a clear structural and molecular interaction-based understanding of the origin of such thermodynamic effects.

The so-called backbone-folding model proposes that the free energy penalty of solvating the side chains is not the only driving force of folding but rather the energetics of backbone hydrogen bonds dominate the folding process (Rose et al. 2006). This implies strong preorganization in the unfolded state (Canchi and Garcia 2011). In its purest form, the backbone-folding model is extreme as a hypothesis. That side chain transfer free energy differences do not demonstrate the expected Tanford driving force for folding and unfolding induced by cosolvents is certainly borne out by experiment (see Figures 12.1 and 12.2) (Jha et al. 2005; Auton, Holthauzen, and

Bolen 2007; Bolen and Rose 2008). However, if oligo glycine may then be taken as a good model of the protein backbone energetics, we do not find much in the way of preorganization in the unfolded state (Hu et al. 2010b) even though we posit that the thermodynamics of the oligo glycine backbone collapse follows that of proteins with definite folds. However, questions remain.

The idea that the protein backbone is as much responsible for the collapse of the protein polymer in water as the side chains challenges 50 years of dogma on the hydrophobic effect (Kauzmann 1959; Nozaki and Tanford 1963; Tanford 1964). Our recent work on hydrophobicity demonstrated the nontrivial nature of the van der Waals attractions on describing whether a compound or surface was *wet* or *dry* (Choudhury and Pettitt 2005a,b, 2006a, 2007; Choudhury 2006). We found that the well-parameterized atomic models of side chains all have attractive forces that demonstrate a *hydrophobic effect* with the correct entropy signature but lack the drying transition (Choudhury and Pettitt 2007). Thus, there is an aspect of the backbone-folding model, which is consistent with the picture that the classical hydrophobic effect does not produce drying in biopolymers. To what extent this is true deserves quantitative testing and modeling. What is clear is that the surface scientist's *dewetting* and the biochemist's ideas of *hydrophobic interactions* are not describing the same phenomena.

The experiments of Tanford (Nozaki and Tanford 1963), now taken as dogma, seemed to show that the transfer free energy of proteins from water to urea solutions was preferentially stabilized by the side chains. This implied, among other things, that urea affected the hydrophobic effect (Kauzmann 1959). To Tanford, the hydrophobic effect was the "dominant energetic factor that leads to the folding of polypeptide chains into compact globular entities" (Tanford 1997). The experiments on transfer free energies of Bolen and coworkers both confirmed and corrected Tanford's experimental results (Auton and Bolen 2004; Auton, Holthauzen, and Bolen 2007). The corrections to the changes in free energies involving activity coefficients, however, when decomposed into side chain and backbone parts, as Tanford had also done, showed that the effect of cosolvents was largely on the backbone and not just on the side chains (Auton, Holthauzen, and Bolen 2007). Recent fluorescence correlation spectroscopy experimental studies of glycine-rich peptides show the same trends of generic backbone interactions sufficient to cause collapse (Jha et al. 2005).

We next consider theoretical and simulation methods which frame our thinking. Density fluctuation correlations are a powerful tool for considering the nature of multicomponent solution systems. The data reviewed above demands a consistent picture for interpretation and extrapolation.

12.2 THEORETICAL FRAMEWORK

The subject of the thermodynamic stability of mixtures of chemical species, for example, water, proteins, osmolytes, and electrolytes, has an old and clear theoretical framework (Callen 1962). The connection to intermolecular interactions similarly can be viewed from the statistical thermodynamics of McMillan and Mayer (MM) (McMillan and Mayer 1945; Friedman 1962) or equally rigorously from the method of Kirkwood and Buff (KB) (1951). For the former, one requires correlations or potentials of mean force at infinite dilution, and then one typically uses a cluster theory to obtain the thermodynamics at the desired finite concentration (Friedman 1962; Rasaiah and Friedman 1968; Pettitt and Rossky 1986). With the latter case, the density–density fluctuation correlations are required at the desired concentration(s). Both have strengths and difficulties in the multicomponent solution context but KB is more in keeping with the theme of this volume and we restrict our attention to that statistical thermodynamic framework.

We start with the change in Helmholtz free energy per particle, $df = d(F/N)$, which can be written for a three-component mixture of solvent (1), protein (2), and other (3) as

$$df = -pdv - sdT + (\mu_2 - \mu_1)dx_2 + (\mu_3 - \mu_1)dx_3 \tag{12.1}$$

Here, the Helmholtz free energy F, volume, and entropy divided by the total number of particles N are represented with lower case f, v, and s, respectively. We shall consider the fluctuations in number of each species and its effect on the thermodynamic potential. We then partition the system into two subsystems, a and b, each defined by a set of variables, $[N_1^a, N_2^a, N_3^a, V^a, T]$ and $[N_1^b, N_2^b, N_3^b, V^b, T]$. We now apply a small perturbation away from equilibrium and the state variables correspondingly readjust to new values $[N_1^a + \delta N_1^a, N_2^a + \delta N_2^a, N_3^a + \delta N_3^a, V^a, T]$ and $[N_1^b + \delta N_1^b, N_2^b + \delta N_2^b, N_3^b + \delta N_3^b, V^b, T]$. One can now write the total change in the free energy, δF, for this process as

$$\delta F = \left(N^a + \delta N^a\right)\left(f + \delta f^a\right) + \left(N^b + \delta N^b\right)\left(f + \delta f^b\right) - \left(N^a + N^b\right)f \tag{12.2}$$

If we substitute the Taylor expansion of f_a and f_b with respect to v, x_2, x_3, and apply the conservation relations,

$$\delta N_1^a + \delta N_1^b = \delta N_2^a + \delta N_2^b = \delta N_3^a + \delta N_3^b = \delta V^a + \delta V^b = 0 \tag{12.3}$$

the first-order terms cancel and we find the variation in F to second order,

$$\delta F \cong \frac{N}{2} \frac{N^b + \delta N^b}{N^a + \delta N^a} \left[\left(\frac{\partial^2 f}{\partial v^2}\right)_{T, x_2, x_3} \left(\delta v^b\right)^2 + \left(\frac{\partial^2 f}{\partial x_2^2}\right)_{T, v, x_3} \left(\delta x_2^b\right)^2 + \left(\frac{\partial^2 f}{\partial x_3^2}\right)_{T, v, x_2} \left(\delta x_3^b\right)^2 + \right.$$

$$2\left(\frac{\partial^2 f}{\partial v \partial x_2}\right)_{T, x_3} \left(\delta v^b\right)\left(\delta x_2^b\right) + 2\left(\frac{\partial^2 f}{\partial v \partial x_3}\right)_{T, x_2} \left(\delta v^b\right)\left(\delta x_3^b\right) +$$

$$\left. 2\left(\frac{\partial^2 f}{\partial x_2 \partial x_3}\right)_{T, v} \left(\delta x_2^b\right)\left(\delta x_3^b\right) \right] \tag{12.4}$$

For stability, we then have $\delta F > 0$, which taken at lowest order is thus a constraint on the second derivatives. To find the constraints one commonly carries out a series of linear transformations on the independent variables, which reduces the quadratic to a sum of squares in the transformed variables. From the resulting relations, by elimination of variables and Legendre transformation, the canonical, all the various semigrand canonical, and grand canonical relations can be derived (Valdeavella, Perkyns, and Pettitt 1994). This yields equations that are expressible in terms of the compressibility, the partial molar volumes, and derivatives of the chemical potential, which are directly calculable from the cofactors of a density weighting of the matrix of zeroth moments of the distribution,

$$G_{ij} = 4\pi \int_0^\infty h_{ij}(r) r^2 dr \tag{12.5}$$

where $h_{ij}(r) = g_{ij}(r) - 1$ is the total radial correlation function in the solution between species i and j (1, 2, or 3). Following the classic derivations of Kirkwood and Buff in a two-component system (Kirkwood and Buff 1951), we have for the other or third component, in this case the osmolyte,

$$\beta\left(\frac{\partial \mu_3}{\partial c_3}\right)_{T,p} = \frac{1}{c_3} + \frac{G_{13} - G_{33}}{1 - (G_{13} - G_{33})c_3} \tag{12.6}$$

We notice that the water–water correlations do not enter directly into the formula. This is not to say that they are not important in influencing G_{13} and G_{33}, but rather that water self-correlations are not a direct measure of the molar concentration dependence of the osmolyte thermodynamics. We will return to this theme later.

The choice of which ensemble to pick and the related choice of concentration scale is important for obtaining practical results (Rösgen, Pettitt, and Bolen 2004; Rösgen et al. 2004). Hill noted that when considering ensembles simultaneously choosing pressure and chemical potential do not have to cause a thermodynamic degeneracy if one uses a semigrand ensemble and the component for which chemical potential is held constant is nonvolatile (Hill 1957). This makes a convenient choice for osmotic pressure or activity coefficient problems (Hill 1959). For simplicity, we first derive an activity expansion for a two-component system in this framework (Rösgen et al. 2004),

$$Z(N_1, \mu_2, p, T) = \sum_{N_2=0}^{\infty} Y_N e^{\beta \mu_2 N_2} = Y_0 + Y_1 f_2 + Y_2 f_2^2 \ldots \tag{12.7}$$

Here, N_1 is the number of waters, μ_2 is the chemical potential, and f_2 the absolute activity of the nonvolatile solute species (such as a protein) at temperature T and pressure p. The Y_N are then partition functions for N_2 solutes in the solvent of N_1 waters at T and p. From this point of view, the standard thermodynamic derivatives yield the various observables required such as,

$$\left(\frac{\partial \ln Z}{\partial \beta \mu_2}\right)_{p,\beta,N_1} = \langle N_2 \rangle \approx \frac{Y_1}{Y_0} f_2 + \dots$$

$$\left(\frac{\partial \ln Z}{\partial \beta p}\right)_{\beta,f_2,N_1} = -\langle V \rangle \approx V_0 + \dots \tag{12.8}$$

where the approximation is good for dilute solutions. Considering the volume in terms of the relative activity $a_2 = f_2 \, V_1/V_0$ provides,

$$\langle V \rangle = \frac{V_0 + a_2 V_1 + \dots}{1 + a_2 + \dots} \tag{12.9}$$

Here, we find V_1 as the partial molar volume. We also find the osmotic coefficient in this scheme as

$$\varphi = \frac{1}{\langle N_2 \rangle} \ln \frac{Z}{Y_0} \tag{12.10}$$

where Y_0 represents the pure solvent system where as Y_1 is a system with one solute and in Y_2 the two solutes may interact. In general, if Y_0 is for pure water we may either calculate the ratios Y_N/Y_0 (see Section 12.3, "Simulation Results" below) or use them as parameters to be fit to experiment and thus solve for f_2. Thus, to second order,

$$\langle N_2 \rangle = \frac{Y_1}{Y_0} f_2 + 2 \frac{Y_2}{Y_0} f_2^2 + \dots \tag{12.11}$$

and all that is left is to solve a quadratic for f_2. In this picture, we may think of Y_N/Y_0 as a measure of how readily the solvent will accept N_2 solutes. Notice $-k_B T \ln (Y_1/Y_0)$ is just the Widom insertion chemical potential (Widom 1963). This formula yields essentially quantitative accuracy when fit to various nonionic cosolvents in water (see Figure 12.3).

Now let us consider the chemical potential up to terms of only first order. Then we have

$$\mu_3 = \mu_3^\circ + k_B T \ln \left(\frac{c_3}{1 - V_1 c_3}\right) \tag{12.12}$$

where μ_3° is the standard state chemical potential. Using the derivative of the chemical potential with respect to the concentration of the osmolyte we get

$$\beta \left(\frac{\partial \mu_3}{\partial c_3}\right)_{T,p} = \frac{1}{c_3} + \frac{V_1}{1 - V_1 c_3} \tag{12.13}$$

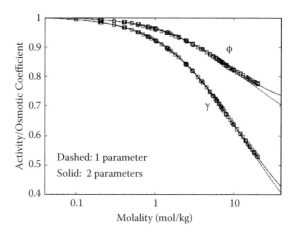

FIGURE 12.3 Activity coefficient and osmotic coefficient versus molality for urea in water (From J. Rösgen, B. M. Pettitt, and D. W. Bolen, 2004, Uncovering the Basis for Nonideal Behavior of Biological Molecules, *Biochemistry*, 43, 14472; J. Rösgen, B. M. Pettitt, J. Perkyns, and D. W. Bolen, 2004, Statistical Thermodynamic Approach to the Chemical Activities in Two-Component Solutions, *Journal of Physical Chemistry B*, 108, 2048.)

which now allows comparison with the result above in the KB framework. We now see that, to first order, the hydrated partial molar volume of the osmolyte is $V_1 = G_{13} - G_{33}$. We have found that V_1 is rather constant in this expression for chemical potential and so the difference $G_{13} - G_{33}$ is not very concentration dependent (Rösgen, Pettitt, and Bolen 2004). Many such structural examples exist when we take experimental data and consider the KB integrals or their differences. Application of KB theory in this inverse sense yields information on protein hydration changes and osmolyte solvation differences separately (Rösgen, Pettitt, and Bolen 2005).

This semigrand partition function approach as phrased in the context of KB theory thus allows one to obtain average structural information from thermodynamic data. We have made some series truncation approximations and so we must ask to what extent the success of the method depends on our assumptions. To consider this we turn to simulation and calculate some of the quantities we have derived. Simulation is, of course, always model dependent but this gives an otherwise independent view of the success of our low-order activity series theory. In particular, by using the ratio of partition functions as parameters in an activity series expansion for quantities such as $\langle N_2 \rangle$ we must address whether these are the real coefficients or effective coefficients, which have considerable contributions or take up the slack from higher-order terms in the series. So for instance, at infinite dilution we can write

$$\langle N_2 \rangle = \frac{Y_1}{Y_0} f_2 \qquad (12.14)$$

implying

$$\mu_2 = -k_B T \ln Y_1 + k_B T \ln Y_0 + k_B T \ln \langle N_2 \rangle \qquad (12.15)$$

For small but finite concentrations, we can easily extend this to include terms up to Y_2. For urea in water, we find the excess chemical potential is nearly constant and much of the variation comes from the ideal part for a dilute solution.

We have used the two-component formulas as a demonstration vehicle here. One can use the same framework for three components. An easy case is when the protein is dilute (Rösgen, Pettitt, and Bolen 2005). We next consider the link through computer simulations.

12.3 SIMULATION RESULTS

Consider simulating the free energy components for the two-component problem of the chemical potential of urea in water at various concentrations. In a previous study, we examined both the simulation method for obtaining the excess free energy (Widom test particle, thermodynamic integration, Bennett reweighting, etc.) as well as the force field dependence (CHARMM, Amber, OPLS) (Kokubo et al. 2007). In that study, we indeed found a rather strong force field dependence, but we also found that some force fields, in particular the Kirkwood–Buff Force Field (KBFF) of Smith and Weerasinghe (Weerasinghe and Smith 2003c) do quite well at representing the experimental activity data. With the force field controlled, and the simulation method found not to be an issue, one can then ask about the reason for the excellent fits of our low-order activity series shown above.

For the seemingly reasonable fits of the one-parameter model, the experimental data implies a hydration number for urea of 0.17 waters. Thus, we see that the shape of the curve fits well but there is considerable renormalization of the higher-order terms such that the fitted coefficients are not the *bare* ones. To test this we used the simulated chemical potentials over a considerable range and directly measured data on the $Y_{0/1}$ as well as the entire effect on the finite concentration system. The one-parameter fit to the simulations gave the hydration number for urea of 2.36, which is more reasonable when we consider that the proportion of urea to water is 1:2.665 at the solubility limit. In the end, we interpret the small improvement of the fitting by using two parameters rather than one to be due to the effective inclusion of much higher-order term effects. The curves had shapes described by the low-order terms, but the coefficients were best given as a numerical (nonphysical) compromise with experiment.

In three-component systems, simulation becomes more important for deconvoluting the effects of fitting with low-order theory. Simulation, however, gives a physical model for which interactions are dominant in different regimes. In investigating the influence of cosolvents on backbone constituents modeled as Gly_{2-5}, we noticed some interesting trends in our calculations. First, the free energy of solvation decreased with each additional monomer (Hu et al. 2010a). The free energy of solvation (sol) being defined as

$$\text{protein}_{(\text{gas})} \xleftrightarrow{\Delta G_{\text{sol}}} \text{protein}_{(\text{aq})} \qquad (12.16)$$

When we add a cosolvent osmolyte to the system, we obtain a free energy difference for solvation in osmolyte solution,

$$\text{protein}_{(\text{gas})} \xleftrightarrow{\Delta G_{\text{os}}} \text{protein}_{(\text{os})} \qquad (12.17)$$

The difference is the transfer free energy from aqueous to osmolyte solution or $\Delta\Delta G_{\text{tr}} = \Delta G_{\text{os}} - \Delta G_{\text{sol}}$.

We have found that we can obtain ΔG_{sol} to a precision of about a quarter of a kcal/mol and can detect slopes in $\Delta\Delta G_{\text{tr}}$ on the order of 10s of cal/mol for entire peptides as a function of conformation. We have results that demonstrate no calculational difficulties from pentapeptides (Hu et al. 2010b) to decapeptides.

With this level of precision, what remains in accuracy is therefore the underlying atomic model choice (Brooks, Karplus, and Pettitt 1988). Using our λ-replica exchange, free energy method, one computed model peptide backbone unit transfer free energy to trimethylamine N-oxide (TMAO) solution of −54 cal/mol/monomer (Figure 12.4c) compares quite favorably with −43 cal/mol/monomer determined experimentally (Auton and Bolen 2004).

While the actual ΔGs reflect the idiosyncrasies of the choice of model, $\Delta\Delta G$ is measurably better behaved in this scheme. In Figure 12.4, we show the decomposition of $\Delta\Delta G$ into (a) vdW, (b) electrostatics, and (c) total for a series of $\text{Gly}_{2\text{-}5}$ peptides. While computational estimates of free energy are clearly important for a variety of reasons, precise computer simulation estimates are cumbersome at best and more likely tiresome for any given system. The use of solutions to analytical theoretical techniques to consider free energy changes in solution has a long history from continuum methods (Gouy 1910; Chapman 1913) to the more modern ideas, which include explicit solvent correlations (Andersen and Chandler 1972).

However, experiment also shows that the solubility defined as precipitation (precip) or liquid–liquid separation (sep) decreases with the addition of each monomer (Auton and Bolen 2005). These are not the processes above but refer to the phase transitions between the solvent phase and the precipitate or liquid phase-separated system,

$$\text{protein}_{(\text{sol})} \xleftrightarrow{\Delta G_{\text{precip}}} \text{protein}_{(\text{precip})} \qquad (12.18)$$

$$\text{protein}_{(\text{sol})} \xleftrightarrow{\Delta G_{\text{sep}}} \text{protein}_{(\text{sep})} \qquad (12.19)$$

Thus, even though oligo glycine has a lower free energy of solvation or favorable thermodynamic interaction with water with the addition of each monomer, the solubility limit lowers; and therefore it phase separates as the concentration increases.

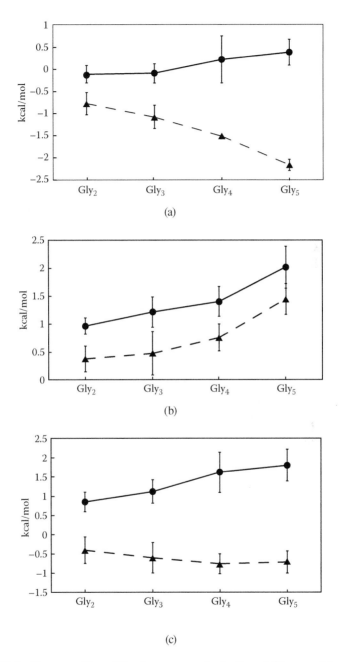

FIGURE 12.4 Decomposition of the free energy of solvation of Gly_{2-5} peptides into (a) a vdW component, (b) an electrostatic component, and (c) the total free energy in 2M aqueous solutions of TMAO (solid) and urea (dashed).

Short oligomers have relatively simple conformational possibilities. Those calculations therefore did not require extensive conformational sampling.

One can then ask what would longer oligomers of Gly_n that have nontrivial conformational possibilities do. We initially chose Gly_{15}, which had been used as a model for denatured states in water and urea solutions (Tran, Mao, and Pappu 2008). We asked whether TMAO can compact Gly_{15}. We found that Gly_{15} is compact in water but the radius of gyration and its probability distribution shrinks in TMAO (Hu et al. 2010b). Given this and the expansion in urea solution (Tran, Mao, and Pappu 2008) we have our first hypothesis from comparing the mechanism of folding of Gly_{15} to the mechanism of solubility of Gly_5. We hypothesize the solvation free energy driving force causing the compaction of Gly_{15} in water and TMAO is mechanistically related to the low solubility limit of Gly_5 through ΔG_{precip} or ΔG_{sep}.

We (Hu et al. 2010b) and others (Tran, Mao, and Pappu 2008) have found that hydrogen bonding within the peptide solute, in this case oligo glycine, is not sufficient to explain the collapse and the correct response to urea denaturation and TMAO compaction. Dipolar solute-induced phase separation (Kirkwood 1934) has been implicated before for peptide conformationally-dependent free energy landscapes (Perkyns, Wang, and Pettitt 1996).

12.4 CONCLUSIONS

Classically, the effects of varying solvent composition are decomposed into electrostatic and hydrophobic effects. Recent reexamination of the classic Tanford experiments on contributions to change in solvation free energy differences (Auton, Holthauzen, and Bolen 2007), along with consideration of what constitutes hydrophobicity (Chandler 2005) brings into question this taxonomy. Given the work reviewed above, we prefer to use a more natural set of variables, electrostatic and van der Waals interactions, which are conveniently part of the underlying intermolecular interaction models, to show the mechanistic trends of solution (de)stabilization.

The experimental measurements have been analyzed both with (Auton, Holthauzen, and Bolen 2007) and without (Auton and Bolen 2005) activity corrections to obtain changes in free energy and apparent free energy changes. What remains is our interpretation in terms of what the molecules are doing and why. By using models, whose fluctuations are more straightforward to explore and decompose than nature, we can test various results of experiment and theories.

Others have postulated that the free energy penalties associated with cavity formation or the vdW component increases in a nontrivial manner with chain length (Tran, Mao, and Pappu 2008). This is interesting when compared to our detailed free energy calculation results, which were specifically decomposed into the vdW and electrostatic components. A central problem which has been revealed recently is that our understanding of cavity hydration in terms of the hydrophobic effect (Pratt and Chandler 1977; Chandler 2005) has changed in light of the nontrivial considerations of the relatively weak van der Waals attractive forces' effects on water structure and whether a completely dry cavity is a reasonable model of biopolymer–water interfacial hydrophobicity (Choudhury and Pettitt 2005b, 2007). Taking the non-electrostatic attractive forces into account produces a qualitatively different picture of the free energy trends of electrostatics and nonpolar contributions (Choudhury and Pettitt 2006b). We find a quite linear trend in the cavity term for relatively short oligomers of glycine (Hu et al. 2010a). In fact, we find the small vdW attractions compensate for the cavity *penalty* and leave a small but favorable contribution for the completely uncharged peptide.

This is not anticipated from cavity hydrophobicity considerations of nonpolar solvation. We need to use our new pictures/theories of the mechanisms to explain this, as clearly the traditional *hydrophobicity plus electrostatics* picture is not quantitatively helpful in explaining the experimental data and simulations at this level.

In addition, the decomposable free energy simulation techniques we have recently developed allow for the calculation of the solvation and transfer free energies of model peptide solutions to be evaluated with excellent reproducibility and precision. We use a variant of replica exchange we have created (Hu et al. 2010a), which has advantages similar to other new methods (Jiang, Hodoscek, and Roux 2009) that allows an essentially automatic coverage of the free energy charging coordinate in a thermodynamic calculation. Previous methods allowed the calculation of chemical potentials of solvent or cosolvent with good results. This method, however, allows the calculation for larger more complex molecular components in the system. Adequate sampling of the fluctuations and their correlations is of critical importance to success in these applications.

Using the new simulation methods together with a rigorous theory gives a different picture of the landscape of protein folding with which to interpret experiment. We find while electrostatics is important, it remains so even when explicit hydrogen bonding patterns do not explain the stability trend. We find that ill-defined concepts like hydrophobicity and its myriad of related scales obfuscate, and are no longer required, as we can rigorously use and define the repulsive and attractive aspects of the van der Waals interaction in solution.

ACKNOWLEDGMENTS

The author thanks D. W. Bolen for many years of stimulating conversations. Partial support is acknowledged from the National Institutes of Health GM037657, and the R. A. Welch Foundation.

References

Abernethy, G. M. and M. J. Gillan. 1980. New method of solving the HNC equation for ionic liquids. *Molecular Physics.* 39, 839.

Abildskov, J., M. D. Ellegaard, and J. P. O'Connell. 2009. Correlation of phase equilibria and liquid densities for gases with ionic liquids. *Fluid Phase Equilibria.* 286, 95.

Abildskov, J., M. D. Ellegaard, and J. P. O'Connell. 2010a. Densities and isothermal compressibilities of ionic liquids-modeling and application. *Fluid Phase Equilibria.* 295, 215.

Abildskov, J., M. D. Ellegaard, and J. P. O'Connell. 2010b. Phase behavior of mixtures of ionic liquids and organic solvents. *Journal of Supercritical Fluids.* 55, 833.

Abramowitz, M. and I. A. Stegun. 1970. *Handbook of Mathematical Functions.* New York: Dover Publications, Inc.

Acree Jr., W. E. 1984. *Thermodynamic Properties of Nonelectrolyte Solutions.* Orlando, FL: Academic Press, Inc.

Adschiri, T., Y. W. Lee, M. Goto, and S. Takami. 2011. Green materials synthesis with supercritical water. *Green Chemistry.* 13, 1380.

Allison, J. R., M. Bergeler, N. Hansen, and W. F. van Gunsteren. 2011. Current computer modeling cannot explain why two highly similar sequences fold into different structures. *Biochemistry.* 50, 10965.

Allison, S. K., J. P. Fox, R. Hargreaves, and S. P. Bates. 2005. Clustering and microimmiscibility in alcohol-water mixtures: Evidence from molecular-dynamics simulations. *Physical Review B.* 71, 024201.

Almasy, L., G. Jancso, and L. Cser. 2002. Application of SANS to the determination of Kirkwood–Buff integrals in liquid mixtures. *Applied Physics A-Materials Science and Processing.* 74, S1376.

Alves, W. A. and P. S. Santos. 2007. Using Raman spectroscopy to investigate donor–acceptor reactions in the formamide/dimethylsulfoxide/acetonitrile system. *Journal of Raman Spectroscopy.* 38, 1332.

Andersen, H. C. and D. Chandler. 1972. Optimized cluster expansions for classical fluids. I. General theory and variational formulation of the mean spherical model and hard sphere Percus–Yevick equations. *Journal of Chemical Physics.* 57, 1918.

Anderson, C. F., E. S. Courtenay, and M. T. Record. 2002. Thermodynamic expressions relating different types of preferential interaction coefficients in solutions containing two solute components. *Journal of Physical Chemistry B.* 106, 418.

Anfinsen, C. B. 1973. Principles that govern the folding of protein chains. *Science.* 181, 223.

Aparicio, S., R. Alcalde, B. García, and J. M. Leal. 2008. Structure-composition relationships in ternary solvents containing methylbenzoate. *Journal of Physical Chemistry B.* 112, 3420.

Aparicio, S., R. Alcalde, J. M. Leal, and B. García. 2005. Characterization and preferential solvation of the hexane/hexan-1-ol/methylbenzoate ternary solvent. *Journal of Physical Chemistry B.* 109, 6375.

Arakawa, T., R. Bhat, and S. N. Timasheff. 1990. Preferential interactions determine protein solubility in 3-component solutions—The $MgCl_2$ system. *Biochemistry.* 29, 1914.

Arakawa, T. and S. N. Timasheff. 1982. Preferential interactions of proteins with salts in concentrated solutions. *Biochemistry.* 21, 6545.

Arakawa, T. and S. N. Timasheff. 1984. Mechanism of protein salting in and salting out by divalent-cation salts—Balance between hydration and salt binding. *Biochemistry.* 23, 5912.

Arakawa, T. and S. N. Timasheff. 1985a. The stabilization of proteins by osmolytes. *Biophysical Journal.* 47, 411.

Arakawa, T. and S. N. Timasheff. 1985b. Theory of protein solubility. *Methods in Enzymology.* 114, 49.

Arakawa, T. and S. N. Timasheff. 1987. Abnormal solubility behavior of β-lactoglobulin—Salting-in by glycine and NaCl. *Biochemistry.* 26, 5147.

Archer, A. J. and N. B. Wilding. 2007. Phase behavior of a fluid with competing attractive and repulsive interactions. *Physical Review E.* 76, 031501.

Attard, P. 1990. Integral-equations and closure relations for the bridge function and for the triplet correlation-function. *Journal of Chemical Physics.* 93, 7301.

Attard, P., D. R. Berard, C. P. Ursenbach, and G. N. Patey. 1991. Interaction free-energy between planar walls in dense fluids—An Ornstein–Zernike approach with results for hard-sphere, Lennard–Jones and dipolar systems. *Physical Review A.* 44, 8224.

Auton, M. and D. W. Bolen. 2004. Additive transfer free energies of the peptide backbone unit that are independent of the model compound and the choice of concentration scale. *Biochemistry.* 43, 1329.

Auton, M. and D. W. Bolen. 2005. Predicting the energetics of osmolyte-induced protein folding/unfolding. *Proceedings of the National Academy of Sciences of the United States of America.* 102, 15065.

Auton, M. and D. W. Bolen. 2007. Application of the transfer model to understand how naturally occurring osmolytes affect protein stability. *Methods in Enzymology.* 428, 397.

Auton, M., D. W. Bolen and J. Rösgen. 2008. Structural thermodynamics of protein preferential solvation: Osmolyte solvation of proteins, aminoacids, and peptides. *Proteins: Structure, Function, and Bioinformatics.* 73, 802.

Auton, M., L. M. F. Holthauzen, and D. W. Bolen. 2007. Anatomy of energetic changes accompanying urea-induced protein denaturation. *Proceedings of the National Academy of Sciences of the United States of America.* 104, 15317.

Auton, M., J. Rösgen, M. Sinev, L. M. F. Holthauzen, and D. W. Bolen. 2011. Osmolyte effects on protein stability and solubility: A balancing act between backbone and side-chains. *Biophysical Chemistry.* 159, 90.

Baldwin, R. L. 1996. How Hofmeister ion interactions affect protein stability. *Biophysical Journal.* 71, 2056.

Ball, P. 2008. Water—An enduring mystery. *Nature.* 452, 291.

Baynes, B. M. and B. L. Trout. 2003. Proteins in mixed solvents: A molecular-level perspective. *Journal of Physical Chemistry B.* 107, 14058.

Behera, R. 1998. On the calculation of thermodynamic properties of electrolyte solutions from Kirkwood–Buff theory. *Journal of Chemical Physics.* 108, 3373.

Ben-Naim, A. 1975. Solute and solvent effects on chemical-equilibria. *Journal of Chemical Physics.* 63, 2064.

Ben-Naim, A. 1977. Inversion of Kirkwood–Buff theory of solutions—Application to water-ethanol system. *Journal of Chemical Physics.* 67, 4884.

Ben-Naim, A. 1987. *Solvation Thermodynamics.* New York: Plenum Press.

Ben-Naim, A. 1988. Theory of preferential solvation of nonelectrolytes. *Cell Biophysics.* 12, 255.

Ben-Naim, A. 1989. Preferential solvation in 2-component systems. *Journal of Physical Chemistry.* 93, 3809.

Ben-Naim, A. 1990a. Inversion of Kirkwood–Buff theory of solutions and its applications. In *Fluctuation Theory of Mixtures*, edited by E. Matteoli and G. A. Mansoori. New York: Taylor & Francis, p. 211.

Ben-Naim, A. 1990b. Preferential solvation in 2-component and in 3-component systems. *Pure and Applied Chemistry.* 62, 25.

Ben-Naim, A. 1992. *Statistical Thermodynamics for Chemists and Biochemists.* New York: Plenum Press.

Ben-Naim, A. 2006. *Molecular Theory of Solutions.* New York: Oxford University Press.

Ben-Naim, A. 2007a. A critique of some recent suggestions to correct the Kirkwood–Buff integrals. *Journal of Physical Chemistry B*. 111, 2896.

Ben-Naim, A. 2007b. Reply to "Comment on 'A critique of some recent suggestions to correct the Kirkwood–Buff integrals.'" *Journal of Physical Chemistry B*. 111, 3072.

Ben-Naim, A. 2008. Comment on "The Kirkwood–Buff theory of solutions and the local composition of liquid mixtures." *Journal of Physical Chemistry B*. 112, 5874.

Ben-Naim, A. 2009. *Molecular Theory of Water and Aqueous Solutions. Part 1: Understanding Water*. Singapore: World Scientific Publishing Co.

Ben-Naim, A. 2011. *Molecular Theory of Water and Aqueous Solutions. Part 2: The Role of Water in Protein Folding, Self-Assembly and Molecular Recognition*. Singapore: World Scientific Publishing Co.

Ben-Naim, A. and A. Santos. 2009. Local and global properties of mixtures in one-dimensional systems. II. Exact results for the Kirkwood–Buff integrals. *Journal of Chemical Physics*. 131, 164512.

Bennion, B. J. and V. Daggett. 2003. The molecular basis for the chemical denaturation of proteins by urea. *Proceedings of the National Academy of Sciences of the United States of America*. 100, 5142.

Bentenitis, N., N. R. Cox, and P. E. Smith. 2009. A Kirkwood–Buff derived force field for thiols, sulfides, and disulfides. *Journal of Physical Chemistry B*. 113, 12306.

Berendsen, H. J. C., J. R. Grigera, and T. P. Straatsma. 1987. The missing term in effective pair potentials. *Journal of Physical Chemistry*. 91, 6269.

Berendsen, H. J. C., J. P. M. Postma, W. F. van Gunsteren, and J. Hermans. 1981. Interaction models for water in relation to protein hydration. In *Intermolecular Forces*, edited by B. Pullman. Dordrecht: D. Reidel, p. 331.

Beutler, T. C., A. E. Mark, R. C. van Schaik, P. R. Gerber, and W. F. van Gunsteren. 1994. Avoiding singularities and numerical instabilities in free-energy calculations based on molecular simulations. *Chemical Physics Letters*. 222, 529.

Biben, T. and J. P. Hansen. 1991. Phase-separation of asymmetric binary hard-sphere fluids. *Physical Review Letters*. 66, 2215.

Bjelkmar, P., P. Larsson, M. A. Cuendet, B. Hess, and E. Lindahl. 2010. Implementation of the CHARMM force field in GROMACS: Analysis of protein stability effects from correction maps, virtual interaction sites, and water models. *Journal of Chemical Theory and Computation*. 6, 459.

Blanco, M. A., E. Sahin, Y. Li, and C. J. Roberts. 2011. Reexamining protein-protein and protein-solvent interactions from Kirkwood–Buff analysis of light scattering in multicomponent solutions. *Journal of Chemical Physics*. 134, 225103.

Blum, L. and A. J. Torruella. 1972. Invariant expansion for 2-body correlations—Thermodynamic functions, scattering, and Ornstein–Zernike equation. *Journal of Chemical Physics*. 56, 303.

Bolen, D. W. 2004. Effects of naturally occurring osmolytes on protein stability and solubility: Issues important in protein crystallization. *Methods*. 34, 312.

Bolen, D. W. and I. V. Baskakov. 2001. The osmophobic effect: Natural selection of a thermodynamic force in protein folding. *Journal of Molecular Biology*. 310, 955.

Bolen, D. W. and G. D. Rose. 2008. Structure and energetics of the hydrogen-bonded backbone in protein folding. *Annual Review of Biochemistry*. 77, 339.

Bondi, A. 1964. Van der Waals volumes and radii. *Journal of Physical Chemistry*. 68, 441.

Bonnet, P. and R. A. Bryce. 2004. Molecular dynamics and free energy analysis of neuraminidase-ligand interactions. *Protein Science*. 13, 946.

Bouillot, B., S. Teychene, and B. Biscans. 2011. An evaluation of thermodynamic models for the prediction of drug and drug-like molecule solubility in organic solvents. *Fluid Phase Equilibria*. 309, 36.

Bowman, G. R., V. A. Voelz, and V. S. Pande. 2011. Taming the complexity of protein folding. *Current Opinion in Structural Biology*. 21, 4.

Branco, R. J. F., M. Graber, V. Denis, and J. Pleiss. 2009. Molecular mechanism of the hydration of *Candida antarctica* lipase B in the gas phase: Water adsorption isotherms and molecular dynamics simulations. *ChemBioChem.* 10, 2913.

Brelvi, S. W. and J. P. O'Connell. 1972. Corresponding states correlations for liquid compressibility and partial molal volumes of gases at infinite dilution in liquids. *American Institute of Chemical Engineers Journal.* 18, 1239.

Brelvi, S. W. and J. P. O'Connell. 1975a. Generalized isothermal equation of state for dense liquids. *American Institute of Chemical Engineers Journal.* 21, 171.

Brelvi, S. W. and J. P. O'Connell. 1975b. Generalized prediction of isothermal compressibilities and an isothermal equation of state for liquid-mixtures. *American Institute of Chemical Engineers Journal.* 21, 1024.

Brelvi, S. W. and J. P. O'Connell. 1975c. Prediction of unsymmetric convention liquidphase activity-coefficients of hydrogen and methane. *American Institute of Chemical Engineers Journal.* 21, 157.

Brennecke, J. F. 1993. Spectroscopic investigation of reactions in supercritical fluids. A review. In *Supercritical Fluid Engineering Science. Fundamentals and Applications*, edited by E. Kiran and J. F. Brennecke. Washington, DC: American Chemical Society, p. 201.

Brennecke, J. F., P. G. Debenedetti, C. A. Eckert, and K. P. Johnston. 1990. Letter to the editor. *American Institute of Chemical Engineers Journal.* 36, 1927.

Brennecke, J. F. and C. A. Eckert. 1989. Phase-equilibria for supercritical fluid process design. *American Institute of Chemical Engineers Journal.* 35, 1409.

Brinkman, H. C. and J. J. Hermans. 1949. The effect of non-homogeneity of molecular weight on the scattering of light by high polymer solutions. *Journal of Chemical Physics.* 17, 574.

Broccio, M., D. Costa, Y. Liu, and S. H. Chen. 2006. The structural properties of a two-Yukawa fluid: Simulation and analytical results. *Journal of Chemical Physics.* 124, 084501.

Brooks, C. L., M. Karplus, and B. M. Pettitt. 1988. *Proteins: A Theoretical Perspective of Dynamics, Structure and Thermodynamics.* Vol. 71, *Advances in Chemical Physics*: Hoboken, NJ: Wiley.

Brunner, G. 1983. Selectivity of supercritical compounds and entrainers with respect to model substances. *Fluid Phase Equilibria.* 10, 289.

Brunner, G. 2005. Supercritical fluids: Technology and application to food processing. *Journal of Food Engineering.* 67, 21.

Brunner, G., ed. 2004. *Supercritical Fluids as Solvents and Reaction Media.* Amsterdam: Elsevier B.V.

Bruno, T. J. and J. F. Ely. 1991. *Supercritical Fluid Technology Reviews in Modern Theory Application.* Boca Raton: CRC Press.

Bryngelson, J. D., J. N. Onuchic, N. D. Socci, and P. G. Wolynes. 1995. Funnels, pathways, and the energy landscape of protein folding: A synthesis. *Proteins: Structure, Function, and Bioinformatics.* 21, 167.

Buff, F. P. and R. Brout. 1955. Molecular formulation of thermodynamic functions encountered in solution theory. *Journal of Chemical Physics.* 23, 458.

Buff, F. P. and F. M. Schindler. 1958. Small perturbations in solution theory. *Journal of Chemical Physics.* 29, 1075.

Bustamante, C. and P. Bustamante. 1996. Nonlinear enthalpy-entropy compensation for the solubility of phenacetin in dioxane-water solvent mixtures. *Journal of Pharmaceutical Sciences.* 85, 1109.

Bustamante, P., B. Escalera, A. Martin, and E. Selles. 1993. A modification of the extended Hildebrand approach to predict the solubility of structurally related drugs in solvent mixtures. *Journal of Pharmacy and Pharmacology.* 45, 253.

Cabezas, H. and J. P. O'Connell. 1986. A fluctuation theory model of strong electrolytes. *Fluid Phase Equilibria.* 30, 213.

Cabezas, H. and J. P. O'Connell. 1993. Some uses and misuses of thermodynamic models for dilute liquid solutions. *Industrial and Engineering Chemistry Research.* 32, 2892.

Cagin, T. and B. M. Pettitt. 1991. Molecular-dynamics with a variable number of molecules. *Molecular Physics.* 72, 169.

Callen, H. B. 1962. *Thermodynamics: An Introduction to the Physical Theories of Equilibrium Thermostatics and Irreversible Thermodynamics.* New York: Wiley.

Campanella, E. A., P. M. Mathias, and J. P. O'Connell. 1987. Equilibrium properties of liquids containing supercritical substances. *American Institute of Chemical Engineers Journal.* 33, 2057.

Canchi, D. R. and A. E. García. 2011. Backbone and side-chain contributions in protein denaturation by urea. *Biophysical Journal.* 100, 1526.

Carter, R. W. 1992. Ph.D. Thesis, University of Delaware.

Casassa, E. F. and H. Eisenberg. 1964. Thermodynamic analysis of multicomponent solutions. *Advances in Protein Chemistry.* 19, 287.

Case, F. H., J. Brennan, A. Chaka, K. D. Dobbs, D. G. Friend, D. Frurip, P. A. Gordon, J. Moore, R. D. Mountain, J. Olson, R. B. Ross, M. Schiller, and V. K. Shen. 2007. The third industrial fluid properties simulation challenge. *Fluid Phase Equilibria.* 260, 153.

Celinski, S. A. and J. M. Scholtz. 2002. Osmolyte effects on helix formation in peptides and the stability of coiled-coils. *Protein Science.* 11, 2048.

Ceperley, D. M. and G. V. Chester. 1977. Perturbation approach to classical one-component plasma. *Physical Review A.* 15, 756.

Chaikin, P. M. and T. C. Lubensky. 2000. *Principles of Condensed Matter Physics.* Cambridge, UK: Cambridge University Press.

Chalikian, T. V. 2003. Volumetric properties of proteins. *Annual Review of Biophysics and Biomolecular Structure.* 32, 207.

Chalikian, T. V. 2011. Volumetric measurements in binary solvents: Theory to experiment. *Biophysical Chemistry.* 156, 3.

Chalikian, T. V. and R. Filfil. 2003. How large are the volume changes accompanying protein transitions and binding? *Biophysical Chemistry.* 104, 489.

Chandler, D. 2005. Interfaces and the driving force of hydrophobic assembly. *Nature.* 437, 640.

Chapman, D. L. 1913. A contribution to the theory of electrocapillarity. LI. *The London, Edinburgh, and Dublin Philisophical Magazine and Journal of Science.* 25, 475.

Cheatham, T. E. and B. R. Brooks. 1998. Recent advances in molecular dynamics simulation toward the realistic representation of biomolecules in solution. *Theoretical Chemistry Accounts.* 99, 279.

Chialvo, A. A. 1990a. An alternative approach to modeling excess Gibbs free energy in terms of Kirkwood–Buff integrals. In *Fluctuation Theory of Mixtures,* edited by E. Matteoli and G. A. Mansoori. New York: Taylor & Francis, p. 131.

Chialvo, A. A. 1990b. Determination of excess Gibbs free-energy from computer-simulation via the single charging-integral approach. I. Theory. *Journal of Chemical Physics.* 92, 673.

Chialvo, A. A. 1991. Excess properties of liquid-mixtures from computer-simulation—A coupling-parameter approach to the determination of their dependence on molecular asymmetry. *Molecular Physics.* 73, 127.

Chialvo, A. A. 1993a. Accurate calculation of excess thermal, infinite dilution, and related properties of liquid mixtures via molecular-based simulation. *Fluid Phase Equilibria.* 83, 23.

Chialvo, A. A. 1993b. Solute-solute and solute-solvent correlations in dilute near-critical ternary mixtures: Mixed-solute and entrainer effects. *Journal of Physical Chemistry.* 97, 2740.

Chialvo, A. A., S. Chialvo, J. M. Simonson, and Y. V. Kalyuzhnyi. 2008. Solvation phenomena in dilute multicomponent solutions. I. Formal results and molecular outlook. *Journal of Chemical Physics.* 128, 214512.

Chialvo, A. A. and P. T. Cummings. 1994. Solute-induced effects on the structure and the thermodynamics of infinitely dilute mixtures. *American Institute of Chemical Engineers Journal*. 40, 1558.

Chialvo, A. A. and P. T. Cummings. 1995. Comment on "Near critical phase behavior of dilute mixtures." *Molecular Physics*. 84, 41.

Chialvo, A. A. and P. T. Cummings. 1999. Molecular-based modeling of water and aqueous solutions at supercritical conditions. In *Advances in Chemical Physics*, edited by S. A. Rice. New York: Wiley and Sons.

Chialvo, A. A., P. T. Cummings, and Y. V. Kalyuzhnyi. 1998. Solvation effect on kinetic rate constant of reactions in supercritical solvents. *American Institute of Chemical Engineers Journal*. 44, 667.

Chialvo, A. A., P. T. Cummings, J. M. Simonson, and R. E. Mesmer. 1999. Solvation in high-temperature electrolyte solutions. II. Some formal results. *Journal of Chemical Physics*. 110, 1075.

Chialvo, A. A., P. T. Cummings, J. M. Simonson, and R. E. Mesmer. 2000a. Solvation in high-temperature aqueous electrolyte solutions. *Journal of Molecular Liquids*. 87, 233.

Chialvo, A. A. and P. G. Debenedetti. 1992. Molecular dynamics study of solute-solute micro-structure in attractive and repulsive supercritical mixtures. *Industrial and Engineering Chemistry Research*. 31, 1391.

Chialvo, A. A., Y. V. Kalyuzhnyi, and P. T. Cummings. 1996. Solvation thermodynamics of gas solubility at sub- and near-critical conditions. *American Institute of Chemical Engineers Journal*. 42, 571.

Chialvo, A. A., P. G. Kusalik, P. T. Cummings, and J. M. Simonson. 2001. Solvation in high-temperature electrolyte solutions. III. Integral equation calculations and interpretation of experimental data. *Journal of Chemical Physics*. 114, 3575.

Chialvo, A. A., P. G. Kusalik, P. T. Cummings, J. M. Simonson, and R. E. Mesmer. 2000b. Molecular approach to high-temperature solvation. Formal, integral equations, and experimental results. *Journal of Physics-Condensed Matter*.

Chialvo, A. A., P. G. Kusalik, Y. V. Kalyuzhnyi, and P. T. Cummings. 2000c. Applications of integral equations calculations to high-temperature solvation phenomena. *Journal of Statistical Physics*. 100, 167.

Chiti, F. and C. M. Dobson. 2006. Protein misfolding, functional amyloid, and human disease. *Annual Review of Biochemistry*. 75, 333.

Chitra, R. and P. E. Smith. 2000. Molecular dynamics simulations of the properties of cosolvent solutions. *Journal of Physical Chemistry B*. 104, 5854.

Chitra, R. and P. E. Smith. 2001a. A comparison of the properties of 2,2,2-trifluoroethanol and 2,2,2-trifluoroethanol/water mixtures using different force fields. *Journal of Chemical Physics*. 115, 5521.

Chitra, R. and P. E. Smith. 2001b. Preferential interactions of cosolvents with hydrophobic solutes. *Journal of Physical Chemistry B*. 105, 11513.

Chitra, R. and P. E. Smith. 2001c. Properties of 2,2,2-trifluoroethanol and water mixtures. *Journal of Chemical Physics*. 114, 426.

Chitra, R. and P. E. Smith. 2002. Molecular association in solution: A Kirkwood–Buff analysis of sodium chloride, ammonium sulfate, guanidinium chloride, urea, and 2,2,2-trifluoro-ethanol in water. *Journal of Physical Chemistry B*. 106, 1491.

Cho, S. S., Y. Levy, and P. G. Wolynes. 2009. Quantitative criteria for native energetic het-erogeneity influences in the prediction of protein folding kinetics. *Proceedings of the National Academy of Sciences of the United States of America*. 106, 434.

Choudhury, N. 2006. A molecular dynamics simulation study of buckyballs in water: Atomistic versus coarse-grained models of C_{60}. *Journal of Chemical Physics*. 125, 034502.

Choudhury, N. and B. M. Pettitt. 2005a. Local density profiles are coupled to solute size and attractive potential for nanoscopic hydrophobic solutes. *Molecular Simulation*. 31, 457.

Choudhury, N. and B. M. Pettitt. 2005b. On the mechanism of hydrophobic association of nanoscopic solutes. *Journal of the American Chemical Society.* 127, 3556.

Choudhury, N. and B. M. Pettitt. 2006a. Enthalpy-entropy contributions to the potential of mean force of nanoscopic hydrophobic solutes. *Journal of Physical Chemistry B.* 110, 8459.

Choudhury, N. and B. M. Pettitt. 2006b. The role of attractive forces on the dewetting of large hydrophobic solutes. In *Modelling Molecular Structure and Reactivity in Biological Systems*, edited by K. J. Naidoo. Cambridge, UK: Royal Society of Chemistry, p. 49.

Choudhury, N. and B. M. Pettitt. 2007. The dewetting transition and the hydrophobic effect. *Journal of the American Chemical Society.* 129, 4847.

Christensen, S., G. H. Peters, F. Y. Hansen, and J. Abildskov. 2007a. Thermodynamic models from fluctuation solution theory analysis of molecular simulations. *Fluid Phase Equilibria.* 261, 185.

Christensen, S., G. H. Peters, F. Y. Hansen, J. P. O'Connell, and J. Abildskov. 2007b. Generation of thermodynamic data for organic liquid mixtures from molecular simulations. *Molecular Simulation.* 33, 449.

Christensen, S., G. H. Peters, F. Y. Hansen, J. P. O'Connell, and J. Abildskov. 2007c. State conditions transferability of vapor-liquid equilibria via fluctuation solution theory with correlation function integrals from molecular dynamics simulation. *Fluid Phase Equilibria.* 260, 169.

Ciach, A. and W. T. Gozdz. 2001. Nonelectrolyte solutions exhibiting structure on the nanoscale. *Annual Reports on the Progress of Chemistry, Section "C" (Physical Chemistry).* 97, 269.

Clifford, A. A. 1999. *Fundamentals of Supercritical Fluids.* New York: Oxford University Press.

Clifford, A. A., K. Pople, W. J. Gaskill, K. D. Bartle, and C. M. Rayner. 1998. Potential tuning and reaction control in the Diels–Alder reaction between cyclopentadiene and methyl acrylate in supercritical carbon dioxide. *Journal of the Chemical Society, Faraday Transactions.* 94, 1451.

Cochran, H. D., L. L. Lee, and D. M. Pfund. 1990. Structure and properties of supercritical fluid mixtures from Kirkwood–Buff fluctuation theory and integral equation methods. In *Fluctuation Theory of Mixtures*, edited by E. Matteoli and G. A. Mansoori. New York: Taylor & Francis, p. 69.

Cohn, E. J. and J. T. Edsall. 1943. *Proteins, Amino Acids and Peptides as Ions and Dipolar Ions.* New York: Reinhold Publishing Corporation.

Conti, G., P. Gianni, L. Lepori, and E. Matteoli. 2003. Volumetric study of (2-methoxyethanol + tetrahydrofuran + cyclohexane) at 298.15 K. *Journal of Chemical Thermodynamics.* 35, 503.

Cooney, W. R. and J. P. O'Connell. 1987. Correlation of partial molar volumes at infinite dilution of salts in water. *Chemical Engineering Communications.* 56, 341.

Cornell, W. D., P. Cieplak, C. I. Bayly, I. R. Gould, K. M. Merz, D. M. Ferguson, D. C. Spellmeyer, T. Fox, J. W. Caldwell, and P. A. Kollman. 1995. A 2nd generation forcefield for the simulation of proteins, nucleic-acids, and organic-molecules. *Journal of the American Chemical Society.* 117, 5179.

Courtenay, E. S., M. W. Capp, C. F. Anderson, and M. T. Record. 2000. Vapor pressure osmometry studies of osmolyte-protein interactions: Implications for the action of osmoprotectants *in vivo* and for the interpretation of "osmotic stress" experiments *in vitro. Biochemistry.* 39, 4455.

Courtenay, E. S., M. W. Capp, and M. T. Record. 2001. Thermodynamics of interactions of urea and guanidinium salts with protein surface: Relationship between solute effects on protein processes and changes in water-accessible surface area. *Protein Science.* 10, 2485.

Coussaert, T. and M. Baus. 1997. Virial approach to hard-sphere demixing. *Physical Review Letters.* 79, 1881.

Curtiss, C. F. and R. B. Bird. 1999. Multicomponent diffusion. *Industrial and Engineering Chemistry Research.* 38, 2515.

D'Arrigo, G., R. Giordano, and J. Teixeira. 2009. Temperature and concentration dependence of SANS spectra of aqueous solutions of short-chain amphiphiles. *European Physical Journal E.* 29, 37.

Dai, S., S. Weerasinghe, and P. E. Smith. 2012. A Kirkwood–Buff derived force field for amines and carboxylic acids. (In preparation, consulted October 2012.)

Davidson, N. R. 1962. *Statistical Mechanics.* New York: McGraw-Hill.

Davis-Searles, P. R., A. J. Saunders, D. A. Erie, D. J. Winzor, and G. J. Pielak. 2001. Interpreting the effects of small uncharged solutes on protein-folding equilibria. *Annual Review of Biophysics and Biomolecular Structure.* 30, 271.

Davis, M. I. 1993. Thermodynamic and related studies of amphiphile plus water-systems. *Chemical Society Reviews.* 22, 127.

Day, R., D. Paschek, and A. E. García. 2010. Microsecond simulations of the folding/unfolding thermodynamics of the Trp-cage miniprotein. *Proteins: Structure, Function, and Bioinformatics.* 78, 1889.

De Boer, J. 1940. Contribution to the Theory of Compressed Gases. Ph.D. Thesis, University of Amsterdam, Amsterdam.

De Boer, J. 1949. Molecular distribution and equation of state of gases. *Reports on Progress in Physics.* 12, 305.

De Gennes, P. G. 1977. *Microscopic Structure and Dynamics of Liquids.* New York: Plenum Press.

Debenedetti, P. G. 1988. Fluctuation-based computer calculation of partial molar properties. 2. A numerically accurate method for the determination of partial molar energies and enthalpies. *Journal of Chemical Physics.* 88, 2681.

Debenedetti, P. G. and A. A. Chialvo. 1992. Solute-solute correlations in infinitely dilute supercritical mixtures. *Journal of Chemical Physics.* 97, 504.

Debenedetti, P. G. and S. K. Kumar. 1986. Infinite dilution fugacity coefficients and the general behavior of dilute binary systems. *American Institute of Chemical Engineers Journal.* 32, 1253.

Debenedetti, P. G. and S. K. Kumar. 1988. The molecular basis of temperature effects in super-critical extraction. *American Institute of Chemical Engineers Journal.* 34, 645.

Debenedetti, P. G. and R. S. Mohamed. 1989. Attractive, weakly attractive and repulsive near-critical systems. *Journal of Chemical Physics.* 90, 4528.

Derr, E. L. and C. H. Deal, Jr. 1969 (September 8–10). Analytical solutions of groups. Correlation of activity coefficients through structural group parameters. Paper read at *Proceedings of the International Symposium on Distillation*, Brighton, England.

Dickhut, R. M., A. W. Andren, and D. E. Armstrong. 1989. Naphthalene solubility in selected organic solvent-water mixtures. *Journal of Chemical and Engineering Data.* 34, 438.

Diedrichs, A. and J. Gmehling. 2011. Solubility calculation of active pharmaceutical ingredients in alkanes, alcohols, water and their mixtures using various activity coefficient models. *Industrial and Engineering Chemistry Research.* 50, 1757.

Dill, K. A. 1990. Dominant forces in protein folding. *Biochemistry.* 29, 7133.

Dill, K. A., K. Ghosh, and J. D. Schmit. 2011. Physical limits of cells and proteomes. *Proceedings of the National Academy of Sciences of the United States of America.* 108, 17876.

Dixit, S., J. Crain, W. C. K. Poon, J. L. Finney, and A. K. Soper. 2002. Molecular segregation observed in a concentrated alcohol-water solution. *Nature.* 416, 829.

Dixon, M. and P. Hutchinson. 1977. Method for extrapolation of pair distribution functions. *Molecular Physics.* 33, 1663.

Dougan, L., S. P. Bates, R. Hargreaves, J. P. Fox, J. Crain, J. L. Finney, V. Reat, and A. K. Soper. 2004. Methanol-water solutions: A bi-percolating liquid mixture. *Journal of Chemical Physics.* 121, 6456.

Dyson, H. J. and P. E. Wright. 2002. Coupling of folding and binding for unstructured proteins. *Current Opinion in Structural Biology.* 12, 54.

Dyson, H. J. and P. E. Wright. 2005. Intrinsically unstructured proteins and their functions. *Nature Reviews Molecular Cell Biology*. 6, 197.

Economou, I. G. 2002. Statistical associating fluid theory: A successful model for the calculation of thermodynamic and phase equilibrium properties of complex fluid mixtures. *Industrial and Engineering Chemistry Research*. 41, 953.

Economou, I. G. and M. D. Donohue. 1990. Mean field calculation of thermodynamic properties of supercritical fluids. *American Institute of Chemical Engineers Journal*. 36, 1920.

Einstein, A. 1910. Theorie der opaleszenz von homogenen flüssigkeiten und flüssigkeitsgemischen in der nähe des kritischen zustandes. *Annalen der Physik*. 33, 1275.

Eisenberg, H. 1976. *Biological Macromolecules and Polyelectrolytes in Solution*. Oxford: Clarendon Press.

Ellegaard, M. D. 2011. Molecular Thermodynamics Using Fluctuation Solution Theory. Ph.D. Thesis, Department of Chemical and Biochemical Engineering, Technical University of Denmark.

Ellegaard, M. D., J. Abildskov, and J. P. O'Connell. 2009. Method for predicting solubilities of solids in mixed solvents. *American Institute of Chemical Engineers Journal*. 55, 1256.

Ellegaard, M. D., J. Abildskov, and J. P. O'Connell. 2010. Molecular thermodynamic modeling of mixed solvent solubility. *Industrial and Engineering Chemistry Research*. 49, 11620.

Ellegaard, M. D., J. Abildskov, and J. P. O'Connell. 2011. Solubilities of gases in ionic liquids using a corresponding-states approach to Kirkwood–Buff solution theory. *Fluid Phase Equilibria*. 302, 93.

Ellington, J. B., K. M. Park, and J. F. Brennecke. 1994. Effect of local composition enhancements on the esterification of phtalic-anhydride with methanol in supercritical carbondioxide. *Industrial and Engineering Chemistry Research*. 33, 965.

Enciso, E. 1985. Numerical-solution of the SSOZ equation by extension of Gillan method to nonhomonuclear molecular fluids. *Molecular Physics*. 56, 129.

Felitsky, D. J. and M. T. Record. 2004. Application of the local-bulk partitioning and competitive binding models to interpret preferential interactions of glycine betaine and urea with protein surface. *Biochemistry*. 43, 9276.

Feng, J. W. A., J. Kao, and G. R. Marshall. 2009. A second look at mini-protein stability: Analysis of FSD-1 using circular dichroism, differential scanning calorimetry, and simulations. *Biophysical Journal*. 97, 2803.

Ferrario, M., M. Haughney, I. R. McDonald, and M. L. Klein. 1990. Molecular-dynamics simulation of aqueous mixtures—Methanol, acetone, and ammonia. *Journal of Chemical Physics*. 93, 5156.

Fisher, M. E. 1964. Correlation functions + critical region of simple fluids. *Journal of Mathematical Physics*. 5, 944.

Foffi, G., C. de Michele, F. Sciortino, and P. Tartaglia. 2005. Arrested phase separation in a short-ranged attractive colloidal system: A numerical study. *Journal of Chemical Physics*. 122, 224903.

Foiles, S. M., N. W. Ashcroft, and L. Reatto. 1984. Structure factor and direct correlation-function of a fluid from finite-range simulation data. *Journal of Chemical Physics*. 81, 6140.

Forsythe, E. L., R. A. Judge, and M. L. Pusey. 1999. Tetragonal chicken egg white lysozyme solubility in sodium chloride solutions. *Journal of Chemical and Engineering Data*. 44, 637.

Fowler, R. H. and E. A. Guggenheim. 1939. *Statistical Thermodynamics: A Version of Statistical Mechanics for Students of Physics and Chemistry*. Cambridge, England: The University Press.

Frank, H. S. and F. Franks. 1968. Structural approach to solvent power of water for hydrocarbons—Urea as a structure breaker. *Journal of Chemical Physics*. 48, 4746.

Franks, F. and D. J. G. Ives. 1966. The structural properties of alcohol-water mixtures. *Quarterly Reviews, Chemical Society*. 20, 1.

Freddolino, P. L., C. B. Harrison, Y. X. Liu, and K. Schulten. 2010. Challenges in protein-folding simulations. *Nature Physics.* 6, 751.

Fredenslund, A., J. Gmehling, and P. Rasmussen. 1977. *Vapor-Liquid Equilibria Using UNIFAC: A Group-Contribution Method.* Amsterdam: Elsevier.

Fredenslund, A., R. L. Jones, and J. M. Prausnitz. 1975. Group-contribution estimation of activity-coefficients in nonideal liquid-mixtures. *American Institute of Chemical Engineers Journal.* 21, 1086.

French, H. T. 1987. Vapour pressures and activity coefficients of (acetonitrile + water) at 308.15 K. *Journal of Chemical Thermodynamics.* 19, 1155.

Friedman, H. L. 1962. *Ionic Solution Theory: Based on Cluster Expansion Methods, Monographs in Statistical Physics and Thermodynamics.* New York: Interscience Publishers.

Friedman, H. L. 1986. Methods to determine structure in water and aqueous solutions. In *Methods in Enzymology,* edited by P. Lester. Orlando, FL: Academic Press, p. 3.

Friedman, H. L. and P. S. Ramanathan. 1970. Theory of mixed electrolyte solutions and application to a model for aqueous lithium chloride-cesium chloride. *Journal of Physical Chemistry.* 74, 3756.

Fries, P. H. and G. N. Patey. 1985. The solution of the hypernetted-chain approximation for fluids of nonspherical particles—A general-method with application to dipolar hard-spheres. *Journal of Chemical Physics.* 82, 429.

Frisch, H. L. and J. L. Lebowitz, eds. 1964. *The Equilibrium Theory of Classical Fluids: A Lecture Note and Reprint Volume.* New York: W. A. Benjamin.

Galam, S. and J. P. Hansen. 1976. Statistical-mechanics of dense ionized matter. 6. Electron screening corrections to thermodynamic properties of one-component plasma. *Physical Review A.* 14, 816.

García, B., S. Aparicio, R. Alcalde, and J. M. Leal. 2003. Preferential solvation in ternary solutions containing methylbenzoate. A Kirkwood–Buff fluctuation theory study. *Journal of Physical Chemistry B.* 107, 13478.

Gardas, R. L., M. G. Freire, P. J. Carvalho, I. M. Marrucho, I. M. A. Fonseca, A. G. M. Ferreira, and J. A. P. Coutinho. 2007. P ρ T measurements of imidazolium-based ionic liquids. *Journal of Chemical and Engineering Data.* 52, 1881.

Gee, M. B. 2010. Computer Simulation and Theory of Amino Acid Interactions in Solution, Chemistry, Ph.D. thesis, Kansas State University, Manhattan, Kansas.

Gee, M. B., N. R. Cox, Y. Jiao, N. Bentenitis, S. Weerasinghe, and P. E. Smith. 2011. A Kirkwood–Buff derived force field for aqueous alkali halides. *Journal of Chemical Theory and Computation.* 7, 1369.

Gee, M. B. and P. E. Smith. 2009. Kirkwood–Buff theory of molecular and protein association, aggregation, and cellular crowding. *Journal of Chemical Physics.* 131, 165101.

Gee, M. B. and P. E. Smith. 2012. A Kirkwood–Buff derived force field for amine salts and carboxylates. (In preparation, consulted October 2012.)

Gekko, K. and S. N. Timasheff. 1981. Thermodynamic and kinetic examination of protein stabilization by glycerol. *Biochemistry.* 20, 4677.

Gianni, P., L. Lepori, and E. Matteoli. 2010. Excess Gibbs energies and volumes of the ternary system chloroform + tetrahydrofuran + cyclohexane at 298.15 K. *Fluid Phase Equilibria.* 297, 52.

Gibbs, J. W. 1902. *Elementary Principles in Statistical Mechanics: Developed with Special Reference to the Rational Foundation of Thermodynamics.* New York: Charles Scribner's & Sons.

Gibbs, J. W. 1948. Elementary principles in statistical mechanics developed with special reference to the rational foundation of thermodynamics. In *The Collected Works of J. Willard Gibbs.* New Haven, CT: Yale University Press.

Gillan, M. J. 1979. New method of solving the liquid structure integral-equations. *Molecular Physics.* 38, 1781.

Girardeau, M. D. and R. M. Mazo. 1973. Variational methods in statistical mechanics. In *Advances in Chemical Physics*, edited by I. Prigogine and S. Rice. Hoboken, NJ: John Wiley and Sons, Inc., p. 187.

Gmehling, J. 1995. From UNIFAC to modified UNIFAC to PSRK with the help of DDB. *Fluid Phase Equilibria*. 107, 1.

Gouy, M. 1910. Sur la constitution de la charge électrique à la surface d'un électrolyte. *Journal de Physique Théorique et Appliquée*. 9, 457.

Gray, C. G. and K. E. Gubbins. 1984. *Theory of Molecular Fluids*. Vol. 1: *Fundamentals*. New York: Oxford University Press.

Greene, R. F. and C. N. Pace. 1974. Urea and guanidine-hydrochloride denaturation of ribonuclease, lysozyme, α-chymotrypsin, and β-lactoglobulin. *Journal of Biological Chemistry*. 249, 5388.

Gross, J. and G. Sadowski. 2001. Perturbed-chain SAFT: An equation of state based on a perturbation theory for chain molecules. *Industrial and Engineering Chemistry Research*. 40, 1244.

Gross, J. and J. Vrabec. 2006. An equation-of-state contribution for polar components: Dipolar molecules. *American Institute of Chemical Engineers Journal*. 52, 1194.

Gruszkiewicz, M. S. and R. H. Wood. 1997. Conductance of dilute LiCl, NaCl, NaBr, and CsBr solutions in supercritical water using a flow conductance cell. *Journal of Physical Chemistry B*. 101, 6549.

Gubbins, K. E. and J. P. O'Connell. 1974. Isothermal compressibility and partial molal volume for polyatomic liquids. *Journal of Chemical Physics*. 60, 3449.

Guo, B., S. Kao, H. McDonald, A. Asanov, L. L. Combs, and W. W. Wilson. 1999. Correlation of second virial coefficients and solubilities useful in protein crystal growth. *Journal of Crystal Growth*. 196, 424.

Guo, J. H., Y. Luo, A. Augustsson, S. Kashtanov, J. E. Rubensson, D. K. Shuh, H. Agren, and J. Nordgren. 2003. Molecular structure of alcohol-water mixtures. *Physical Review Letters*. 91, 157401.

Hall, D. G. 1971. Kirkwood–Buff theory of solutions—Alternative derivation of part of it and some applications. *Transactions of the Faraday Society*. 67, 2516.

Hall, D. G. 1983. Thermodynamics of micellar solutions. In *Aggregation Processes in Solution*, edited by J. Gormally and E. Wyn-Jones. Amsterdam: Elsevier Scientific Publishing Co.

Hamer, W. J. and Y. C. Wu. 1972. Osmotic coefficients and mean activity coefficients of uni-univalent electrolytes in water at 25 degrees. *Journal of Physical Chemistry Reference Data*. 1, 1047.

Handa, Y. P. and G. C. Benson. 1979. Volume changes on mixing 2 liquids—Review of the experimental-techniques and the literature data. *Fluid Phase Equilibria*. 3, 185.

Hansen, J. P. and I. R. McDonald. 1986. *Theory of Simple Liquids*, 2nd. ed. New York: Academic Press.

Hansen, J. P. and I. R. McDonald. 2006. *Theory of Simple Liquids*, 3rd. ed. Amsterdam: Academic Press.

Henderson, D. 1974. Theory of simple mixtures. *Annual Review of Physical Chemistry*. 25, 461.

Henry, W. 1803. Experiments on the quantity of gases absorbed by water, at different temperatures, and under different pressures. *Philosophical Transactions of the Royal Society of London*. 93, 29.

Herczenik, E. and M. F. B. G. Gebbink. 2008. Molecular and cellular aspects of protein misfolding diseases. *The FASEB Journal*. 22, 2115.

Hess, B., C. Kutzner, D. van der Spoel, and E. Lindahl. 2008. GROMACS 4: Algorithms for highly efficient, load-balanced, and scalable molecular simulation. *Journal of Chemical Theory and Computation*. 4, 435.

Hess, B. and N. F. A. van der Vegt. 2009. Cation specific binding with protein surface charges. *Proceedings of the National Academy of Sciences of the United States of America.* 106, 13296.

Hildebrand, J. H. 1924. *Solubility.* New York: Chemical Catalog Company.

Hill, T. L. 1956. *Statistical Mechanics: Principles and Selected Applications.* New York: McGraw-Hill.

Hill, T. L. 1957. Theory of solutions. I. *Journal of the American Chemical Society.* 79, 4885.

Hill, T. L. 1959. Theory of solutions. II. Osmotic pressure virial expansion and light scattering in two component solutions. *Journal of Chemical Physics.* 30, 93.

Hnědkovský, L., V. Majer, and R. H. Wood. 1995. Volumes and heat-capacities of $H_3BO_3(aq)$ at temperatures from 298.15 K to 705 K and at pressures to 35 MPa. *Journal of Chemical Thermodynamics.* 27, 801.

Hnědkovský, L., R. H. Wood, and V. Majer. 1996. Volumes of aqueous solutions of CH_4, CO_2, H_2S, and NH_3 at temperatures from 298.15 K to 705 K and pressures to 35 MPa. *Journal of Chemical Thermodynamics.* 28, 125.

Horinek, D. and R. R. Netz. 2011. Can simulations quantitatively predict peptide transfer free energies to urea solutions? Thermodynamic concepts and force field limitations. *Journal of Physical Chemistry A.* 115, 6125.

Hornak, V., R. Abel, A. Okur, B. Strockbine, A. Roitberg, and C. Simmerling. 2006. Comparison of multiple Amber force fields and development of improved protein backbone parameters. *Proteins: Structure, Function, and Bioinformatics.* 65, 712.

Horta, B. A. C., P. F. J. Fuchs, W. F. van Gunsteren, and P. H. Hünenberger. 2011. New interaction parameters for oxygen compounds in the GROMOS force field: Improved pure-liquid and solvation properties for alcohols, ethers, aldehydes, ketones, carboxylic acids, and esters. *Journal of Chemical Theory and Computation.* 7, 1016.

Hu, C. Y., H. Kokubo, G. C. Lynch, D. W. Bolen, and B. M. Pettitt. 2010a. Backbone additivity in the transfer model of protein solvation. *Protein Science.* 19, 1011.

Hu, C. Y., G. C. Lynch, H. Kokubo, and B. M. Pettitt. 2010b. Trimethylamine N-oxide influence on the backbone of proteins: An oligoglycine model. *Proteins: Structure, Function, and Bioinformatics.* 78, 695.

Hu, C. Y., B. M. Pettitt, and J. Rösgen. 2009. Osmolyte solutions and protein folding. *F1000 Biology Reports.* 1, 1.

Hu, Y. 1997. Fluctuation Solution Theory for Strong Electrolyte Solutions. Ph.D. Thesis, University of Virginia.

Hua, L., R. Zhou, D. Thirumalai, and B. J. Berne. 2008. Urea denaturation by stronger dispersion interactions with proteins than water implies a 2-stage unfolding. *Proceedings of the National Academy of Sciences of the United States of America.* 105, 16928.

Huang, Y. H. and J. P. O'Connell. 1987. Corresponding states correlation for the volumetric properties of compressed liquids and liquid-mixtures. *Fluid Phase Equilibria.* 37, 75.

Hutchenson, K. W., A. M. Scurto, and B. Subramaniam, eds. 2009. *Gas-Expanded Liquids and Near-Critical Media: Green Chemistry and Engineering.* Vol. 1006, *ACS Symposium Series*, Washington, DC.

Hynes, J. T. 1985. The theory of reactions in solution. In *Theory of Chemical Reaction Dynamics*, edited by M. Baer. Boca Raton: CRC Press, p. 171.

Jacquemin, J., P. Husson, V. Mayer, and I. Cibulka. 2007. High-pressure volumetric properties of imidazolium-based ionic liquids: Effect of the anion. *Journal of Chemical and Engineering Data.* 52, 2204.

Jagla, E. A. 1999. Core-softened potentials and the anomalous properties of water. *Journal of Chemical Physics.* 111, 8980.

Jakob, A., H. Grensemann, J. Lohmann, and J. Gmehling. 2006. Further development of modified UNIFAC (Dortmund): Revision and extension 5. *Industrial and Engineering Chemistry Research.* 45, 7924.

Janz, G. J., B. G. Oliver, G. R. Lakshmin, and G. E. Mayer. 1970. Electrical conductance, diffusion, viscosity, and density of sodium nitrate, sodium perchlorate, and sodium thiocyanate in concentrated aqueous solutions. *Journal of Physical Chemistry.* 74, 1285.

Jha, A. K., A. Colubri, M. H. Zaman, S. Koide, T. R. Sosnick, and K. F. Freed. 2005. Helix, sheet, and polyproline II frequencies and strong nearest neighbor effects in a restricted coil library. *Biochemistry.* 44, 9691.

Jiang, W., M. Hodoscek, and B. Roux. 2009. Computation of absolute hydration and binding free energy with free energy perturbation distributed replica-exchange molecular dynamics (FEP/REMD). *Journal of Chemical Theory and Computation.* 5, 2583.

Jiao, Y., F. Chen, S. Weerasinghe, and P. E. Smith. 2012. A Kirkwood–Buff derived force field for alcohols in water. (In preparation, consulted October 2012.)

Jiao, Y. and P. E. Smith. 2011. Fluctuation theory of molecular association and conformational equilibria. *Journal of Chemical Physics.* 135, 014502.

Jolly, D. L. and R. J. Bearman. 1980. Molecular-dynamics simulation of the mutual and self-diffusion coefficients in Lennard–Jones liquid-mixtures. *Molecular Physics.* 41, 137.

Jolly, D. L., B. C. Freasier, and R. J. Bearman. 1976. Extension of simulation radial-distribution functions to an arbitrary range by Baxter's factorization technique. *Chemical Physics.* 15, 237.

Jonah, D. A. 1983. Investigation of binary-liquid mixtures via the study of infinitely dilute-solutions. *Advances in Chemistry Series.* 395.

Jonah, D. A. 1986. On a method of numerical differentiation and its application to the evaluation of limiting activity-coefficients and their 1st derivatives. *Chemical Engineering Science.* 41, 2261.

Jonah, D. A. and H. D. Cochran. 1994. Chemical-potentials in dilute, multicomponent solutions. *Fluid Phase Equilibria.* 92, 107.

Jorgensen, W. L., J. Chandrasekhar, J. D. Madura, R. W. Impey, and M. L. Klein. 1983. Comparison of simple potential functions for simulating liquid water. *Journal of Chemical Physics.* 79, 926.

Jorgensen, W. L. and J. Tirado-Rives. 1988. The OPLS potential functions for proteins—Energy minimizations for crystals of cyclic-peptides and crambin. *Journal of the American Chemical Society.* 110, 1657.

Jorgensen, W. L. and J. Tirado-Rives. 2005. Potential energy functions for atomic-level simulations of water and organic and biomolecular systems. *Proceedings of the National Academy of Sciences of the United States of America.* 102, 6665.

Joung, I. S. and T. E. Cheatham, III. 2008. Determination of alkali and halide monovalent ion parameters for use in explicitly solvated biomolecular simulations. *Journal of Physical Chemistry B.* 112, 9020.

Joung, I. S. and T. E. Cheatham, III. 2009. Molecular dynamics simulations of the dynamic and energetic properties of alkali and halide ions using water-model-specific ion parameters. *Journal of Physical Chemistry B.* 113, 13279.

Jouyban-Gharamaleki, A., L. Valaee, M. Barzegar-Jalali, B. J. Clark, and W. E. Acree. 1999. Comparison of various cosolvency models for calculating solute solubility in water-cosolvent mixtures. *International Journal of Pharmaceutics.* 177, 93.

Jouyban, A., S. Romero, H. K. Chan, B. J. Clark, and P. Bustamante. 2002. A cosolvency model to predict solubility of drugs at several temperatures from a limited number of solubility measurements. *Chemical and Pharmaceutical Bulletin.* 50, 594.

Juraszek, J. and P. G. Bolhuis. 2008. Rate constant and reaction coordinate of Trp-cage folding in explicit water. *Biophysical Journal.* 95, 4246.

Kaminski, G. A., R. A. Friesner, J. Tirado-Rives, and W. L. Jorgensen. 2001. Evaluation and reparametrization of the OPLS-AA force field for proteins via comparison with accurate quantum chemical calculations on peptides. *Journal of Physical Chemistry B.* 105, 6474.

Kan, A. T. and M. B. Tomson. 1996. UNIFAC prediction of aqueous and nonaqueous solubilities of chemicals with environmental interest. *Environmental Science and Technology.* 30, 1369.

Kang, M. and P. E. Smith. 2006. A Kirkwood–Buff derived force field for amides. *Journal of Computational Chemistry.* 27, 1477.

Kang, M. and P. E. Smith. 2007. Preferential interaction parameters in biological systems by Kirkwood–Buff theory and computer simulation. *Fluid Phase Equilibria.* 256, 14.

Kang, M. and P. E. Smith. 2008. Kirkwood–Buff theory of four and higher component mixtures. *Journal of Chemical Physics.* 128, 244511.

Karunaweera, S., M. B. Gee, S. Weerasinghe, and P. E. Smith. 2012. Theory and simulation of multicomponent osmotic systems. *Journal of Chemical Theory and Computation.* 8, 3493.

Kauzmann, W. 1959. Some factors in the interpretation of protein denaturation. In *Advances in Protein Chemistry*, edited by C. B. Anfinsen, M. L. Anson, K. Bailey and J. T. Edsall. San Diego: Academic Press Inc., p. 1.

Kežić, B. and A. Perera. 2011. Towards a more accurate reference interaction site model integral equation theory for molecular liquids. *Journal of Chemical Physics.* 135, 234104.

Kim, J. I. 1978. A critical study of the Ph_4AsPh_4B assumption for single ion thermodynamics in amphiprotic and dipolar-aprotic solvents; Evaluation of physical parameters relevant to theoretical consideration. *Zeitschrift für Physikalische Chemie.* 113, 129.

Kimura, Y. and Y. Yoshimura. 1992. Chemical-equilibrium in fluids from the gaseous to liquid states: Solvent density dependence of the dimerization equilibrium of 2-methyl-2-nitrosopropane in carbon dioxide, chlorotrifluoromethane, and trifluoromethane. *Journal of Chemical Physics.* 96, 3085.

Kiran, E., P. G. Debenedetti, and C. J. Peters, eds. 2000. *Supercritical Fluids. Fundamentals and Applications.* Vol. E366, *NATO ASI.* Dordrecht: Klüwer Academic Publishers.

Kiran, E. and J. M. H. Levelt Sengers. 1994. *Supercritical Fluids. Fundamentals for Applications.* Vol. E273, *NATO ASI.* Dordrecht: Klüwer Academic Publishers.

Kirkwood, J. G. 1934. Theory of solutions of molecules containing widely separated charges with special application to zwitterions. *Journal of Chemical Physics.* 2, 351.

Kirkwood, J. G. 1935. Statistical mechanics of fluid mixtures. *Journal of Chemical Physics.* 3, 300.

Kirkwood, J. G. 1936. Statistical mechanics of liquid solutions. *Chemical Reviews.* 19, 275.

Kirkwood, J. G. 1947, privately circulated. *Selected topics in statistical mechanics based on a series of lectures.* Lectures delivered at Princeton University, Spring term of 1947, Princeton, NJ. Notes by J. H. Irving.

Kirkwood, J. G. and F. P. Buff. 1951. The statistical mechanical theory of solutions. 1. *Journal of Chemical Physics.* 19, 774.

Kirkwood, J. G. and R. J. Goldberg. 1950. Light scattering arising from composition fluctuations in multi-component systems. *Journal of Chemical Physics.* 18, 54.

Klein-Seetharaman, J., M. Oikawa, S. B. Grimshaw, J. Wirmer, E. Duchardt, T. Ueda, T. Imoto, L. J. Smith, C. M. Dobson, and H. Schwalbe. 2002. Long-range interactions within a nonnative protein. *Science.* 295, 1719.

Knez, E. 2009. Enzymatic reactions in dense gases. *Journal of Supercritical Fluids.* 47, 357.

Koga, Y. 2007. *Solution Thermodynamics and its Application to Aqueous Solutions.* Amsterdam: Elsevier.

Kojima, K. and K. Tochigi. 1979. *Prediction of Vapor-Liquid Equilibria by the ASOG Method.* Tokyo: Elsevier Scientific Publishing Company.

Kojima, K., S. J. Zhang, and T. Hiaki. 1997. Measuring methods of infinite dilution activity coefficients and a database for systems including water. *Fluid Phase Equilibria.* 131, 145.

Kokubo, H., C. Y. Hu, and B. M. Pettitt. 2011. Peptide conformational preferences in osmolyte solutions: Transfer free energies of decaalanine. *Journal of the American Chemical Society.* 133, 1849.

Kokubo, H. and B. Pettitt. 2007. Preferential solvation in urea solutions at different concentrations: Properties from simulation studies. *Journal of Physical Chemistry B.* 111, 5233.

Kokubo, H., J. Rösgen, D. W. Bolen, and B. M. Pettitt. 2007. Molecular basis of the apparent near ideality of urea solutions. *Biophysical Journal.* 93, 3392.

Kontogeorgis, G. M., M. L. Michelsen, G. K. Folas, S. Derawi, N. von Solms, and E. H. Stenby. 2006a. Ten years with the applications of cubic-plus-association EOS to pure organic compounds and self-associating system (cubic-plus-association) equation of state. Part 1. Pure compounds and self-associating systems. *Industrial and Engineering Chemistry Research.* 45, 4855.

Kontogeorgis, G. M., M. L. Michelsen, G. K. Folas, S. Derawi, N. von Solms, and E. H. Stenby. 2006b. Ten years with the CPA (cubic-plus-association) equation of state. Part 2. Cross-associating and multicomponent systems. *Industrial and Engineering Chemistry Research.* 45, 4869.

Kontogeorgis, G. M., E. C. Voutsas, I. V. Yakoumis, and D. P. Tassios. 1996. An equation of state for associating fluids. *Industrial and Engineering Chemistry Research.* 35, 4310.

Kornblatt, J. A. and M. J. Kornblatt. 2002. The effects of osmotic and hydrostatic pressures on macromolecular systems. *Biochimica Et Biophysica Acta-Protein Structure and Molecular Enzymology.* 1595, 30.

Krichevskii, I. R. 1937. Thermodynamics of an infinitely dilute solution in mixed solvents. I. The Henry coefficient in a mixed solvent behaving as an ideal solvent. *Zhurnal Fizicheskoi Khimii.* 9, 41.

Kruse, A. and E. Dinjus. 2007. Hot compressed water as reaction medium and reactant—Properties and synthesis reactions. *Journal of Supercritical Fluids.* 39, 362.

Kulkarni, A. M. and C. F. Zukoski. 2002. Nanoparticle crystal nucleation: Influence of solution conditions. *Langmuir.* 18, 3090.

Kumar, S. K. and K. P. Johnston. 1988. Modelling the solubility of solids in supercritical fluids with density as the independent variable. *Journal of Supercritical Fluids.* 1, 15.

Kumelan, J., A. P. S. Kamps, D. Tuma, and G. Maurer. 2006. Solubility of H_2 in the ionic liquid [bmim][PF_6]. *Journal of Chemical and Engineering Data.* 51, 11.

Kuntz Jr., I. D. and W. Kauzmann. 1974. Hydration of proteins and polypeptides. *Advances in Protein Chemistry.* 28, 239.

Kurnik, R. T. and R. C. Reid. 1982. Solubility of solid mixtures in supercritical fluids. *Fluid Phase Equilibria.* 8, 93.

Kusalik, P. G. and G. N. Patey. 1987. The thermodynamic properties of electrolyte-solutions—Some formal results. *Journal of Chemical Physics.* 86, 5110.

Kusalik, P. G. and G. N. Patey. 1988. On the molecular theory of aqueous-electrolyte solutions. 1. The solution of the RHNC approximation for models at finite concentration. *Journal of Chemical Physics.* 88, 7715.

Kwiatkowski, J., Z. Lisicki, and W. Majewski. 1984. An experimental-method for measuring solubilities of solids in supercritical fluids. *Berichte der Bunsen-Gesellschaft-Physical Chemistry Chemical Physics.* 88, 865.

Lado, F. 1964. Perturbation correction to radial distribution function. *Physical Review A-General Physics.* 135, 1013.

Lazaridis, T. 2002. Binding affinity and specificity from computational studies. *Current Organic Chemistry.* 6, 1319.

Lazaridis, T., A. Masunov, and F. Gandolfo. 2002. Contributions to the binding free energy of ligands to avidin and streptavidin. *Proteins: Structure, Function, and Genetics.* 47, 194.

Lebowitz, J. L. and J. K. Percus. 1961. Long-range correlations in a closed system with applications to nonuniform fluids. *Physical Review.* 122, 1675.

Lebowitz, J. L. and J. K. Percus. 1963. Asymptotic behavior of radial distribution function. *Journal of Mathematical Physics.* 4, 248.

Lee, J. C. and S. N. Timasheff. 1981. The stabilization of proteins by sucrose. *Journal of Biological Chemistry.* 256, 7193.

Lee, M. E. and N. F. A. van der Vegt. 2005. A new force field for atomistic simulations of aqueous tertiary butanol solutions. *Journal of Chemical Physics.* 122, 114509.

Lepori, L. and E. Matteoli. 1997. Excess Gibbs energies of the ternary system ethanol + tetrahydrofuran + cyclohexane at 298.15 K. *Fluid Phase Equilibria.* 134, 113.

Lepori, L. and E. Matteoli. 1998. Excess volumes of the ternary system ethanol + tetrahydrofuran + cyclohexane at 298.15 K. *Fluid Phase Equilibria.* 145, 69.

Lepori, L., E. Matteoli, G. Conti, and P. Gianni. 1998. Excess Gibbs energies of the ternary system ethanol + N,N-dimethylformamide + cyclohexane at 298.15 K. *Fluid Phase Equilibria.* 153, 293.

Lepree, J. M., M. J. Mulski, and K. A. Connors. 1994. Solvent effects on chemical processes. 6. The phenomenological model applied to the solubility of naphthalene and 4-nitroaniline in binary aqueous-organic solvent mixtures. *Journal of the Chemical Society-Perkin Transactions.* 2, 1491.

Levelt Sengers, J. M. H. 1991. Solubility near the solvent's critical point. *Journal of Supercritical Fluids.* 4, 215.

Lewis, G. N. 1900a. A new conception of thermal pressure and a theory of solutions. *Proceedings of the American Academy of Arts and Sciences.* 36, 145.

Lewis, G. N. 1900b. A new view of the thermal pressure and theory of solutions. *Zeitschrift für Physikalische Chemie.* 35, 343.

Lewis, G. N. 1901. The law of physico-chemical change. *Proceedings of the American Academy of Arts and Sciences.* 37, 49.

Li, A., W. J. Doucette, and A. W. Andren. 1994. Estimation of aqueous solubility, octanol/water partition-coefficient, and Henry's law constant for polychlorinated-biphenyls using UNIFAC. *Chemosphere.* 29, 657.

Li, T. W., E. H. Chimowitz, and F. Munoz. 1995. First-order corrections to infinite dilution fugacity coefficients using computer-simulation. *American Institute of Chemical Engineers Journal.* 41, 2300.

Licence, P. and M. Poliakoff. 2008. An introduction to supercritical fluids: From bench scale to commercial plant. In *New Methodologies and Techniques for a Sustainable Organic Chemistry*, edited by A. Mordini and F. Faigl. Dordrecht: Springer, p. 171.

Lim, W. K., J. Rösgen, and S. W. Englander. 2009. Urea, but not guanidinium, destabilizes proteins by forming hydrogen bonds to the peptide group. *Proceedings of the National Academy of Sciences of the United States of America.* 106, 2595.

Lin, T. Y. and S. N. Timasheff. 1994. Why do some organisms use a urea-methylamine mixture as osmolyte? Thermodynamic compensation of urea and trimethylamine n-oxide interactions with protein. *Biochemistry.* 33, 12695.

Lin, C. L. and R. H. Wood. 1996. Prediction of the free energy of dilute aqueous methane, ethane, and propane at temperatures from 600 to 1200 degrees C and densities from 0 to 1 g cm^{-3} using molecular dynamics simulations. *Journal of Physical Chemistry.* 100, 16399.

Liu, H. and J. P. O'Connell. 1998. On the measurement of solute partial molar volumes in near-critical fluids with supercritical fluid chromatography. *Industrial and Engineering Chemistry Research.* 37, 3323.

Liu, Y. and D. Bolen. 1995. The peptide backbone plays a dominant role in protein stabilization by naturally occurring osmolytes. *Biochemistry.* 34, 12884.

Longuet-Higgins, H. C. 1951. The statistical thermodynamics of multicomponent systems. *Proceedings of the Royal Society of London. Series A. Mathematical and Physical Sciences.* 205, 247.

Lupis, C. H. P. 1983. *Chemical Thermodynamics of Materials*. New York: North-Holland.

Mackerell, A. D. 2004. Empirical force fields for biological macromolecules: Overview and issues. *Journal of Computational Chemistry*. 25, 1584.

Mackerell, A. D., D. Bashford, M. Bellott, R. L. Dunbrack, J. D. Evanseck, M. J. Field, S. Fischer, J. Gao, H. Guo, S. Ha, D. J. McCarthy, L. Kuchnir, K. Kuczera, F. T. K. Lau, C. Mattos, S. Michnick, T. Ngo, D. T. Nguyen, B. Prodhom, W. E. Reiher, B. Roux, M. Schlenkrich, J. C. Smith, R. Stote, J. Straub, M. Watanabe, J. Wiorkiewicz-Kuczera, D. Yin, and M. Karplus. 1998. All-atom empirical potential for molecular modeling and dynamics studies of proteins. *Journal of Physical Chemistry B*. 102, 3586.

Mackerell, A. D., M. Feig, and C. L. Brooks. 2004. Extending the treatment of backbone energetics in protein force fields: Limitations of gas-phase quantum mechanics in reproducing protein conformational distributions in molecular dynamics simulations. *Journal of Computational Chemistry*. 25, 1400.

Maerzke, K. A., N. E. Schultz, R. B. Ross, and J. I. Siepmann. 2009. TraPPE-UA force field for acrylates and Monte Carlo simulations for their mixtures with alkanes and alcohols. *Journal of Physical Chemistry B*. 113, 6415.

Majer, V., L. Hui, R. Crovetto, and R. H. Wood. 1991. Volumetric properties of aqueous 1-1 electrolyte-solutions near and above the critical-temperature of water. 1. Densities and apparent molar volumes of NaCl(aq) from 0.0025 mol·kg^{-1} to 3.1 mol·kg^{-1}, 604.4-K to 725.5-K, and 18.5-MPa to 38.0-MPa. *Journal of Chemical Thermodynamics*. 23, 213.

Majer, V., J. Sedlbauer, and G. Bergin. 2008. Henry's law constant and related coefficients for aqueous hydrocarbons, CO_2 and H_2S over a wide range of temperature and pressure. *Fluid Phase Equilibria*. 272, 65.

Mansoori, G. A. and J. F. Ely. 1985. Statistical mechanical theory of local compositions. *Fluid Phase Equilibria*. 22, 253.

Marcus, Y. 1983. A quasi-lattice, quasi-chemical theory of preferential solvation of ions in mixed solvents. *Australian Journal of Chemistry*. 36, 1719.

Marcus, Y. 1989. Preferential solvation. Part 3. Binary solvent mixtures. *Journal of the Chemical Society, Faraday Transactions 1: Physical Chemistry in Condensed Phases*. 85, 381.

Marcus, Y. 1990. Preferential solvation in mixed solvents. Part 5. Binary mixtures of water and organic solvents. *Journal of the Chemical Society, Faraday Transactions*. 86, 2215.

Marcus, Y. 1991. Preferential solvation in mixed solvents. Part 6. Binary mixtures containing methanol, ethanol, acetone or triethylamine and another organic solvent. *Journal of the Chemical Society, Faraday Transactions*. 87, 1843.

Marcus, Y. 1995. Preferential solvation in mixed solvents. Part 7. Binary mixtures of water and alkanolamines. *Journal of the Chemical Society, Faraday Transactions*. 91, 427.

Marcus, Y. 1999. Preferential solvation in mixed solvents. Part 8. Aqueous methanol from sub-ambient to elevated temperatures. *Physical Chemistry Chemical Physics*. 1, 2975.

Marcus, Y. 2000. Some thermodynamic and structural aspects of mixtures of glycerol with water. *Physical Chemistry Chemical Physics*. 2, 4891.

Marcus, Y. 2001. Preferential solvation in mixed solvents. Part 10. Completely miscible aqueous co-solvent binary mixtures at 298.15 K. *Monatshefte für Chemie*. 132, 1387.

Marcus, Y. 2002a. Preferential solvation in mixed solvents. Part 11. Eight additional completely miscible aqueous co-solvent binary mixtures and the relationship between the volume-corrected preferential solvation parameters and the structures of the co-solvents. *Physical Chemistry Chemical Physics*. 4, 4462.

Marcus, Y. 2002b. *Solvent Mixtures: Properties and Selective Solvation*. New York: Marcel Dekker, Inc.

Marcus, Y. 2003. Preferential solvation in mixed solvents. Part 12. Aqueous glycols. *Journal of Molecular Liquids*. 107, 109.

Marcus, Y. 2006a. Preferential solvation in mixed solvents. Part 13. Mixtures of tetrahydro-furan with organic solvents: Kirkwood–Buff. *Journal of Solution Chemistry.* 35, 251.

Marcus, Y. 2006b. Preferential solvation in mixed solvents. Part 14. Mixtures of 1,4-dioxane with organic solvents: Kirkwood–Buff integrals and volume-corrected preferential solvation parameters. *Journal of Molecular Liquids.* 128, 115.

Marcus, Y. 2008. Clustering in liquid mixtures of water and acetonitrile. In *Solution Chemistry Research Progress,* edited by D. V. Bostrelli. New York: Nova Science Publishers, Inc., p. 1.

Marcus, Y. 2011. Water structure enhancement in water-rich binary solvent mixtures. *Journal of Molecular Liquids.* 158, 23.

Marcus, Y. 2012. Water structure enhancement in water-rich binary solvent mixtures. Part 2. The excess partial molar heat capacity of the water. *Journal of Molecular Liquids.* 166, 62.

Marcus, Y. and Y. Migron. 1991. Polarity, hydrogen bonding, and structure of mixtures of water and cyanomethane. *Journal of Physical Chemistry.* 95, 400.

Marlow, G. E. and B. M. Pettitt. 1994. Studies of salt-peptide solutions: Theoretical and experimental approaches. In *Advances in Computational Biology,* edited by H. O. Villar. South San Francisco: Elsevier Science, p. 231.

Martin, W., J. Baross, D. Kelley, and M. J. Russell. 2008. Hydrothermal vents and the origin of life. *Nature Reviews Microbiology.* 6, 805.

Mathias, P. M. and J. P. O'Connell. 1979. A predictive method for *PVT* and phase behavior of liquids containing supercritical components. In *Equations of State in Engineering and Research.* Washington, DC: American Chemical Society.

Mathias, P. M. and J. P. O'Connell. 1981. Molecular thermodynamics of liquids containing supercritical compounds. *Chemical Engineering Science.* 36, 1123.

Matteoli, E. 1997. A study on Kirkwood–Buff integrals and preferential solvation in mixtures with small deviations from ideality and/or with size mismatch of components. Importance of a proper reference system. *Journal of Physical Chemistry B.* 101, 9800.

Matteoli, E., P. Gianni, and L. Lepori. 2010. Excess Gibbs energies of the ternary system 2-methoxyethanol + tetrahydrofuran + cyclohexane and other relevant binaries at 298.15 K. *Journal of Chemical and Engineering Data.* 55, 5441.

Matteoli, E. and L. Lepori. 1984. Solute–solute interactions in water. 2. An analysis through the Kirkwood–Buff integrals for 14 organic solutes. *Journal of Chemical Physics.* 80, 2856.

Matteoli, E. and L. Lepori. 1990. The ternary system water + 1-propanol + urea at 298.15 K. Activity coefficients, partial molal volumes and Kirkwood–Buff integrals. *Journal of Molecular Liquids.* 47, 89.

Matteoli, E. and L. Lepori. 1995. Kirkwood–Buff integrals and preferential solvation in ternary non-electrolyte mixtures. *Journal of the Chemical Society, Faraday Transactions.* 91, 431; Corrigendum: 91, 1885.

Matteoli, E. and L. Lepori. 2007. Comment on "A critique of some recent suggestions to correct the Kirkwood–Buff integrals." *Journal of Physical Chemistry B.* 111, 3069.

Matteoli, E. and L. Lepori. 2008. Reply to "Comment on 'The Kirkwood–Buff theory of solutions and the local composition of liquid mixtures.'" *Journal of Physical Chemistry B.* 112, 5878.

Matteoli, E. and G. A. Mansoori, eds. 1990. *Fluctuation Theory of Mixtures.* Edited by G. A. Mansoori, *Advances in Thermodynamics,* Vol. 2. New York: Taylor & Francis.

Matteoli, E. and G. A. Mansoori. 1995. A simple expression for radial-distribution functions of pure fluids and mixtures. *Journal of Chemical Physics.* 103, 4672.

Mazo, R. M. 1958. Statistical mechanical theory of solutions. *Journal of Chemical Physics.* 29, 1122.

Mazo, R. M. 2006. A fluctuation theory analysis of the salting-out effect. *Journal of Physical Chemistry B.* 110, 24077.

Mazo, R. M. 2007. Salting out near the critical point. *Journal of Physical Chemistry B*. 111, 7288.

McGuigan, D. B. and P. A. Monson. 1990. Analysis of infinite dilution partial molar volumes using distribution function theory. *Fluid Phase Equilibria*. 57, 227.

McMillan, W. G. and J. E. Mayer. 1945. The statistical thermodynamics of multicomponent systems. *Journal of Chemical Physics*. 13, 276.

McNaught, A. D. and A. Wilkinson. 1997. *IUPAC Compendium of Chemical Terminology: The Gold Book*, 2nd ed. Oxford: Blackwell Scientific Publications.

Mecke, M., A. Muller, J. Winkelmann, J. Vrabec, J. Fischer, R. Span, and W. Wagner. 1996. An accurate van der Waals-type equation of state for the Lennard–Jones fluid. *International Journal of Thermophysics*. 17, 391.

Miyawaki, O. 2009. Thermodynamic analysis of protein unfolding in aqueous solutions as a multisite reaction of protein with water and solute molecules. *Biophysical Chemistry*. 144, 46.

Modell, M. and R. C. Reid. 1983. *Thermodynamics and its Applications*, 2nd ed. Englewood Cliffs, NJ: Prentice Hall.

Morris, J. W., P. J. Mulvey, M. M. Abbott, and H. C. van Ness. 1975. Excess thermodynamic functions for ternary systems. I. Acetone-chloroform-methanol at 50 degrees. *Journal of Chemical and Engineering Data*. 20, 403.

Morris, K. R. 1988. Solubility of Aromatic Compounds in Mixed Solvents. Ph.D. Thesis, Pharmaceutical Sciences Graduate College, The University of Arizona.

Mota, F. L., A. J. Queimada, A. E. Andreatta, S. P. Pinho, and E. A. Macedo. 2012. Calculation of drug-like molecules solubility using predictive activity coefficient models. *Fluid Phase Equilibria*. 322, 48.

Mu, Y. G., D. S. Kosov, and G. Stock. 2003. Conformational dynamics of trialanine in water. 2. Comparison of Amber, CHARMM, GROMOS, and OPLS force fields to NMR and infrared experiments. *Journal of Physical Chemistry B*. 107, 5064.

Mukherji, D., N. F. A. van der Vegt, K. Kremer, and L. Delle Site. 2012. Kirkwood–Buff analysis of liquid mixtures in an open boundary simulation. *Journal of Chemical Theory and Computation*. 8, 375.

Müller, E. A. and K. E. Gubbins. 2001. Molecular-based equations of state for associating fluids: A review of SAFT and related approaches. *Industrial and Engineering Chemistry Research*. 40, 2193.

Munoz, F. and E. H. Chimowitz. 1993. Critical phenomena in mixtures. 2. Synergistic and other effects near mixture critical-points. *Journal of Chemical Physics*. 99, 5450.

Munoz, F., T. W. Li, and E. H. Chimowitz. 1995. Henry's law and synergism in dilute near-critical solutions—Theory and simulation. *American Institute of Chemical Engineers Journal*. 41, 389.

Münster, A. 1969. *Statistical Thermodynamics*. Berlin: Springer-Verlag.

Münster, A. 1970. *Classical Thermodynamics*. London: Wiley-Interscience.

Myers, D. B. and R. L. Scott. 1963. Thermodynamic functions for nonelectrolyte solutions. *Industrial and Engineering Chemistry*. 55, 43.

Myers, J. K., C. N. Pace, and J. M. Scholtz. 1995. Denaturant *m*-values and heat-capacity changes—Relation to changes in accessible surface-areas of protein unfolding. *Protein Science*. 4, 2138.

Nagata, I. and K. Miyazaki. 1992. Excess enthalpies of (aniline + propan-2-ol) and of (aniline + propan-2-ol + benzene) at the temperature 298.15 K. *Journal of Chemical Thermodynamics*. 24, 1175.

Newman, K. E. 1994. Kirkwood–Buff solution theory—Derivation and applications. *Chemical Society Reviews*. 23, 31.

Nichols, J. W., S. G. Moore, and D. R. Wheeler. 2009. Improved implementation of Kirkwood–Buff solution theory in periodic molecular simulations. *Physical Review E*. 80, 051203.

Nikolova, P. V., S. J. B. Duff, P. Westh, et al. 2000. A thermodynamic study of aqueous aceto-nitrile: Excess chemical potentials, partial molar enthalpies, entropies and volumes, and fluctuations. *Canadian Journal of Chemistry*. 78, 1553.

Novák, J. P., J. Matouš, and J. Pick. 1987. *Liquid-Liquid Equilibria*. Amsterdam: Elsevier.

Nozaki, Y. and C. Tanford. 1963. The solubility of amino acids and related compounds in aqueous urea solutions. *Journal of Biological Chemistry*. 238, 4074.

Nymeyer, H. 2009. Energy landscape of the Trpzip2 peptide. *Journal of Physical Chemistry B*. 113, 8288.

O'Brien, E. P., G. Ziv, G. Haran, B. R. Brooks, and D. Thirumalai. 2008. Effects of dena-turants and osmolytes on proteins are accurately predicted by the molecular trans-fer model. *Proceedings of the National Academy of Sciences of the United States of America*. 105, 13403.

O'Connell, J. P. 1971a. Molecular thermodynamics of gases in mixed solvents. *American Institute of Chemical Engineers Journal*. 17, 658.

O'Connell, J. P. 1971b. Thermodynamic properties of solutions based on correlation func-tions. *Molecular Physics*. 20, 27.

O'Connell, J. P. 1981. Thermodynamic properties of solutions and the theory of fluctuations. *Fluid Phase Equilibria*. 6, 21.

O'Connell, J. P. 1990. Thermodynamic properties of mixtures from fluctuation solution the-ory. In *Fluctuation Theory of Mixtures*, edited by E. Matteoli and G. A. Mansoori. New York: Taylor & Francis, p. 45.

O'Connell, J. P. 1993. Application of fluctuation solution theory to strong electrolyte-solutions. *Fluid Phase Equilibria*. 83, 233.

O'Connell, J. P. 1994. Thermodynamics and fluctuation solution theory with some applications to systems at near or supercritical conditions. In Part II: Fluctuations, crossover effects, transport properties. In *Supercritical Fluids: Fundamentals for Application*, edited by E. Kiran and J. M. H. Levelt Sengers. Dordrecht, The Netherlands: Klüwer Academic Publishers, p. 191.

O'Connell, J. P. 1995. Application of fluctuation solution theory to thermodynamic properties of solutions. *Fluid Phase Equilibria*. 104, 21.

O'Connell, J. P. and R. Clairmont. 2010. Evaluation of equation of state combining and mix-ing rules with properties from fluctuation solution theory. Poster presented at the 12th International Conference on Properties and Phase Equilibria for Product and Process Design. Suzhou, Jiangsu, China.

O'Connell, J. P. and J. M. Haile. 2005. *Thermodynamics: Fundamentals for Applications*. New York: Cambridge University Press.

O'Connell, J. P., Y. Q. Hu, and K. A. Marshall. 1999. Aqueous strong electrolyte solution activ-ity coefficients and densities from fluctuation solution theory. *Fluid Phase Equilibria*. 158, 583.

O'Connell, J. P. and H. Q. Liu. 1998. Thermodynamic modelling of near-critical solutions. *Fluid Phase Equilibria*. 144, 1.

O'Connell, J. P. and J. M. Prausnitz. 1964. Thermodynamics of gas solubility in mixed sol-vents. *Industrial and Engineering Chemistry Fundamentals*. 3, 347.

O'Connell, J. P., A. V. Sharygin, and R. H. Wood. 1996. Infinite dilution partial molar vol-umes of aqueous solutes over wide ranges of conditions. *Industrial and Engineering Chemistry Research*. 35, 2808.

Okur, A., B. Strockbine, V. Hornak, and C. Simmerling. 2003. Using PC clusters to evalu-ate the transferability of molecular mechanics force fields for proteins. *Journal of Computational Chemistry*. 24, 21.

Oostenbrink, C., A. Villa, A. E. Mark, and W. F. van Gunsteren. 2004. A biomolecular force field based on the free enthalpy of hydration and solvation: The GROMOS force-field parameter sets 53a5 and 53a6. *Journal of Computational Chemistry*. 25, 1656.

Pace, C. N. 1986. Determination and analysis of urea and guanidine hydrochloride denaturation curves. In *Methods in Enzymology. Enzyme Structure Part L*, edited by C. H. W. Hirs and S. N. Timasheff. Orlando, FL: Academic Press, p. 266.

Pace, C. N., S. Trevino, E. Prabhakaran, and J. M. Scholtz. 2004. Protein structure, stability and solubility in water and other solvents. *Philosophical Transactions of the Royal Society of London Series B-Biological Sciences.* 359, 1225.

Paduszynski, K. and U. Domanska. 2012. A new group contribution method for prediction of density of pure ionic liquids over a wide range of temperature and pressure. *Industrial and Engineering Chemistry Research.* 51, 591.

Pappenberger, G., C. Saudan, M. Becker, A. E. Merbach, and T. Kiefhaber. 2000. Denaturant-induced movement of the transition state of protein folding revealed by high-pressure stopped-flow measurements. *Proceedings of the National Academy of Sciences of the United States of America.* 97, 17.

Parsegian, V. A., R. P. Rand, and D. C. Rau. 1995. Macromolecules and water: Probing with osmotic stress. *Energetics of Biological Macromolecules.* 259, 43.

Parsegian, V. A., R. P. Rand, and D. C. Rau. 2000. Osmotic stress, crowding, preferential hydration, and binding: A comparison of perspectives. *Proceedings of the National Academy of Sciences of the United States of America.* 97, 3987.

Paschek, D., R. Day, and A. E. García. 2011. Influence of water-protein hydrogen bonding on the stability of Trp-cage miniprotein. A comparison between the TIP3P and TIP4P-Ew water models. *Physical Chemistry Chemical Physics.* 13, 19840.

Paschek, D., S. Hempel, and A. E. García. 2008. Computing the stability diagram of the Trp-cage miniprotein. *Proceedings of the National Academy of Sciences of the United States of America.* 105, 17754.

Paul, S. and G. Patey. 2007. Structure and interaction in aqueous urea-trimethylamine-N-oxide solutions. *Journal of the American Chemical Society.* 129, 4476.

Pauling, L., R. B. Corey, and H. R. Branson. 1951. The structure of proteins: Two hydrogen-bonded helical configurations of the polypeptide chain. *Proceedings of the National Academy of Sciences of the United States of America.* 37, 205.

Pearson, F. J. and G. S. Rushbrooke. 1957. XX. On the theory of binary fluid mixtures. *Proceedings of the Royal Society of Edinburgh, Section: A Mathematics.* 64, 305.

Peng, D. Y. and D. B. Robinson. 1976. A new two-constant equation of state. *Industrial and Engineering Chemistry Fundamentals.* 15, 59.

Perera, A. 2009. Bridge function and fundamental measure theory: A test in dimension one. *Molecular Physics.* 107, 2251.

Perera, A. 2011. On the microscopic structure of liquid water. *Molecular Physics.* 109, 2433.

Perera, A., B. Kežić, F. Sokolić, and L. Zoranić. 2012. Micro-heterogeneity in complex liquids. In *Molecular Dynamics—Studies of Synthetic and Biological Macromolecules*, edited by L. Wang. Rijeka, Croatia: InTech.

Perera, A., R. Mazighi, and B. Kežić. 2012. Fluctuations and micro-heterogeneity in aqueous mixtures. *Journal of Chemical Physics.* 136, 174516.

Perera, A. and G. N. Patey. 1988. The solution of the hypernetted-chain and Percus–Yevick approximations for fluids of hard spherocylinders. *Journal of Chemical Physics.* 89, 5861.

Perera, A. and F. Sokolić. 2004. Modeling nonionic aqueous solutions: The acetone-water mixture. *Journal of Chemical Physics.* 121, 11272.

Perera, A., F. Sokolić, L. Almasy, P. Westh, and Y. Koga. 2005. On the evaluation of the Kirkwood–Buff integrals of aqueous acetone mixtures. *Journal of Chemical Physics.* 123, 024503.

Perera, A., F. Sokolić, L. Almasy, and Y. Koga. 2006. Kirkwood–Buff integrals of aqueous alcohol binary mixtures. *Journal of Chemical Physics.* 124, 124515.

Perera, A., L. Zoranić, F. Sokolić, and R. Mazighi. 2011. A comparative molecular dynamics study of water-methanol and acetone-methanol mixtures. *Journal of Molecular Liquids.* 159, 52.

Perkyns, J. S., Y. Wang, and B. M. Pettitt. 1996. Salting in peptides: Conformationally dependent solubilities and phase behavior of a tripeptide zwitterion in electrolyte solution. *Journal of the American Chemical Society.* 118, 1164.

Perry, R. L., H. Cabezas, and J. P. O'Connell. 1988. Fluctuation thermodynamic properties of strong electrolyte-solutions. *Molecular Physics.* 63, 189.

Perry, R. L. and J. P. O'Connell. 1984. Fluctuation thermodynamic properties of reactive components from species correlation-function integrals. *Molecular Physics.* 52, 137.

Perry, R. L., J. C. Telotte, and J. P. O'Connell. 1981. Solution thermodynamics for reactive components. *Fluid Phase Equilibria.* 5, 245.

Pettitt, B. M. and P. J. Rossky. 1986. Alkali halides in water: Ion–solvent correlations and ion–ion potentials of mean force at infinite dilution. *Journal of Chemical Physics.* 84, 5836.

Pfund, D. M. and H. D. Cochran. 1993. Chemical potentials in ternary supercritical fluid mixtures. In *Supercritical Engineering Science. Fundamentals Studies and Applications,* edited by E. Kiran and J. F. Brennecke. Washington, DC: American Chemical Society.

Pierce, V., M. Kang, M. Aburi, S. Weerasinghe, and P. E. Smith. 2008. Recent applications of Kirkwood–Buff theory to biological systems. *Cell Biochemistry and Biophysics.* 50, 1.

Pitzer, K. S. 1995. *Thermodynamics.* New York: McGraw-Hill.

Ploetz, E. A., N. Bentenitis, and P. E. Smith. 2010a. Developing force fields from the microscopic structure of solutions. *Fluid Phase Equilibria.* 290, 43.

Ploetz, E. A., N. Bentenitis, and P. E. Smith. 2010b. Kirkwood–Buff integrals for ideal solutions. *Journal of Chemical Physics.* 132, 164501.

Ploetz, E. A. and P. E. Smith. 2011a. A Kirkwood–Buff force field for the aromatic amino acids. *Physical Chemistry Chemical Physics.* 13, 18154.

Ploetz, E. A. and P. E. Smith. 2011b. Local fluctuations in solution mixtures. *Journal of Chemical Physics.* 135, 044506.

Plyasunov, A. V. 2011. Thermodynamics of $B(OH)_3$ in the vapor phase of water: Vapor-liquid and Henry's constants, fugacity and second cross virial coefficients. *Fluid Phase Equilibria.* 305, 212.

Plyasunov, A. V., J. P. O'Connell, and R. H. Wood. 2000. Infinite dilution partial molar properties of aqueous solutions of nonelectrolytes. I. Equations for partial molar volumes at infinite dilution and standard thermodynamic functions of hydration of volatile nonelectrolytes over wide ranges of conditions. *Geochimica et Cosmochimica Acta.* 64, 495.

Plyasunov, A. V., J. P. O'Connell, R. H. Wood, and E. L. Shock. 2000. Infinite dilution partial molar properties of aqueous solutions of nonelectrolytes. II. Equations for the standard thermodynamic functions of hydration of volatile nonelectrolytes over wide ranges of conditions including subcritical temperatures. *Geochimica et Cosmochimica Acta.* 64, 2779.

Plyasunov, A. V., J. P. O'Connell, R. H. Wood, and E. L. Shock. 2001. Semiempirical equation of state for the infinite dilution thermodynamic functions of hydration of nonelectrolytes over wide ranges of temperature and pressure. *Fluid Phase Equilibria.* 183, 133.

Plyasunov, A. V., E. L. Shock, and J. P. O'Connell. 2006. Corresponding-states correlations for estimating partial molar volumes of nonelectrolytes at infinite dilution in water over extended temperature and pressure ranges. *Fluid Phase Equilibria.* 247, 18.

Pohorille, A. and L. R. Pratt. 1990. Cavities in molecular liquids and the theory of hydrophobic solubilities. *Journal of the American Chemical Society.* 112, 5066.

Poland, D. C. and H. A. Scheraga. 1965. Hydrophobic bonding and micelle stability. *Journal of Physical Chemistry.* 69, 2431.

Poling, B. E., J. M. Prausnitz, and J. P. O'Connell. 2000. *The Properties of Gases and Liquids*, 5th ed. New York: McGraw-Hill. Errata at http://www.che.virginia.edu/PGL5/.

Polishuk, I. and A. Mulero. 2011. The numerical challenges of SAFT EoS models. *Reviews in Chemical Engineering*. 27, 241.

Pratt, L. R. and D. Chandler. 1977. Theory of the hydrophobic effect. *Journal of Chemical Physics*. 67, 3683.

Prausnitz, J. M., R. N. Lichtenthaler, and E. Gomes de Azevedo. 1986. *Molecular Thermodynamics of Fluid Phase Equilibria*, 2nd. ed. Englewood Cliffs, NJ: Prentice-Hall.

Prausnitz, J. M., R. N. Lichtenthaler, and E. Gomes de Azevedo. 1999. *Molecular Thermodynamics of Fluid-Phase Equilibria*, 3rd ed., *Prentice Hall International Series in the Physical and Chemical Engineering Sciences*. Upper Saddle River, NJ: Prentice-Hall.

Press, W. H., S. A. Teukolsky, W. T. Vetterling, and B. P. Flannery. 1992. *Numerical Recipes in C: The Art of Scientific Computing*, 2nd ed. Cambridge: Cambridge University Press.

Prigogine, I. with contributions from A. Bellemans and V. Mathot. 1957. *The Molecular Theory of Solutions*. Amsterdam: North-Holland Publishing Co.

Prigogine, I. and R. Dufay. 1954. *Chemical Thermodynamics, Treatise on Thermodynamics*. London; New York: Longmans, Green.

Puliti, G., S. Paolucci, and M. Sen. 2011. Thermodynamic properties of gold-water nanolayer mixtures using molecular dynamics. *Journal of Nanoparticle Research*. 13, 4277.

Qu, Y. X., C. L. Bolen, and D. W. Bolen. 1998. Osmolyte-driven contraction of a random coil protein. *Proceedings of the National Academy of Sciences of the United States of America*. 95, 9268.

Ramirez, R., M. Mareschal, and D. Borgis. 2005. Direct correlation functions and the density functional theory of polar solvents. *Chemical Physics*. 319, 261.

Raoult, F. M. 1878. Vapour-tension and solidifying point of saline solutions. *Comptes Rendus Hebdomadaires des Séances de l'Académie des Sciences*. 87, 167.

Raoult, F. M. 1882. Law of freezing of solvents. *Comptes Rendus Hebdomadaires des Séances de l'Académie des Sciences*. 95, 1030.

Raoult, F. M. 1887. General law of the vapor pressure of solvents. *Comptes Rendus Hebdomadaires des Séances de l'Académie des Sciences*. 104, 1430.

Raoult, F. M. 1888. Vapour-tensions of ethereal solutions. *Zeitschrift für Physikalische Chemie, Stöchiometrie und Verwandtschaftslehre*. 2, 353.

Rasaiah, J. C. and H. L. Friedman. 1968. Integral equation methods in the computation of equilibrium properties of ionic solutions. *Journal of Chemical Physics*. 48, 2742.

Reaves, J. T. and C. B. Roberts. 1999. Subcritical solvent effects on a parallel Diels–Alder reaction network. *Industrial and Engineering Chemistry Research*. 38, 855.

Record, M. T., W. T. Zhang, and C. F. Anderson. 1998. Analysis of effects of salts and uncharged solutes on protein and nucleic acid equilibria and processes: A practical guide to recognizing and interpreting polyelectrolyte effects, Hofmeister effects, and osmotic effects of salts. *Advances in Protein Chemistry*. 51, 281.

Redlich, O. and A. T. Kister. 1948. Thermodynamics of nonelectrolyte solutions—x-y-T relations in a binary system. *Industrial and Engineering Chemistry*. 40, 341.

Ren, P. Y., C. J. Wu, and J. W. Ponder. 2011. Polarizable atomic multipole-based molecular mechanics for organic molecules. *Journal of Chemical Theory and Computation*. 7, 3143.

Retailleau, P., A. Ducruix, and M. Ries-Kautt. 2002. Importance of the nature of anions in lysozyme crystallisation correlated with protein net charge variation. *Acta Crystallographica Section D-Biological Crystallography*. 58, 1576.

Retailleau, P., M. Ries-Kautt, and A. Ducruix. 1997. No salting-in of lysozyme chloride observed at low ionic strength over a large range of pH. *Biophysical Journal*. 73, 2156.

Richards, J. P. 2011. Magmatic to hydrothermal metal fluxes in convergent and collided margins. *Ore Geology Reviews*. 40, 1.

Ries-Kautt, M. M. and A. F. Ducruix. 1989. Relative effectiveness of various ions on the solubility and crystal-growth of lysozyme. *Journal of Biological Chemistry.* 264, 745.

Roberts, C. B., J. F. Brennecke, and J. E. Chateauneuf. 1995. Solvation effects on reactions of triplet benzophenone in supercritical fluids. *American Institute of Chemical Engineers Journal.* 41, 1306.

Roberts, C. B., J. W. Zhang, J. E. Chateauneuf, and J. F. Brennecke. 1995. Laser flash-photolysis and integral-equation theory to investigate reactions of dilute solutes with oxygen in supercritical fluids. *Journal of the American Chemical Society.* 117, 6553.

Robinson, C. R., S. G. Sligar, M. L. Johnson, and G. K. Ackers. 1995. Hydrostatic and osmotic pressure as tools to study macromolecular recognition. *Energetics of Biological Macromolecules.* 259, 395.

Romankiw, L. A. and I. M. Chou. 1983. Densities of aqueous NaCl, KCl, $MgCl_2$, and $CaCl_2$ binary-solutions in the concentration range 0.5-6.1-*m* at 25-degrees-C, 30-degrees-C, 35-degrees-C, 40-degrees-C, and 45-degrees-C. *Journal of Chemical and Engineering Data.* 28, 300.

Romero, S., A. Reillo, B. Escalera, and P. Bustamante. 1996. The behavior of paracetamol in mixtures of amphiprotic and amphiprotic-aprotic solvents. Relationship of solubility curves to specific and nonspecific interactions. *Chemical and Pharmaceutical Bulletin.* 44, 1061.

Rose, G., P. Fleming, J. Banavar, and A. Maritan. 2006. A backbone-based theory of protein folding. *Proceedings of the National Academy of Sciences of the United States of America.* 103, 16623.

Rosenberg, R. M. and W. L. Peticolas. 2004. Henry's law: A retrospective. *Journal of Chemical Education.* 81, 1647.

Rösgen, J., B. M. Pettitt, and D. W. Bolen. 2004. Uncovering the basis for nonideal behavior of biological molecules. *Biochemistry.* 43, 14472.

Rösgen, J., B. M. Pettitt, and D. W. Bolen. 2005. Protein folding, stability, and solvation structure in osmolyte solutions. *Biophysical Journal.* 89, 2988.

Rösgen, J., B. M. Pettitt, J. Perkyns, and D. W. Bolen. 2004. Statistical thermodynamic approach to the chemical activities in two-component solutions. *Journal of Physical Chemistry B.* 108, 2048.

Rossky, P. J. 2008. Protein denaturation by urea: Slash and bond. *Proceedings of the National Academy of Sciences of the United States of America.* 105, 16825.

Rowlinson, J. S. and F. L. Swinton. 1982. *Liquids and Liquid Mixtures,* 3rd ed. London: Butterworth Scientific.

Ruckenstein, E. and I. L. Shulgin. 2001a. Effect of a third component on the interactions in a binary mixture determined from the fluctuation theory of solutions. *Fluid Phase Equilibria.* 180, 281.

Ruckenstein, E. and I. L. Shulgin. 2001b. Entrainer effect in supercritical mixtures. *Fluid Phase Equilibria.* 180, 345.

Ruckenstein, E. and I. Shulgin. 2002a. The solubility of solids in mixtures composed of a supercritical fluid and an entrainer. *Fluid Phase Equilibria.* 200, 53.

Ruckenstein, E. and I. L. Shulgin. 2002b. Salting-out or -in by fluctuation theory. *Industrial and Engineering Chemistry Research.* 41, 4674.

Ruckenstein, E. and I. L. Shulgin. 2003a. Ideal multicomponent liquid solution as a mixed solvent. *Industrial and Engineering Chemistry Research.* 42, 4406.

Ruckenstein, E. and I. L. Shulgin. 2003b. Solubility of drugs in aqueous solutions. Part 1: Ideal mixed solvent approximation. *International Journal of Pharmaceutics.* 258, 193.

Ruckenstein, E. and I. L. Shulgin. 2003c. Solubility of drugs in aqueous solutions. Part 2: Binary nonideal mixed solvent. *International Journal of Pharmaceutics.* 260, 283.

Ruckenstein, E. and I. L. Shulgin. 2003d. Solubility of drugs in aqueous solutions. Part 3: Multicomponent mixed solvent. *International Journal of Pharmaceutics.* 267, 121.

Ruckenstein, E. and I. L. Shulgin. 2004. Solubility of drugs in aqueous solutions. Part 4: Drug solubility by the dilute approximation. *International Journal of Pharmaceutics.* 278, 221.

Ruckenstein, E. and I. L. Shulgin. 2005. Solubility of hydrophobic organic pollutants in binary and multicomponent aqueous solvents. *Environmental Science and Technology.* 39, 1623.

Ruckenstein, E. and I. L. Shulgin. 2006. Effect of salts and organic additives on the solubility of proteins in aqueous solutions. *Advances in Colloid and Interface Science.* 123, 97.

Ruckenstein, E. and I. L. Shulgin, eds. 2009. *Thermodynamics of Solutions: From Gases to Pharmaceutics to Proteins.* Dordrecht: Springer.

Salacuse, J. J., A. R. Denton, and P. A. Egelstaff. 1996. Finite-size effects in molecular dynamics simulations: Static structure factor and compressibility. 1. Theoretical method. *Physical Review E.* 53, 2382.

Sasahara, K., M. Sakurai, and K. Nitta. 1999. The volume and compressibility changes of lysozyme associated with guanidinium chloride and pressure-assisted unfolding. *Journal of Molecular Biology.* 291, 693.

Sasahara, K., M. Sakurai, and K. Nitta. 2001. Pressure effect on denaturant-induced unfolding of hen egg white lysozyme. *Proteins: Structure, Function, and Genetics.* 44, 180.

Savage, P. E., S. Golapan, T. I. Mizan, C. J. Martino, and E. E. Brock. 1995. Reactions at supercritical conditions: Applications and fundamentals. *American Institute of Chemical Engineers Journal.* 41, 1723.

Schachman, H. K. and M. A. Lauffer. 1949. The hydration, size and shape of tobacco mosaic virus. *Journal of the American Chemical Society.* 71, 536.

Schellman, J. A. 1987. Selective binding and solvent denaturation. *Biopolymers.* 26, 549.

Schmid, B. and J. Gmehling. 2012. Revised parameters and typical results of the VTPR group contribution equation of state. *Fluid Phase Equilibria.* 317, 110.

Schmid, N., A. P. Eichenberger, A. Choutko, S. Riniker, M. Winger, A. E. Mark, and W. F. van Gunsteren. 2011. Definition and testing of the GROMOS force-field versions 54A7 and 54B7. *European Biophysics Journal with Biophysics Letters.* 40, 843.

Schnell, S. K., X. Liu, J. M. Simon, A. Bardow, D. Bedeaux, T. J. H. Vlugt, and S. Kjelstrup. 2011. Calculating thermodynamic properties from fluctuations at small scales. *Journal of Physical Chemistry B.* 115, 10911.

Schoen, M. and C. Hoheisel. 1984. The mutual diffusion coefficient-D_{12} in binary-liquid model mixtures—Molecular-dynamics calculations based on Lennard–Jones (12-6) potentials. 1. The method of determination. *Molecular Physics.* 52, 33.

Sedlbauer, J., J. P. O'Connell, and R. H. Wood. 2000. A new equation of state for correlation and prediction of standard molal thermodynamic properties of aqueous species at high temperatures and pressures. *Chemical Geology.* 163, 43.

Sedlbauer, J., E. M. Yezdimer, and R. H. Wood. 1998. Partial molar volumes at infinite dilution in aqueous solutions of NaCl, LiCl, NaBr, and CsBr at temperatures from 550 K to 725 K. *Journal of Chemical Thermodynamics.* 30, 3.

Shimizu, S. 2004. Estimating hydration changes upon biomolecular reactions from osmotic stress, high pressure, and preferential hydration experiments. *Proceedings of the National Academy of Sciences of the United States of America.* 101, 1195.

Shimizu, S. 2011. Molecular origin of the cosolvent-induced changes in the thermal stability of proteins. *Chemical Physics Letters.* 514, 156.

Shimizu, S. and C. L. Boon. 2004. The Kirkwood–Buff theory and the effect of cosolvents on biochemical reactions. *Journal of Chemical Physics.* 121, 9147.

Shimizu, S. and N. Matubayasi. 2006. Preferential hydration of proteins: A Kirkwood–Buff approach. *Chemical Physics Letters.* 420, 518.

Shimizu, S., W. M. McLaren, and N. Matubayasi. 2006. The Hofmeister series and protein-salt interactions. *Journal of Chemical Physics.* 124, 234905.

Shukla, D. and B. L. Trout. 2011. Understanding the synergistic effect of arginine and glutamic acid mixtures on protein solubility. *Journal of Physical Chemistry B.* 115, 11831.

Shulgin, I. and E. Ruckenstein. 2002a. The solubility of binary mixed gases by the fluctuation theory. *Industrial and Engineering Chemistry Research.* 41, 6279.

Shulgin, I. L. and E. Ruckenstein. 2002b. Henry's constant in mixed solvents from binary data. *Industrial and Engineering Chemistry Research.* 41, 1689.

Shulgin, I. L. and E. Ruckenstein. 2003. Prediction of gas solubility in binary polymer plus solvent mixtures. *Polymer.* 44, 901.

Shulgin, I. L. and E. Ruckenstein. 2005a. A protein molecule in an aqueous mixed solvent: Fluctuation theory outlook. *Journal of Chemical Physics.* 123, 054909.

Shulgin, I. L. and E. Ruckenstein. 2005b. Relationship between preferential interaction of a protein in an aqueous mixed solvent and its solubility. *Biophysical Chemistry.* 118, 128.

Shulgin, I. L. and E. Ruckenstein. 2006a. The Kirkwood–Buff theory of solutions and the local composition of liquid mixtures. *Journal of Physical Chemistry B.* 110, 12707.

Shulgin, I. L. and E. Ruckenstein. 2006b. Preferential hydration and solubility of proteins in aqueous solutions of polyethylene glycol. *Biophysical Chemistry.* 120, 188.

Shulgin, I. L. and E. Ruckenstein. 2008a. Excess around a central molecule with application to binary mixtures. *Physical Chemistry Chemical Physics.* 10, 1097.

Shulgin, I. L. and E. Ruckenstein. 2008b. Reply to "Comment on 'The Kirkwood–Buff theory of solutions and the local composition of liquid mixtures.'" *Journal of Physical Chemistry B.* 112, 5876.

Simnick, J. J., H. M. Sebastian, H. M. Lin, and K. C. Chao. 1979a. Gas-liquid equilibrium in mixtures of methane+*m*-xylene, and methane-meta-cresol. *Fluid Phase Equilibria.* 3, 145.

Simnick, J. J., H. M. Sebastian, H. M. Lin, and K. C. Chao. 1979b. Vapor-liquid-equilibrium in methane + quinoline mixtures at elevated-temperatures and pressures. *Journal of Chemical and Engineering Data.* 24, 239.

Smit, B. 1992. Phase-diagrams of Lennard-Jones fluids. *Journal of Chemical Physics.* 96, 8639.

Smith, P. E. 2004. Cosolvent interactions with biomolecules: Relating computer simulation data to experimental thermodynamic data. *Journal of Physical Chemistry B.* 108, 18716.

Smith, P. E. 2006a. Chemical potential derivatives and preferential interaction parameters in biological systems from Kirkwood–Buff theory. *Biophysical Journal.* 91, 849.

Smith, P. E. 2006b. Equilibrium dialysis data and the relationships between preferential interaction parameters for biological systems in terms of Kirkwood–Buff integrals. *Journal of Physical Chemistry B.* 110, 2862.

Smith, P. E. 2008. On the Kirkwood–Buff inversion procedure. *Journal of Chemical Physics.* 129, 124509.

Smith, P. E., G. E. Marlow, and B. M. Pettitt. 1993. Peptides in ionic-solutions—A simulation study of a bis(penicillamine) enkephalin in sodium-acetate solution. *Journal of the American Chemical Society.* 115, 7493.

Smith, P. E. and R. M. Mazo. 2008. On the theory of solute solubility in mixed solvents. *Journal of Physical Chemistry B.* 112, 7875.

Smith, P. E. and B. M. Pettitt. 1991. Effects of salt on the structure and dynamics of the bis(penicillamine) enkephalin zwitterion—A simulation study. *Journal of the American Chemical Society.* 113, 6029.

Smith, P. E. and B. M. Pettitt. 1992. Amino-acid side-chain populations in aqueous and saline solution: Bis-penicillamine enkephalin. *Biopolymers.* 32, 1623.

Soper, A. K., L. Dougan, J. Crain, and J. L. Finney. 2006. Excess entropy in alcohol-water solutions: A simple clustering explanation. *Journal of Physical Chemistry B.* 110, 3472.

Staples, B. R. and R. L. Nuttall. 1977. Activity and osmotic coefficients of aqueous calcium-chloride at 298.15-K. *Journal of Physical and Chemical Reference Data.* 6, 385.

Steinfeld, J. I., J. S. Francisco, and W. L. Hase. 1989. *Chemical Kinetics and Dynamics.* Englewood Cliffs, NJ: Prentice Hall.

Stell, G., G. N. Patey, and J. S. Høye. 1981. Dielectric constants of fluid models: Statistical mechanical theory and its quantitative implementation. *Advances in Chemical Physics.* 48, 183.

Stockmayer, W. H. 1950. Light scattering in multi-component systems. *Journal of Chemical Physics.* 18, 58.

Stradner, A., H. Sedgwick, F. Cardinaux, W. C. K. Poon, S. U. Egelhaaf, and P. Schurtenberger. 2004. Equilibrium cluster formation in concentrated protein solutions and colloids. *Nature.* 432, 492.

Tanford, C. 1964. Isothermal unfolding of globular proteins in aqueous urea solutions. *Journal of American Chemical Society.* 86, 2050.

Tanford, C. 1974. Thermodynamics of micelle formation: Prediction of micelle size and size distribution. *Proceedings of the National Academy of Sciences.* 71, 1811.

Tanford, C. 1997. How protein chemists learned about the hydrophobic factor. *Protein Science.* 6, 1358.

Teubner, M. and R. Strey. 1987. Origin of the scattering peak in microemulsions. *Journal of Chemical Physics.* 87, 3195.

Teufel, D. P., C. M. Johnson, J. K. Lum, and H. Neuweiler. 2011. Backbone-driven collapse in unfolded protein chains. *Journal of Molecular Biology.* 409, 250.

Theodorou, D. N. and U. W. Suter. 1985. Geometrical considerations in model systems with periodic boundaries. *Journal of Chemical Physics.* 82, 955.

Timasheff, S. N. 1992. A physicochemical basis for the selection of osmolytes by nature. In *Water and Life: Comparative Analysis of Water Relationships at the Organismic, Cellular and Molecular Levels,* edited by G. N. Somero, C. B. Osmond, and C. L. Bolis. Berlin: Springer-Verlag, p. 70.

Timasheff, S. N. 1993. The control of protein stability and association by weak-interactions with water—How do solvents affect these processes? *Annual Review of Biophysics and Biomolecular Structure.* 22, 67.

Timasheff, S. N. 1998a. Control of protein stability and reactions by weakly interacting cosolvents: The simplicity of the complicated. *Advances in Protein Chemistry.* 51, 355.

Timasheff, S. N. 1998b. In disperse solution, "osmotic stress" is a restricted case of preferential interactions. *Proceedings of the National Academy of Sciences of the United States of America.* 95, 7363.

Timasheff, S. N. 2002a. Protein-solvent preferential interactions, protein hydration, and the modulation of biochemical reactions by solvent components. *Proceedings of the National Academy of Sciences of the United States of America.* 99, 9721.

Timasheff, S. N. 2002b. Protein hydration, thermodynamic binding, and preferential hydration. *Biochemistry.* 41, 13473.

Timasheff, S. N. and T. Arakawa. 1988. Mechanism of protein precipitation and stabilization by co-solvents. *Journal of Crystal Growth.* 90, 39.

Timasheff, S. N. and T. Arakawa. 1989. Stabilization of protein structure by solvents. In *Protein Structure: A Practical Approach,* edited by T. E. Creighton. Oxford: IRL Press at Oxford University Press, p. 331.

Timasheff, S. N. and G. F. Xie. 2003. Preferential interactions of urea with lysozyme and their linkage to protein denaturation. *Biophysical Chemistry.* 105, 421.

Tokunaga, J. 1975. Solubilities of oxygen, nitrogen, and carbon-dioxide in aqueous alcohol solutions. *Journal of Chemical and Engineering Data.* 20, 41.

Tolmachev, V. V. 1960. Relationship between the statistic variation principle and the method of forming partial sums for diagrams in the thermodynamic perturbation theory for a modified statement of the problem of Bose–Einstein non-ideal system. *Doklady Akademii Nauk USSR.* 134, 1324.

Tran, H. T., A. Mao, and R. V. Pappu. 2008. Role of backbone-solvent interactions in determining conformational equilibria of intrinsically disordered proteins. *Journal of the American Chemical Society.* 130, 7380.

Treece, J. M., R. S. Sheinson, and T. L. McMeekin. 1964. Solubilities of β-lactoglobulins A, B, and AB. *Archives of Biochemistry and Biophysics.* 108, 99.

Treiner, C., P. Tzias, M. Chemla, and G. M. Poltoratskii. 1976. Solvation of tetrabutylammonium bromide in water + acetonitrile mixtures at 298.15 K from vapour pressure measurements of dilute solutions. *Journal of the Chemical Society, Faraday Transactions 1: Physical Chemistry in Condensed Phases.* 72, 2007.

Treszczanowicz, T., A. J. Treszczanowicz, T. Kasprzycka-Guttman, and A. Kulesza. 2001. Solubility of β-carotene in binary solvents formed by some hydrocarbons with ketones. *Journal of Chemical and Engineering Data.* 46, 792.

Uversky, V. N., C. J. Oldfield, and A. K. Dunker. 2005. Showing your ID: Intrinsic disorder as an ID for recognition, regulation and cell signaling. *Journal of Molecular Recognition.* 18, 343.

Valdeavella, C. V., J. S. Perkyns, and B. M. Pettitt. 1994. Investigations into the common ion effect. *Journal of Chemical Physics.* 101, 5093.

Van't Hoff, J. H. 1887. Die rolle des osmotischen druckes in der analogie zwischen lösungen und gasen. *Zeitschrift für Physikalische Chemie, Stöchiometrie und Verwandtschaftslehre.* 1, 481.

Van't Hoff, J. H. 1894. How the theory of solutions arose. *Berichte der Deutschen Chemischen Gesellschaft.* 27, 6.

Van Alsten, J. G. and C. A. Eckert. 1993. Effect of entrainers and of solute size and polarity in supercritical-fluid solutions. *Journal of Chemical and Engineering Data.* 38, 605.

Van Ness, H. C. 1995. Thermodynamics in the treatment of vapor/liquid equilibrium (VLE) data. *Pure and Applied Chemistry.* 67, 859.

Van Ness, H. C. and M. M. Abbott. 1982. *Classical Thermodynamics of Nonelectrolyte Solutions: With Applications to Phase Equilibria.* New York: McGraw Hill.

Vergara, A., L. Paduano, F. Capuano, and R. Sartorio. 2002. Kirkwood–Buff integrals for polymer-solvent mixtures. Preferential solvation and volumetric analysis in aqueous PEG solutions. *Physical Chemistry Chemical Physics.* 4, 4716.

Verlet, L. 1968. Computer experiments on classical fluids. 2. Equilibrium correlation functions. *Physical Review.* 165, 201.

Vlcek, L., A. A. Chialvo, and D. R. Cole. 2011. Optimized unlike-pair interactions for water-carbon dioxide mixtures described by the SPC/E and EPM2 models. *Journal of Physical Chemistry B.* 115, 8775.

Wallas, S. M. 1985. *Phase Equilibria in Chemical Engineering.* Boston: Butterworth-Heinemann.

Wang, J. M., P. Cieplak, and P. A. Kollman. 2000. How well does a restrained electrostatic potential (RESP) model perform in calculating conformational energies of organic and biological molecules? *Journal of Computational Chemistry.* 21, 1049.

Wang, S. S., C. G. Gray, P. A. Egelstaff, and K. E. Gubbins. 1973. Monte-Carlo study of pair correlation-function for a liquid with non-central forces. *Chemical Physics Letters.* 21, 123.

Wang, W., O. Donini, C. M. Reyes, and P. A. Kollman. 2001. Biomolecular simulations: Recent developments in force fields, simulations of enzyme catalysis, protein-ligand, protein-protein, and protein-nucleic acid noncovalent interactions. *Annual Review of Biophysics and Biomolecular Structure.* 30, 211.

Wedberg, N. H. R. I. 2011. Molecular Modeling of Enzyme Dynamics Toward Understanding Solvent Effects. Ph.D. Thesis, DTU Chemical Engineering, Technical University of Denmark, Kongens Lyngby.

Wedberg, R., J. Abildskov, and G. H. Peters. 2012. Protein dynamics in organic media at varying water activity studied by molecular dynamics simulation. *Journal of Physical Chemistry B.* 116, 2575.

Wedberg, R., J. P. O'Connell, G. H. Peters, and J. Abildskov. 2010. Accurate Kirkwood–Buff integrals from molecular simulations. *Molecular Simulation.* 36, 1243.

Wedberg, R., J. P. O'Connell, G. H. Peters, and J. Abildskov. 2011a. Pair correlation function integrals: Computation and use. *Journal of Chemical Physics*. 135, 084113.

Wedberg, R., J. P. O'Connell, G. H. Peters, and J. Abildskov. 2011b. Total and direct correlation function integrals from molecular simulation of binary systems. *Fluid Phase Equilibria*. 302, 32.

Wedberg, R., G. H. Peters, and J. Abildskov. 2008. Total correlation function integrals and isothermal compressibilities from molecular simulations. *Fluid Phase Equilibria*. 273, 1.

Weerasinghe, S., M. B. Gee, M. Kang, N. Bentenitis, and P. E. Smith. 2010. Developing force fields from the microscopic structure of solutions: The Kirkwood–Buff approach. In *Modeling Solvent Environments: Applications to Simulations of Biomolecules*, edited by M. Feig. Weinheim: Wiley-VCH Verlag GmbH and Co. KGaA.

Weerasinghe, S. and B. M. Pettitt. 1994. Ideal chemical-potential contribution in moleculardynamics simulations of the grand-canonical ensemble. *Molecular Physics*. 82, 897.

Weerasinghe, S. and P. E. Smith. 2003a. Cavity formation and preferential interactions in urea solutions: Dependence on urea aggregation. *Journal of Chemical Physics*. 118, 5901.

Weerasinghe, S. and P. E. Smith. 2003b. Kirkwood–Buff derived force field for mixtures of acetone and water. *Journal of Chemical Physics*. 118, 10663.

Weerasinghe, S. and P. E. Smith. 2003c. A Kirkwood–Buff derived force field for mixtures of urea and water. *Journal of Physical Chemistry B*. 107, 3891.

Weerasinghe, S. and P. E. Smith. 2003d. A Kirkwood–Buff derived force field for sodium chloride in water. *Journal of Chemical Physics*. 119, 11342.

Weerasinghe, S. and P. E. Smith. 2004. A Kirkwood–Buff derived force field for the simulation of aqueous guanidinium chloride solutions. *Journal of Chemical Physics*. 121, 2180.

Weerasinghe, S. and P. E. Smith. 2005. A Kirkwood–Buff derived force field for methanol and aqueous methanol solutions. *Journal of Physical Chemistry B*. 109, 15080.

Weinreb, P. H., W. Zhen, A. Poon, K. A. Conway, and P. T. Lansbury. 1996. NACP, a protein implicated in Alzheimer's disease and learning, is natively unfolded. *Biochemistry*. 35, 13709.

Weinstein, R. D., A. R. Renslo, R. L. Danheiser, J. G. Harris, and J. W. Tester. 1996. Kinetic correlation of Diels–Alder reactions in supercritical carbon dioxide. *Journal of Physical Chemistry*. 100, 12337.

Wen, X. G. 2004. *Quantum Field Theory of Many-Body Systems: From the Origin of Sound to an Origin of Light and Electrons*. Oxford: Oxford University Press.

Wheeler, D. R. and J. Newman. 2004a. Molecular dynamics simulations of multicomponent diffusion. 1. Equilibrium method. *Journal of Physical Chemistry B*. 108, 18353.

Wheeler, D. R. and J. Newman. 2004b. Molecular dynamics simulations of multicomponent diffusion. 2. Nonequilibrium method. *Journal of Physical Chemistry B*. 108, 18362.

Widom, B. 1963. Some topics in the theory of fluids. *Journal of Chemical Physics*. 39, 2808.

Wienke, G. and J. Gmehling. 1998. Prediction of octanol-water partition coefficients, Henry coefficients, and water solubilities using UNIFAC. *Toxicological and Environmental Chemistry*. 65, 57.

Wiggins, P. 2008. Life depends upon two kinds of water. *PLoS One*. 3, e1406.

Wilczek-Vera, G. and J. H. Vera. 2011. The activity of individual ions. A conceptual discussion of the relation between the theory and the experimentally measured values. *Fluid Phase Equilibria*. 312, 79.

Wilhelm, E., R. Battino, and R. J. Wilcock. 1977. Low-pressure solubility of gases in liquid water. *Chemical Reviews*. 77, 219.

Williams, J. R. and A. A. Clifford, eds. 2000. *Supercritical Fluid Methods and Protocols. Methods in Biotechnology* series, series editor: J. M. Waker. Totowa, New Jersey: Humana Press Inc.

Wilson, G. M. 1964. Vapor-liquid equilibrium. XI. A new expression for the excess free energy of mixing. *Journal of the American Chemical Society*. 86, 127.

Wong, D. S. H. and S. I. Sandler. 1992. A theoretically correct mixing rule for cubic equations of state. *American Institute of Chemical Engineers Journal.* 38, 671.

Wooley, R. J. and J. P. O'Connell. 1991. A database of fluctuation thermodynamic properties and molecular correlation-function integrals for a variety of binary liquids. *Fluid Phase Equilibria.* 66, 233.

Wrabl, J. and D. Shortle. 1999. A model of the changes in denatured state structure underlying m value effects in staphylococcal nuclease. *Nature Structural Biology.* 6, 876.

Wright, P. E. and H. J. Dyson. 1999. Intrinsically unstructured proteins: Re-assessing the protein structure-function paradigm. *Journal of Molecular Biology.* 293, 321.

Wyman, J. 1964. Linked functions and reciprocal effects in hemoglobin—A 2nd look. *Advances in Protein Chemistry.* 19, 223.

Yalkowsky, S. H. and T. J. Roseman. 1981. Solubilization of drugs by cosolvents. In *Techniques of Solubilization of Drugs*, edited by S. H. Yalkowsky. New York: Marcel Dekker, p. 91.

Yancey, P., M. Clark, S. Hand, R. Bowlus, and G. Somero. 1982. Living with water stress: Evolution of osmolyte systems. *Science.* 217, 1214.

Yesodharan, S. 2002. Supercritical water oxidation: An environmentally safe method for the disposal of organic wastes. *Current Science.* 82, 1112.

Zeck, S. and H. Knapp. 1985. Solubilities of ethylene, ethane, and carbon-dioxide in mixed-solvents consisting of methanol, acetone, and water. *International Journal of Thermophysics.* 6, 643.

Zhang, W. T., M. W. Capp, J. P. Bond, C. F. Anderson, and M. T. Record. 1996. Thermodynamic characterization of interactions of native bovine serum albumin with highly excluded (glycine betaine) and moderately accumulated (urea) solutes by a novel application of vapor pressure osmometry. *Biochemistry.* 35, 10506.

Zhang, C. and J. P. Ma. 2010. Enhanced sampling and applications in protein folding in explicit solvent. *Journal of Chemical Physics.* 132, 16.

Zhong, Y. and S. Patel. 2010. Nonadditive empirical force fields for short-chain linear alcohols: Methanol to butanol. Hydration free energetics and Kirkwood–Buff analysis using charge equilibration models. *Journal of Physical Chemistry B.* 114, 11076.

Zhu, X., P. E. M. Lopes, and A. D. Mackerell, Jr. 2012. Recent developments and applications of the CHARMM force fields. *Wiley Interdisciplinary Reviews-Computational Molecular Science.* 2, 167.

Zielkiewicz, J. 1995a. Kirkwood–Buff integrals in the binary and ternary mixtures containing heptane and aliphatic alcohol. *Journal of Physical Chemistry.* 99, 3357.

Zielkiewicz, J. 1995b. Solvation of DMF in the *N,N*-dimethylformamide + alcohol + water mixtures investigated by means of the Kirkwood–Buff integrals. *Journal of Physical Chemistry.* 99, 4787.

Zoranić, L., R. Mazighi, F. Sokolić, and A. Perera. 2009. Concentration fluctuations and microheterogeneity in aqueous amide mixtures. *Journal of Chemical Physics.* 130, 124315.

Zou, Q., B. J. Bennion, V. Daggett, and K. P. Murphy. 2002. The molecular mechanism of stabilization of proteins by TMAO and its ability to counteract the effects of urea. *Journal of the American Chemical Society.* 124, 1192.

Index